SPECTROSCOPIC DATA

Volume 1
Heteronuclear Diatomic Molecules

SPECTROSCOPIC DATA

Volume 1
Heteronuclear Diatomic Molecules

Edited by S. N. Suchard
The Aerospace Corporation
Los Angeles, California

Part B

IFI/PLENUM • NEW YORK-WASHINGTON-LONDON

Library of Congress Cataloging in Publication Data

Suchard, S N
 Spectroscopic data.

 Includes bibliographical references.
 CONTENTS: v. 1. Heteronuclear diatomic molecules.
 1. Spectrum analysis — Tables, etc. I. Title.
 QC453.S85   535'.84'0212   74-34288
 ISBN 0-306-68311-3

Published in 1975 by IFI/Plenum Data Company
A Division of Plenum Publishing Corporation
227 West 17th Street, New York, N.Y. 10011

United Kingdom edition published by Plenum Press, London
A Division of Plenum Publishing Company, Ltd.
4a Lower John Street, London W1R 3PD, England

All rights reserved

No part of this book may be reproduced, stored in a retrieval system, or transmitted, in any form or by any means, electronic, mechanical, photocopying, microfilming, recording, or otherwise, without written permission from the Publisher

Printed in the United States of America

Preface

The origin, rationale, and organization of this compendium are discussed in the introductory text on pages ix to xviii. Because the number of pages is so large that a single volume would be difficult to handle, Volume 1 has been published in two parts. To avoid inconvenience to the user, the introductory text and the spectroscopic information summary on pages xxi to xxix, to which the user may frequently wish to refer, have been printed at the front of each part.

During the preparation of this compilation, many people contributed; the compiler wishes to thank all of them. In particular he appreciated the moral support of his wife, Phyllis, and of his children, and the efforts of V. Gilbertson and F. Plotnik, the manuscript typists; J. A. Kiley, K. C. Bregand, and W. H. McPherson, who made valuable editorial suggestions; and especially J. Melzer, without whose constant enthusiasm and hard work this report would not have been published. In addition, he extends his gratitude to Mr. R. S. Bradford and Professor H. P. Broida of the University of California at Santa Barbara, who made available to him their personal spectroscopic information collections on metal oxides and halides, and to Dr. L. Wilson of the Air Force Weapons Laboratory, who gave the initial impetus to this project. Finally, the author would like to thank the Advanced Research Projects Agency of the Department of Defense for their support during the compilation of this compendium.

Contents

I.	Introduction	ix
II.	Organization of the Spectroscopic Table	x
	Methods of Production and Experimental Techniques	x
	Band Systems	x
	Spectroscopic Constants	xi
	Perturbations and General Information	xii
	Bibliography	xii
III.	Notation and Notational Conversion Formulas	xiii
IV.	Conclusions on the Availability of Spectroscopic Information	xvi
	References	xix
	Spectroscopic Information Summary	xxi

Part A

A	1
B	96
C	252
F	451
E	456
G	474
H	543
I	557

Part B

L	611
M	645
N	693
O	786
P	811
R	907
S	925
T	1125
U	1184
V	1190
W	1199
Y	1202
Z	1218

I. INTRODUCTION

In recent years the need for a complete collection of information relevant to diatomic molecules has become evident. Several excellent collections of this type of information have been available for many years (Refs. 1-3); however, the state of our collective knowledge has been considerably expanded since their publication. At present, if recent information concerning a specific molecular species is desired, it is necessary to institute a time-consuming library search to collect the required information from the large number of sources in which it may have been presented. If information concerning an entire isoelectric molecular series is required, the time involved in collecting the information can be considerable.

This compilation was assembled in the hopes of solving this time problem. We have attempted to gather a complete collection of spectroscopic information relevant to selected molecular systems. When this project was initiated, all diatomic molecules were to be included; however, after further consideration, the decision was made to restrict our collection to heteronuclear diatomic molecules, which, from thermodynamic reasoning, could be produced in an electronically excited state by a chemical reaction between a ground state atom and a ground state molecule. This restriction was not strictly followed in several cases for the sake of completeness. In general, however, only molecules with stable electronically excited states lying below the molecular dissociation energy of the ground state species are included.

The organization of the material has been patterned after the compendium of Rosen (Ref. 1). We believe that his form of presentation displays the spectroscopic information in a manner that is amenable to efficient retrieval. The material itself was located with the help of several excellent earlier compendia (Refs. 1-3) and from the continuously updated Berkeley Newsletters collected by J. G. Phillips and S. P. Davis of the University of California at Berkeley (Ref. 4).

II. ORGANIZATION OF THE SPECTROSCOPIC TABLE

The information that we are presenting deals primarily with the electronic spectra of selected diatomic systems. In general, the spectroscopic constants have been derived from the interpretation of electronic spectra; however, when the amount of data was insufficient, the information was taken from alternative sources.

For simplicity, the molecules are presented in alphabetical order. The information for most of the molecules is broken into five separate sections, with several of the sections broken further into several subsections. Information is presented according to the following format.

METHODS OF PRODUCTION AND EXPERIMENTAL TECHNIQUES

The most favorable sources for the production of the molecule of interest and important experimental techniques used for studying the molecule.

BAND SYSTEMS

This section is broken into two subsections. In the first subsection, a general description of the molecular transition of each system or group is presented; the description is divided into eight headings.

1. System numbers or common designations for the system, or both (e.g., β band of NO)

2. Transition. Conventional or quantum designation for the states involved. The signs →, ←, and ⇄ refer to systems observed in emission, in absorption, or in both emission and absorption.

3. Sources. The most favorable sources for producing the particular system

4. Wavelength Limits. Spectral range of the system (Å)

5. Degrading. Direction of band head shading (R ≡ red, V ≡ violet)

6. Band Head ν_{00} or Characteristic Bands λ. Characteristic spectral bands free from overlapping bands or with sharp heads

7. Remarks. Additional information that is useful in characterizing the particular system
8. Bibliography. Listing of references that concern themselves with the particular system

The second subsystem presents a more detailed analysis of the system. Wavelengths (Å) of band heads or origins, intensities, and vibrational classifications are presented where available. Other available and relevant information is presented to characterize the system.

SPECTROSCOPIC CONSTANTS

The molecular constants that totally define the electronic states of the molecule are presented under ten headings. If not specifically mentioned, these constants refer to molecules made of the most abundant isotopes.

1. State. Quantum specification of the electronic state
2. T_e. Electronic energy above ground state (cm^{-1})
3. ω_e. Vibrational spacing (cm^{-1})
4. $x_e \omega_e$. Anharmonic correction to vibrational spacing (cm^{-1})
5. B_e. Rigid rotator rotational spacing (cm^{-1})
6. α_e. Nonrigid rotator correction to B_e (cm^{-1})
7. D_e. Anharmonic correction to rotational spacing (cm^{-1})
8. r_e. Equilibrium internuclear distance (Å)
9. Remarks. Additional relevant information
10. Bibliography. Important references

Other molecular constants or information, if known, are given as footnotes. The dissociation energy D_0^0 or D_{298}^0 is given in cm^{-1}, kcal/mole, and eV. The greater majority of the values for the molecular dissociation energies have been adopted from the book, <u>Dissociation Energies and Spectra of Diatomic Molecules</u>, by Gaydon (Ref. 5). When molecular values that are more recent than Gaydon's for the dissociation energy exist, the appropriate reference is cited. Dissociation energies taken directly from Gaydon, it should be noted, are not followed by a reference.

PERTURBATIONS AND GENERAL INFORMATION

This section encompasses all other information that would be useful for a complete understanding of the specific molecule. The information, where available, includes predissociations, perturbations, dipole moments, ionization potentials, potential energy curves, Franck-Condon factors, spontaneous lifetimes, rates of production, and deactivation and branching ratios. We hope that this section, in conjunction with the other information set forth, presents a complete description of the physical parameters associated with each molecular species.

BIBLIOGRAPHY

Following the format of Rosen (Ref. 1), in addition to presenting the references used to gather information, a short description of the important points of the paper is given. The bibliography takes into account most papers published through 1972.

Our referencing system is similar to that employed by Rosen in that the references are presented in terms of two numbers. The first number refers to the year of publication, and the second number refers to a running count of the references for each specific molecule. For references that were located after typing had begun, we have employed the designation L (e.g., (72.L1)).

III. NOTATION AND NOTATIONAL CONVERSION FORMULAS

The total energy of a given state of a diatomic molecule is given by the formula

$$T = T_e + G + F \tag{1}$$

where T_e is electronic energy, G is vibrational energy, and F is rotational energy. Further breaking down these different forms of energy, the electronic energy T_e is given by

$$T_e = T_0 + A\Lambda\Sigma \tag{2}$$

where T_0 is the electronic energy if spin is neglected, A is spin-orbit coupling, Λ is the electronic orbital angular momentum quantum number about the internuclear axis, and Σ is the component of the resulting spin. The vibrational energy G is given by

$$G = \omega_e(v + 1/2) - x_e \omega_e(v + 1/2)^2 + y_e \omega_e(v + 1/2)^3 + \ldots \tag{3}$$

where v is the vibrational quantum number, ω_e is the harmonic oscillator vibrational spacing, $x_e \omega_e$ is the first anharmonic correction to the vibrational spacing, and $y_e \omega_e$ is the second anharmonic correction. The rotational energy F is given by

$$F = B_v J(J + 1) - D_v J^2(J + 1)^2 + H_v J^3(J + 1)^3 + \ldots \tag{4}$$

where J is the rotational quantum number, B_v is the rigid rotator rotational spacing, D_v is the first anharmonic correction to the rotational spacing, and H_v is the second anharmonic correction. In addition, there are nonrigid rotator corrections to both B_v and D_v. These corrections are given by

$$B_v = B_e - \alpha_e(v + 1/2) + \gamma_e(v + 1/2)^2 + \ldots \tag{5}$$

and

$$D_v = D_e + \beta_e(v + 1/2)^2 + \ldots \tag{6}$$

where B_e is $\hbar^2/2\mu r_e^2$, μ is the reduced mass, r_e is the equilibrium internuclear distance, α_e and β_e are the first anharmonic corrections, and γ_e is the second anharmonic correction.

Using these formulas, a transition from state 1 at energy T_1, to state 2 at energy T_2 will be at an energy (cm^{-1}) of

$$v = T_1 - T_2 = (T_{e_1} - T_{e_2}) + (G_1 - G_2) + (F_1 - F_2) \tag{7}$$

Since, in general, the rotational energy changes are much smaller than either the vibrational or electronic changes, neglecting rotation,

$$\begin{aligned}v_{v', v''} = &\ v_e + \omega_e'(v + 1/2) - x_e'\omega_e'(v + 1/2)^2 + y_e'\omega_e'(v + 1/2)^3 + \ldots \\ &- [\omega_e''(v + 1/2) - x_e''\omega_e''(v + 1/2)^2 + y_e''\omega_e''(v + 1/2)^3 + \ldots]\end{aligned} \tag{8}$$

Assuming, as is often the case in absorption, $v' = v'' = 0$, and substituting in Eq. (8),

$$\begin{aligned}v_{v', v''} = &\ v_{00} + \omega_0'v' - x_0'\omega_0'v'^2 + y_0'\omega_0'v'^3 + \ldots \\ &- (\omega_0''v'' - x_0''\omega_0''v''^2 + y_0''\omega_0''v''^2 + \ldots)\end{aligned} \tag{9}$$

where

$$\omega_0 = \omega_e - x_e \omega_e + 3/4 y_e \omega_e + \ldots$$
$$x_0 \omega_0 = x_e \omega_e - 3/2 y_e \omega_e + \ldots$$
$$y_0 \omega_0 = y_e \omega_e + \ldots$$

A final quantity that is also reported is $\Delta G_{1/2}$. This quantity corresponds to the energy difference between vibrational levels $v = 0$ and $v = 1$, neglecting $y_0 \omega_0$, and is represented by

$$\Delta G_{1/2} = \omega_0 - x_0 \omega_0 = \omega_e - 2x_e \omega_e$$

Several other molecular constants that are reported for several molecules are represented in the following list:

f = oscillator strength (f-value)
λ, γ = spin-coupling constants for multiplet Σ states (cm^{-1})
q, p = Λ doubling constants (cm^{-1})
μ = electronic dipole moment (D)
R_b = branching ratio

IV. CONCLUSIONS ON THE AVAILABILITY OF SPECTROSCOPIC INFORMATION

It was our hope, when this search was initiated, to find sufficient information in the literature from which to draw definite conclusions regarding the feasibility of the production of an electronic transition chemical laser. As can be seen in the following table, our hopes were not fulfilled. At the present time there does not appear to be sufficient information about any diatomic molecule that would lead one to believe that molecule would definitely produce a chemically pumped electronic transition laser.

We have charted how much is known about the molecular systems we have researched. In general, very little information is available for any given system. The charts immediately precede the detailed information for the collected systems in each volume.

Sufficient information is available for the identification of the emitting species from potential laser systems. If, however, the experiment is to measure the partitioning of energy between the various accessible electronic levels of the molecule, molecular lifetimes and Franck-Condon factors must be known in order to interpret the data. To ascertain the feasibility of a specific molecule for a laser entails knowing not only the branching ratio, but also the lifetime, pumping rates, and deactivation rates. We do not yet have all of this information for a single molecule.

Knowing that, at present, there is almost a complete lack of the necessary information for producing a chemically pumped electronic transition laser, we realize that we are in the same situation as experimenters who were trying to produce vibrational lasers during the early 1960's. At that time, pumping rates, vibrational distributions, and deactivation rates were all unknown, as is now the case for electronic transition lasers. Also in analogy with vibrational lasers, however, the ultimate demonstration of a chemically pumped electronic transition laser should be possible.

The question then arises: What information is required and is there any ordering in the importance of the information? The answer becomes obvious when you remember that an inversion must be created before laser action can occur. Therefore, a knowledge of the branching ratio is absolutely necessary before predictions can be made as to the suitability of a molecule as a potential laser. Even if the lifetimes and pumping rates appear suitable, an inversion must exist before the system can lase. As previously mentioned, the interpretation of spectral intensities to determine the branching ratio for a particular transition requires a knowledge of both the Franck-Condon factors and radiative lifetime for the transition. The Franck-Condon factors, however, cannot be accurately calculated without accurate knowledge of the spectroscopic constants defining the two electronic states between which the transition takes place. So, at least from the experimental point of view, it is imperative that accurate spectroscopic constants and radiative lifetimes be known for the determination of branching ratios. From a theoretical viewpoint it may only be necessary to have accurate spectroscopic constants.

Whereas a sufficiently large branching ratio may be a necessary condition for a system to reach threshold, it is not sufficient. Even with a sufficiently large branching ratio, if the electronic state of interest is not produced in a time that is fast as compared to the rate at which it is being quenched, either by spontaneous emission or de-excitation, the required inversion cannot be produced. Consequently, once a particular set of reactants have been shown to produce a molecule with a sufficiently large branching ratio, either by experiment, theory, or spin conservation rules, deactivation studies of the molecule should be actively pursued as well as similar studies on other molecules of the same family (e.g., BaF, BaCl, SrF, and SrCl, or NF, NCl, PF, and PCl). Once a suitable branching ratio is found for a system and the important quenching rates measured, kinetic calculations can be made to determine proper operating conditions for the production of an inversion and laser action.

The above approach for determining the feasibility of chemically pumped electronic transition is sound. Whereas the actual production of the laser may prove to be quite difficult, the probability of finding one approaches unity. If one were to assign priorities to the information needed to assist the researcher, they would be as follows:

1. <u>Branching ratios</u>. To ascertain the maximum possible inversion

2. <u>Vibrational level distributions of the product molecules</u>. A total inversion may not exist between the entire upper and lower electronic levels, but it still can exist between specific vibration-rotation levels of the two states.

3. <u>Radiative lifetimes</u>. Necessary for the experimental determination of branching ratios and also for the dictation of minimum reaction rates

4. <u>Spectroscopic constants and dissociation energies</u>. Needed to calculate Franck-Condon factors

5. <u>Pumping and quenching rates</u>. Even if the inversion exists, it must do so on the proper time scale.

Lastly, while the straightforward research approach to the production of a laser may be aesthetically pleasing, the intuitive "shotgun" approach should not be discounted. The largest boost that could accelerate the discovery of new lasers is the discovery and understanding of the first chemically pumped electronic transition laser system.

Since the beginning of the keeping of sports statistics, no human being had run a four-minute mile. Once this record had been broken, however, it was not long before many had done so. Possibly this analogy will hold here.

REFERENCES

1. B. Rosen, ed., <u>Selected Constants - Spectroscopic Data Relative to Diatomic Molecules</u>, Pergamon Press, Oxford (1970).

2. G. Herzberg, <u>Molecular Spectra and Molecular Structure. I. Spectra of Diatomic Molecules</u>, D. Van Nostrand Co., Inc., Princeton (1950).

3. R. W. B. Pearse and A. G. Gaydon, <u>The Identification of Molecular Spectra</u>, Chapman and Hall Ltd., London (1965).

4. J. G. Phillips and S. P. Davis, <u>Berkeley Newsletters</u>, University of California, Berkeley (1960-present).

5. A. G. Gaydon, <u>Dissociation Energies and Spectra of Diatomic Molecules</u>, Chapman and Hall Ltd., London (1968).

SPECTROSCOPIC INFORMATION SUMMARY

MOLECULE	VIBRA-TIONAL CONSTANTS	ROTA-TIONAL CONSTANTS	VIBRATIONAL LEVEL DISTRIBU-TIONS	DISSO-CIATION ENERGY	LIFE-TIMES	FRANCK-CONDON FACTORS	BRANCH-ING RATIOS	QUENCH-ING	LASER ACTION OBSERVED VIBRA-TIONAL	LASER ACTION OBSERVED ELEC-TRONIC
AlBr	X	X		X						
AlCl	P	X		X						
AlF	P	X		X		X				X
AlH	P	P		X						
AlI	P	P		X						
AlO	P	X		X	P	P				
AlP				X						
AlS	P	X		X						
AlSe	X			X						
AsF	P	P								
AsN	P	X		X						
AsO	P	P		X		P				
AsP	X									
AsS	X	X								
BBr	X	X		X	X					
BC				X						
BCl	X	X		X	P	X				
BF	P	X		X	P	P				
BH	P	P		X						
BN	P	P		X		P				
BO	X	X		X		X				
BS	P	X		X						
BaBr	X		P	X	P		P			
BaCl	X		P	X	P		X			
BaF	X	P	P	X	P	P	X			
BaH	P	X		X						
BaI	X		P	X	P					
BaO	X	P	P	X	P	P	X	P		
BaS	P	X		X						
BeBr	X			X						
BeCl	X	X		X						

X=SUBSTANTIAL INFORMATION; P=SKETCHY INFORMATION; NO NOTATION=NO INFORMATION

MOLECULE	VIBRA-TIONAL CONSTANTS	ROTA-TIONAL CONSTANTS	VIBRATIONAL LEVEL DISTRIBUTIONS	DISSO-CIATION ENERGY	LIFE-TIMES	FRANCK-CONDON FACTORS	BRANCH-ING RATIOS	QUENCH-ING	LASER ACTION OBSERVED VIBRA-TIONAL	ELEC-TRONIC
BeF	P	P		X		P				
BeI	X	X								
BeO	X	X		X		P				
BeP				X						
BeS	P	P		X						
BiBr	X			X						
BiCl	P	P		X						
BiO	P	P		X						
BrCl	P	P		X						
CBr	P	P		X						
CCl	P	P		X						
CF	X	X		X	X	X			X	
CH	P	X		X	P	P				
CN	P	X		X	P	P		P	X	X
CO	P	P		X	P				X	X
CP	X	X		X		X				
CS	P	P		X	P	P				
CSe	P	X		X						
CaBr	X			X	P					
CaCl	X	P		X	P		P			
CaF	X	P		X	P		X			
CaI	X			X	P					
CaO	X	X		X		P				
CaS	X	X		X						
CeB				X						
CeN				X						
CeO		X		X						
CeS				X						
ClF	X	X		X						
CoBr	X									

X=SUBSTANTIAL INFORMATION; P=SKETCHY INFORMATION; NO NOTATION=NO INFORMATION

MOLECULE	VIBRA-TIONAL CONSTANTS	ROTA-TIONAL CONSTANTS	VIBRATIONAL LEVEL DISTRIBU-TIONS	DISSO-CIATION ENERGY	LIFE-TIMES	FRANCK-CONDON FACTORS	BRANCH-ING RATIOS	QUENCH-ING	LASER ACTION OBSERVED VIBRA-TIONAL	ELEC-TRONIC
CoCl	P			X						
CoF				X						
CoH		X								
CoO	P			X						
CoS				X						
CoSi				X						
CrBr				X						
CrCl	P			X						
CrF				X						
CrH	P	X		X						
CrI				X						
CrO	X	X		X		X				
CrS	P			X						
EuF				X						
EuO				X						
EuS				X						
EuSe				X						
EuTe				X						
FO				X						
FeBr	P			X						
FeCl	P			X						
FeF									X	
FeO	P	P		X	P					
FeS				X						
FeSi										
GaBr	P	P		X						
GaCl	P	P		X						
GaF	X	X		X						
GaH	P	X		X						
GaI	P	P		X						
GaP				X						

X=SUBSTANTIAL INFORMATION; P=SKETCHY INFORMATION; NO NOTATION=NO INFORMATION

MOLECULE	VIBRA-TIONAL CONSTANTS	ROTA-TIONAL CONSTANTS	VIBRATIONAL LEVEL DISTRIBU-TIONS	DISSO-CIATION ENERGY	LIFE-TIMES	FRANCK-CONDON FACTORS	BRANCH-ING RATIOS	QUENCH-ING	LASER ACTION OBSERVED VIBRA-TIONAL	ELEC-TRONIC
GeBr	P			X						
GeC				X						
GeCl	X	P		X						
GeF	X	P		X						
GeH	P	X		X						
GeI	P									
GeO	P	P		X		P				
GeS	P	P		X		P				
GeSe	X	P		X						
GeTe	X	P		X						
HfI										
HfO	X	P		X		P				
HoO				X						
HoS				X						
HoSe				X						
IBr	P	P		X						
ICl	P	P		X						
InBr	X			X						
InCl	P	X		X						
InF	P	P		X		P				
InH	P	X		X		P				
InI	P	P		X						
InO	X			X						
IrB				X						
IrC	P	P		X						
IrO				X						
IrSi				X						
LaF		X								
LaO	P	P		X		P				
LaS	X	X		X						
LaSe				X						

X=SUBSTANTIAL INFORMATION; P=SKETCHY INFORMATION; NO NOTATION=NO INFORMATION

MOLECULE	VIBRA-TIONAL CONSTANTS	ROTA-TIONAL CONSTANTS	VIBRATIONAL LEVEL DISTRIBU-TIONS	DISSO-CIATION ENERGY	LIFE-TIMES	FRANCK-CONDON FACTORS	BRANCH-ING RATIOS	QUENCH-ING	LASER ACTION OBSERVED VIBRA-TIONAL	ELEC-TRONIC
LaTe				X						
LuF	X									
LuH	P	P		X						
LuO	P	P		X						
LuS				X						
LuSe				X						
LuTe				X						
MgBr	X	P		X	P					
MgF	X	P		X	P	P	X		X	
MgO	X	P		X		P				
MnCl	P			X						
MnF	P			X						
MnH	P	X		X						
MnO	P			X						
MnS	X			X						
NBr	X	X		X						
NCl	P	P		X						
NF	P	X		X	P			P		
NH	P	X		X	P	X		P		
NI				X						
NO	X	X		X	P	X		P	X	X
NS	X	X		X		P				
NSe	P	X		X						
NbO	X	P		X		P				
NdF				X						
NdO				X						
NdS				X						
NdSe				X						
NdTe				X						
NiBr	P			X						
NiCl	P	P		X						

X=SUBSTANTIAL INFORMATION; P=SKETCHY INFORMATION; NO NOTATION=NO INFORMATION

MOLECULE	VIBRA- TIONAL CONSTANTS	ROTA- TIONAL CONSTANTS	VIBRA- TIONAL LEVEL DISTRIBU- TIONS	DISSO- CIATION ENERGY	LIFE- TIMES	FRANCK- CONDON FACTORS	BRANCH- ING RATIOS	QUENCH- ING	LASER ACTION OBSERVED VIBRA- TIONAL	ELEC- TRONIC
NiF				X					X	
NiH	P	X		X						
NiI				X						
NiO	P			X						
NiS				X						
NiSi				X						
OH	X	X		X	P	P		P	X	
PF	P	X		X						
PH		X		X	P					
PN	X	X		X		P				
PO	P	P		X		P				
PS	X	P		X						
PbCl	X	P		X						
PbF	P	P		X						
PbO	P	P		X	P	P				
PbS	P	P		X						
PbSe	X	P		X						
PdB				X						
PdH										
PdO				X						
PdSi				X						
PmS				X						
PrO				X						
PrS				X						
PtB				X						
PtC				X						
PtF									X	
PtH	P	P		X						
PtO	X	P		X						
PtSi				X						
PuF				X						

X=SUBSTANTIAL INFORMATION; P=SKETCHY INFORMATION; NO NOTATION=NO INFORMATION

MOLECULE	VIBRA-TIONAL CONSTANTS	ROTA-TIONAL CONSTANTS	VIBRATIONAL LEVEL DISTRIBUTIONS	DISSO-CIATION ENERGY	LIFE-TIMES	FRANCK-CONDON FACTORS	BRANCH-ING RATIOS	QUENCH-ING	LASER ACTION OBSERVED VIBRATIONAL	ELEC-TRONIC
ReO										
RhB				X						
RhC	P	P		X		P				
RhO				X						
RhSi				X						
RuB				X						
RuC	P	P		X						
RuO	P	X		X						
RuSi										
SF	P	P		X						
SO	P	X		X	P	P				
SbBr	X			X						
SbCl	X			X						
SbF	P	P		X						
SbO	P	P		X						
ScCl	P	P		X						
ScF	P	P		X	P					
ScO	X	X		X		P				
ScS				X						
SeO	P	P		X						
SiBr	P	P		X						
SiC				X						
SiCl	P	P		X		P				
SiF	P	P		X		P				
SiI	P	P		X						
SiN	P	P		X		P				
SiO	P	P		X	P	P				
SiS	X	X		X	P					
SiSe	X	P		X						
SiTe	X			X						
SmF				X						

X=SUBSTANTIAL INFORMATION; P=SKETCHY INFORMATION; NO NOTATION=NO INFORMATION

MOLECULE	VIBRATIONAL CONSTANTS	ROTATIONAL CONSTANTS	VIBRATIONAL LEVEL DISTRIBUTIONS	DISSOCIATION ENERGY	LIFETIMES	FRANCK-CONDON FACTORS	BRANCHING RATIOS	QUENCHING	LASER ACTION OBSERVED VIBRATIONAL	ELECTRONIC
SmO				X						
SmS				X						
SnCl	X			X						
SnF	P	P		X			P			
SnO	P	P		X	P		P			
SnS	P	P		X	P					
SnSe	P	P		X						
SnTe	X	P		X						
SrBr	P			X	P					
SrCl	X			X	P					
SrF	X	P		X	P		X			
SrI	X			X	P					
SrO	P	X		X		P				
SrS	X	X		X						
TaO	P	P		X						
TeO	P	P		X						
ThO	P	X		X		X				
ThP				X						
TiBr										
TiC				X						
TiCl	P			X						
TiF				X					X	
TiH										
TiI										
TiN	P	X		X						
TiO	P	P		X	P	P			X	
TiS	P	X		X						
TlF	P	P		X						
UB				X						
UF									X	
UN				X						

X=SUBSTANTIAL INFORMATION; P=SKETCHY INFORMATION; NO NOTATION=NO INFORMATION

MOLECULE	VIBRA-TIONAL CONSTANTS	ROTA-TIONAL CONSTANTS	VIBRATIONAL LEVEL DISTRIBUTIONS	DISSOCIATION ENERGY	LIFE-TIMES	FRANCK-CONDON FACTORS	BRANCHING RATIOS	QUENCHING	LASER ACTION OBSERVED VIBRATIONAL	ELECTRONIC
UO				X					X	
US				X						
VCl										
VO	X	X		X	P	P				
VS				X						
WO	P			X						
YCl	X	X		X						
YF	P	X		X						
YO	X	X		X		X				
YS				X						
ZrBr										
ZrCl										
ZrI										
ZrN				X						
ZrO	P	P		X		P				
ZrS				X						

X=SUBSTANTIAL INFORMATION; P=SKETCHY INFORMATION; NO NOTATION=NO INFORMATION

LaF

Methods of Production and Experimental Technique

Absorption (La + AlF$_3$) in a King furnace.
Thermal emission.

Band Systems

In the region $9000\,\text{Å} > \lambda > 4000\,\text{Å}$ a large number of red degrading bands are observed. Two system series are observed in absorption at 2000°C: singlets based on the $X^1\Sigma^+$ state, and triplets based on the $a^3\Delta$ state.

Characteristic band heads, (0, 0), λ:

Transition	R Head	Q Head	Transition	R Head
$^3\Phi_2 - {}^3\Delta_1$	5461.22	5464.22	$^1\Sigma^+ - X^1\Sigma^+$	8570.32
$^3\Phi_3 - {}^3\Delta_2$	5395.57	5398.10	$^1\Pi - X^1\Sigma^+$	6175.84
$^3\Phi_4 - {}^3\Delta_3$	5312.00	5315.01	$^1\Pi - X^1\Sigma^+$	4768.03
			$^1\Sigma^+ - X^1\Sigma^+$	4446.09
			$^1\Sigma^+ - X^1\Sigma^+$	4428.59

Emission bands in the region $3625\,\text{Å} > \lambda > 3200\,\text{Å}$ have also been observed (65.1).

LaF

SPECTROSCOPIC CONSTANTS

State	T_{oo}	$\Delta G_{1/2}$	B_o	$\alpha_e \times 10^3$	$D_e \times 10^7$	r_e	Remarks	Bibliography
	Triplet States							
$^3\phi_4$	22340.20 + a_3	-	0.2302	-	1.7	-		(n. p. 3)
$^3\phi_3$	22149.20 + a_2	-	0.2294	-	2.0	-		(n. p. 3)
$^3\phi_2$	21834.46 + a_1	-	0.2277	-	1.7	-		(n. p. 3)
$^3\phi_4$	18809.40 + a_3	483	0.2336	-	2.2	-		(n. p. 3, n. p. 4)
$^3\phi_3$	18519.90 + a_2	473.2	0.2321	-	2.3	-		(n. p. 3, n. p. 4)
$^3\phi_2$	18295.80 + a_1	484.6	0.2322	-	1.9	-		(n. p. 3, n. p. 4)
$^3\Delta_2$	13597.27 + a_2	-	0.2191	-	1.9	-		(n. p. 3, 67.2)
$^3\Delta_1$	13567.00 + a_1	-	0.2189	-	2.0	-		(n. p. 3, 67.2)
$^3\Delta_2$	13316.3 + a_2	-	0.2216	-	2.9	-		(n. p. 3)
$a_3\,^3\Delta_3$	a_3	537	0.23875	-	1.87	-		(n. p. 3, n. p. 4, 67.2)
$a_2\,^3\Delta_2$	a_2	537.6	0.23842	1.21	1.88	-		(n. p. 3, n. p. 4, 67.2)
$a_1\,^3\Delta_1$	a_1	537.1	0.23771	1.20	1.79	2.057		(n. p. 3, n. p. 4, 67.2)

LaF

SPECTROSCOPIC CONSTANTS

State	T_{oo}	$\Delta G_{1/2}$	B_o	$\alpha_e \times 10^3$	$D_e \times 10^7$	r_e	Remarks	Bibliography
	Singlet States							
$^1\Sigma^+$	(22574.22)	(421)	0.229	–	–	–	Strong perturbations between these states.	(n.p. 3)
$^1\Sigma^+$	(22485.35)	–	0.242	–	–	–		(n.p. 3)
$^1\Pi$	20959.83	549	0.2374	–	1.6	–	Λ doubling	(n.p. 3, n.p. 4)
$^1\Pi$	16184.00	474	0.2293	–	2.7	–	Λ doubling	(n.p. 3, n.p. 4)
$^1\Sigma^+$	11661.9	489.4	0.2273	1.1	2.1	–		(n.p. 3, 67.2)
$X^1\Sigma^+$	0	570	0.2456	–	1.7	2.0254		(n.p. 3, n.p. 4, 67.2)

Dissociation energy is unknown.

BIBLIOGRAPHY

(65. 1) Emission,
E. A. Shenyavskaya, L. V. Gurvich, and A. A. Mal'stev,
<u>Vestnik Moskov. Univ., Ser. II: Khim.</u> <u>20</u>, No. 4, 10-3

(67. 2) Preliminary Rotational Analysis,
R. F. Barrow, M. W. Bastin, D. L. G. Moore, and C. J. Pott,
<u>Nature</u> <u>215</u>, 1072-3

(n. p. 3) Rotational Analysis,
R. F. Barrow, L. Lee, and D. J. Partridge,
(unpublished)

(n. p. 4) Vibrational Analysis,
R. Hauge,
(unpublished)

LaO

Methods of Production and Experimental Technique

Absorption in a furnace from a carbon tube (t ~ 2000°C): in a rare gas matrix at T° = 4°K (67.21).

Emission from arcs and flames; carbon flame (57.12).

Astrophysics: stellar absorption.

Emission from an La anode and Cu cathode in an O_2 atmosphere (71.34, 71.35).

Emission from a King furnace (69.27).

Emission from a discharge in Lanthanum acetate (72.36, 70.29).

Low pressure arc.

Hollow cathode lamp.

BAND SYSTEMS

System	Transition	Sources	Wavelength Limits	Degrading	Characteristic Bands, λ	Remarks	Bibliography
I	$A^2\Pi \rightleftarrows X^2\Sigma^+$	Carbon flame	9730-6850	R	(0, 0) 7877.2(R_1) 7910.5(Q_1) 7379.8($^2R_{21}$) 7403.5($^RQ_{21}$)		(67.21, 62.14, 57.12, 31.4, 31.5)
II	$B^2\Sigma^+ \rightleftarrows X^2\Sigma^+$	Carbon flame	6450-5015	R	5602.4 5599.9 ∣ (0, 0)		(70.29, 67.21, 62.14, 57.12, 31.4, 31.5)
III	$C^2\Pi_r \rightleftarrows X^2\Sigma^+$	Carbon flame	4625-4350	R	(0, 0) 4418.2(Q_1) 4372.0($^RQ_{21}$)		(71.31, 71.34, 69.27, 67.21, 59.13, 57.12, 31.4, 31.5)
IV	$D^2\Sigma^+ \rightarrow X^2\Sigma^+$	Discharge, flame	3715-3510	V [a]	3604.4(0, 0) 3608.1(1, 1)		(71.32, 70.28, 59.13, 57.12, 31.5)
V	$F^2\Sigma^+ \rightarrow X^2\Sigma^+$	Discharge, flame	3670-3450	V	3566.1 3565.8 ∣ (0, 0)		(71.32, 59.13, 57.12)
VI	$C^2\Pi \rightarrow A'^2\Delta$	Discharge	6825-6430	V	6607.7 6605.1 6821.5		(71.34, 71.35, 59.13, 57.12)
VII	H - ?	Flame	~3710	V	3709.7		(59.13, 57.12)

[a] Vibrational structure degrades R.

LaO

I. $A^2\Pi \rightleftarrows X^2\Sigma^+$ System

Bands of greatest intensity (31.4):

$$^2\Pi_{1/2} \rightarrow {}^2\Sigma_{1/2}$$

v', v''	3, 3	2, 2	1, 1	0, 0	0, 0
Branches	Q_1	Q_1	Q_1	Q_1	R_1
λ	8014.78	7979.68	7944.93	7910.50	7877.18
Intensity	6	8	10	15	10

$$^2\Pi_{3/2} \rightarrow {}^2\Sigma_{1/2}$$

v', v''	3, 3	3, 3	2, 2	2, 2	1, 1	0, 0	0, 0
Branches	$^RQ_{21}$	$^SR_{21}$	$^RQ_{21}$	$^SR_{21}$	$^RQ_{21}$	$^RQ_{21}$	$^SR_{21}$
λ	7496.50	7474.83	7465.25	7442.92	7434.28	7403.52	7379.84
Intensity	6	5	8	6	10	20	15

II. $B^2\Sigma^+ \rightleftarrows X^2\Sigma^+$ System

Bands of greatest intensity (31.4):

v', v''	λ	Intensity	v', v''	λ	Intensity
3, 4	5951.26	5	1, 1	5628.60	10
				5626.03	10
2, 3	5923.97	7			
	5920.84	5		5602.50	20
			0, 0	5602.36	10
1, 2	5896.67	8		5600.02	10
	5893.57	6		5599.89	5
0, 1	5869.50	5	1, 0	5380.4	
	5866.3				
2, 2	5654.82	5			
	5652.34	5			

III. $C^2\Pi \rightleftarrows X^2\Sigma^+$ System

Two intense (0,0) sequences characterize the system. C state exhibits large Λ doubling (71.31).

IV. $D^2\Sigma^+ \rightleftarrows X^2\Sigma^+$ System

Band heads (70.28, 59.13, 57.12):

(v', v'')	(0, 1)	(1, 1)	(0, 0)	(1, 0)
λ	3713.0	3608.1	3604.4	3505.7

V. $F^2\Sigma^+ \rightarrow X^2\Sigma^+$ System

Band heads (71.32, 59.13, 57.12):

(v', v'')	(0, 0)	(1, 1)	(1, 0)
λ	3566.1 3565.8	3561.0	3460.9

VI. $C^2\Pi \rightarrow A'^2\Delta$ System

Band heads of greatest intensity:

$C^2\Pi_{1/2} \rightarrow A'^2\Delta_{3/2}$

(v', v'')	(0, 0)	(1, 1)	(2, 2)
λ	6607.8	6597.1	6586.4

$C^2\Pi_{3/2} \rightarrow A'^2\Delta_{5/2}$

(v', v'')	(0, 0)	(1, 1)	(2, 2)
λ	6821.5	6808.9	6796.3

LaO

SPECTROSCOPIC CONSTANTS

State	T_e	ω_e	$\omega_e x_e$	B_e	$\alpha_e \times 10^3$	$D_e \times 10^6$	r_e	Remarks	Bibliography
H	26950+x	–	–	–	–	–	–		(59.13, 57.12)
h	x	–	–	–	–	–	–		(59.13, 57.12)
$F^2\Sigma^+$	28049.2[a]	~850	–	0.3618	2.4	0.259	1.802		(71.32, 59.13)
$D^2\Sigma^+$	26960.0[a]	~786	–	0.3643	1.9	0.310	1.796		(71.32, 70.28, 59.13)
$C^2\Pi_{3/2}$	22847.7	801.14	3.291	0.3471	2.3	0.275	1.84	$y_e\omega_e = 0.0231$ cm^{-1}	(71.31, 71.34, 69.27)
$C^2\Pi_{1/2}$	22631.3	792.37	2.919	0.3405	1.7	–		$y_e\omega_e = 0.0187$ cm^{-1}	(71.31, 71.34, 69.27)
$B^2\Sigma^+$	17884.9	696.0	1.83				1.855		(70.29, 67.21)
$A^2\Pi_{3/2}$	13497.63	757.2[b]	2.07	0.3463	1.7	–	1.84		(67.21, 62.14, 57.12)
$A^2\Pi_{1/2}$	12635.65		2.18						(67.21, 62.14, 57.12)
$A'^2\Delta_{5/2}$	8190.1	773.87	3.229	–	–	–	–	$y_e\omega_e = 0.0131$ cm^{-1}	(71.34, 71.35)
$A'^2\Delta_{3/2}$	7493.4	768.20	2.990	–	–	–	–	$y_e\omega_e = 0.0296$ cm^{-1}	(71.34, 71.35)
$X^2\Sigma^+$	0	817.26	3.097	0.3519[c]	1.4	–	1.825	$y_e\omega_e = 0.0406$ cm^{-1}	(71.34, 67.21)

[a] T_0, [b] $\Delta G_{1/2}$, [c] B_0.

Dissociation energy = 8.25 ± 0.11 eV, 190.3 kcal/mole, 66542 cm^{-1} (67.22, 67.23).

Perturbations and General Information

Potential energy curves — RKR potential (70.30):

State	v	r_{min} (Å)	r_{max} (Å)
$B^2\Sigma^+$	0	1.801	1.915
	1	1.764	1.961
	2	1.740	1.995
	3	1.722	2.023
	4	1.706	2.049
	5	1.692	2.072
	6	1.680	2.094
	7	1.669	2.115
	8	1.660	2.135
	9	1.650	2.154
	10	1.642	2.173

Franck-Condon factors — Morse potential (64.16):

$A^2\Pi_{3/2} - X^2\Sigma$

v', v''	0	1	2	3	4	5	6
0	0.96210	0.03643	0.00145	0.00002	0.00000	0.00000	0.00000
1	0.03762	0.88808	0.06994	0.00427	0.00009	0.00000	0.00000
2	0.00027	0.07454	0.81609	0.10050	0.00838	0.00022	0.00001
3	0.00001	0.00094	0.11045	0.74639	0.12809	0.01371	0.00043
4	0.00000	0.00002	0.00201	0.14505	0.57924	0.15274	0.02017
5	0.00000	0.00001	0.00004	0.00365	0.17807	0.61487	0.17444
6	0.00000	0.00000	0.00003	0.00007	0.00592	0.20930	0.55413
7	0.00000	0.00000	0.00000	0.00006	0.00009	0.00896	0.24405

$A^2\Pi_{1/2} - X^2\Sigma$

v', v''	0	1	2	3	4	5	6
0	0.96173	0.03686	0.00138	0.00002	0.00000	0.00000	0.00000
1	0.03794	0.88639	0.07145	0.00413	0.00009	0.00000	0.00000
2	0.00033	0.07564	0.81200	0.10357	0.00822	0.00024	0.00001
3	0.00000	0.00109	0.11271	0.73903	0.13306	0.01361	0.00048
4	0.00000	0.00001	0.00242	0.14875	0.66792	0.15977	0.02024
5	0.00000	0.00001	0.00002	0.00443	0.18336	0.59909	0.18359
6	0.00000	0.00000	0.00002	0.00002	0.00724	0.21621	0.53347
7	0.00000	0.00000	0.00000	0.00003	0.00002	0.01107	0.25461

LaO

$B^2\Sigma - X^2\Sigma$

v', v''	0	1	2	3	4	5	6
0	0.86268	0.12295	0.01340	0.00092	0.00005	0.00000	0.00000
1	0.13074	0.62559	0.20435	0.03573	0.00335	0.00024	0.00001
2	0.00654	0.23181	0.43796	0.23210	0.06327	0.00762	0.00066
3	0.00003	0.01937	0.30557	0.29329	0.27332	0.09304	0.01385
4	0.00001	0.00019	0.03789	0.35484	0.18526	0.27432	0.12264
5	0.00000	0.00006	0.00063	0.06118	0.38274	0.10783	0.26062
6	0.00000	0.00002	0.00013	0.00155	0.08813	0.39250	0.05522
7	0.00000	0.00000	0.00007	0.00021	0.00320	0.11715	0.37768
8	0.00000	0.00000	0.00001	0.00016	0.00032	0.00449	-

BIBLIOGRAPHY

(29.1) Preliminary Note,
G. Piccardi,
Nature 124, 129

(29.2) All the Systems,
W. Jevons,
Proc. Phys. Soc. 41, 520-45

(30.3) Preliminary Note,
J. Querbach,
Z. Physik 60, 109-24

(31.4) Spectral Reproduction,
W. F. Meggers and J. A. Wheeler,
Bur. Stand. J. Res. 6, 239-75

(31.5) Interpretation,
L. W. Johnson and R. C. Johnson,
Proc. Roy. Soc. A 133, 207-19

(32.6) Comparison with Analogous Molecules,
F. A. Jenkins and A. Harvey,
Phys. Rev. 39, 922-31

(33.7) Excitation in Flame,
G. Piccardi,
Gazz. Chim. 63, 127-38

(38. 8)   W. W. Watson,
          Phys. Rev. 53, 678

(39. 9)   Spectral Analysis,
          G. Piccardi,
          Spectrochim. Acta 1, 249-60

(42. 10) Spectral Reproduction,
          A. Gatterer,
          Ricerche Spettroscop. Lab. Astrofis. Specola Vaticana 1, 153-79

(48. 11) Stellar Absorption,
          P. C. Keenan,
          Astrophys. J. 107, 420-1

(57. 12) Vibrational Analysis. Spectral Reproduction,
          A. Gatterer, J. Junkes, E. W. Salpeter, and B. Rosen,
          "Molecular Spectra of Metallic Oxides,"
          Ed. Specola Vaticana, 1957

(59. 13) Vibrational Analysis,
          S. Hautecler and B. Rosen,
          Bull. Cl. Sci. Acad. Roy. 45, 790-803

(62. 14) Systems A, B-X, Rotational Analysis,
          L. Akerlind,
          Arkiv Fysik 22, 65-93

(62. 15) A, B-X Systems, Rotational Analysis,
          L. Akerlind,
          "Rotational Analysis of the Band Systems of LaO,"
          Naturwissenschaften 49, 7

(64. 16) Franck-Condon Factors, A, B-X Systems,
          F. S. Ortenberg, V. B. Glasko, and A. I. Dmitriev,
          "Vibrational Transition Probabilities for Some Band Systems
          of Diatomic Molecules II,"
          Sov. Astron-AJ 8, 258-61

(65. 17) Electric Deflection of Molecular Beam, Relation of Pure
          Precision Between $A^2\pi$ and $B^2\Sigma^+$,
          R. A. Berg, L. Wharton, W. Klemperer, A. Büchler and
          J. L. Stauffer,
          J. Chem. Phys. 43, 2416-21

(65. 18) Fluorescence in Molecular Beam, Ground State of LaO,
          L. Brewer and R. M. Walsh,
          J. Chem. Phys. 42, 4055

LaO

(65.19) P. H. Rasai and W. Weltner, Jr.,
"Ground States and Hyperfine-Structure Separations of ScO, YO and LaO from ESR Spectra at $4°K$,"
J. Chem. Phys. 43, 2553

(65.20) Exhaustive Study of all Systems, Lifetime,
R. M. Walsh,
"Molecular-Beam Studies of Lanthanum Monoxide,"
Ph.D. Thesis, U. C. Berkeley, April 1965

(67.21) Optical Absorption Spectrum and Electron Spin Resonance in Matricies of Rare Gases at $4°K$,
W. Weltner Jr., D. MacLeod Jr., and P. H. Kasai,
J. Chem. Phys. 46, 3172-84

(67.22) Dissociation Energy,
L. L. Ames, P. N. Walsh, and O. White,
J. Phys. Chem. 71, 2707-18

(67.23) Dissociation Energy,
P. Coppens, S. Smoes, and J. Drowart,
Trans. Faraday Soc. 63, 2140-8

(67.24) I. V. Veitz and L. V. Gurvich,
"Absorption Spectra of Molecules of Substance not Readily Volatile and of Radicals in Shock Waves,
Dokl. Akad. Nank. S.S.S.R. 173, 1325-7

(68.25) R. Bacis,
"Very High Resolution Study of Molecular Spectra with the aid of a Fabry-Perot Photoelectric Spectrometer and a Hollow Cathode Lamp Made of an Alloy: Case of the Oxide LaO,"
C. R. Acad. Sci. 266B, 1071-4

(69.26) N. S. Murthy and B. N. Murthy,
"Integrated Intensities of the Green-Yellow System of Lanthanum Monoxide Bands,"
Nature 223, 181-2

(69.27) C-X System Analysis, Possibly Incorrect,
P. Carette and J. Blondeau,
"Rotational Analysis of the Indigo System Emission Spectrum of the LaO Molecule,"
C. R. Acad. Sci. 268B, 1743-5

LaO

(70.28) D, F → X Systems, Vibrational Analysis,
P. Caratte and R. Houdart,
"High Resolution Study of the Vibrational Structure of the Ultraviolet Emission System of LaO,"
C. R. Acad. Sci. 271B, 110-2

(70.29) B-X System, Vibrational Analysis,
C. B. Suarez,
"Vibrational Analysis of the Green System of $La^{18}O$,"
J. Phys. B 3, 729-31

(70.30) RKRV Potential Energy Curves - $B^2\Sigma^+$,
N. S. Murthy and B. N. Murthy,
"True Potential Energy Curves for LaO, VO and CP,"
J. Phys. B (Atom Mol. Phys.) 3, L16-L18

(71.31) C-X System, Rotational-Vibrational Analysis,
D. W. Green,
"Rotational Analysis of the $C^2\pi \rightarrow X^2\Sigma^+$ Electronic Transition of LaO,"
Can. J. Phys. 49, 2552-64

(71.32) P. Carette and R. Houdart,
"Rotational Analysis of the D and F Ultraviolet Emission Systems of the LaO Molecule,"
C. R. Acad. Sci. 272B, 595-8

(71.33) Comparison to Isoelectronic Species,
D. W. Green,
"Low-Lying Electronic States of the ScO, YO and LaO Molecules,"
J. Phys. Chem. 75, 3103-6

(71.34) D. W. Green,
"Vibrational Constants of the $C^2\pi$, $A'\,^2\Delta$, and $X^2\Sigma^+$ Electronic States of LaO,"
J. Mol. Spectrosc. 40, 501-10

(71.35) D. W. Green,
"Rotational Analysis of the $C^2\pi_r \rightarrow A'\,^2\Delta_r$ Electronic Transition in LaO,"
J. Mol. Spectrosc. 38, 155-74

LaO

(72.36) Franck-Condon Factors, r-centroids, B-X,
N. S. Murthy and B. N. Murthy,
"Intensity Measurements and Relative Band Strengths of the Bands of (B-X) System of LaO,"
J. Phys. B. (Atom. Mol. Phys.) 5, 714

(72.L1) C. B. Suarez,
"Rotational Constants of the B-X System of La^{18}O Band Spectrum,"
Chem. Phys. Letters 16, 515-6

LaS

Methods of Production and Experimental Technique

High temperature carbon tube furnace with La and ZnS (70.5).

BAND SYSTEMS

System	Transition	Sources	Wavelength Limits	Degrading	Band Head, $\nu_{0,0}$	Remarks	Bibliography
I	$B^2\Sigma^+ \to X^2\Sigma^+$	Furnace	8500-7000	-	14182.3(R_{23}) (1,0)		(70.5)
II	$? \to X^2\Sigma^2$	Furnace	4500-4200	-			(70.5)

LaS

I. $B^2\Sigma^+ \to X^2\Sigma^+$ System

Four R and four P heads. Bands of greatest intensity:

v', v''	Head	λ(Å)	v', v''	Head	λ(Å)
1, 0	R_{23}	7051.04	0, 0	R_{13}	7260.99
	R_{24}	7051.23		R_{14}	7261.24
	R_{13}	7051.89			
	R_{14}	7052.12	0, 1	R_{23}	7507.92
2, 1	R_{23}	7075.20		R_{24}	7508.17
	R_{24}	7075.44		R_{13}	7509.06
	R_{13}	7076.05		R_{14}	7509.32
	R_{14}	7076.29	1, 2	R_{23}	7533.19
0, 0	R_{23}	7260.01		R_{24}	7533.44
	R_{24}	7260.25		R_{13}	7534.31
				R_{14}	7534.56

SPECTROSCOPIC CONSTANTS

State	T_e	ω_e	$x_e\omega_e$	B_o	$\alpha_e \times 10^3$	$D_o \times 10^8$	r_e	Remarks	Bibliography
$B^2\Sigma^+$	13766.86	410.07	0.94	0.11099	–	3.1	2.417	$Y_o = -0.09627$ cm^{-1}	(70.5)
$X^2\Sigma^+$	0	456.7	0.96	0.11693	–	3.0	2.355		(70.5)

Dissociation energy = 5.90 ± 0.12 eV, 136.0 kcal/mole, 47588 cm^{-1} (68.4).

BIBLIOGRAPHY

(65.1) Dissociation Energy,
E. D. Cater, T. E. Lee, E. W. Johnson, E. G. Rauh, and H. A. Eick,
J. Phys. Chem. <u>69</u>, 2684-9

(67.2) Dissociation Energy,
P. Coppens, S. Smoes, and J. Drowart,
Trans. Faraday Soc. <u>63</u>, 2140-8

(68.3) Dissociation Energy,
E. D. Cater and R. P. Steiger,
J. Phys. Chem. <u>72</u>, 2231-3

(68.4) Dissociation Energy,
R. P. Steiger,
"Mass Spectrometric Investigation of the Vaporization Thermodynamics and Dissociation Energies of LaS, ScS, YS, ZrS, and UO,"
Disc. Abstr. <u>29B</u>, 2009

(70.5) M. Marcano and R. F. Barrow,
"Analysis of the Transition $B^2\Sigma^+ - X^2\Sigma^+$ in Gaseous LaS,"
J. Phys. B (Atom. Mol. Phys.) <u>3</u>, L121-3

LaSe

Dissociation energy = 4.89 ± 0.15 eV, 39440 cm^{-1} (72.1).

BIBLIOGRAPHY

(72.1)   M. Shafi, C. L. Beckel, and R. Engelke,
"Diatomic Molecule Ground State Dissociation Energies,"
J. Mol. Spectrosc. 42, 578-581

LaTe

Dissociation energy = 3.92 ± 0.15 eV, 31618 cm^{-1} (72.1).

BIBLIOGRAPHY

(72.1) M. Shafi, C. L. Beckel, and R. Engelke,
"Diatomic Molecule Ground State Dissociation Energies,"
J. Mol. Spectrosc. 42, 578-581

LuF

Methods of Production and Experimental Technique

Emission from a hollow cathode lamp lined with an alloy of LuF_3 + Ca (72.1).

BAND SYSTEMS

System	Transition	Sources	Wavelength Limits	Degrading	Band Head, $\nu_{0,0}$	Remarks	Bibliography
I	$A^1\Sigma \to X^1\Sigma$	Hollow cathode	-	R	16169.4(0,0)	Bands observed from 8000 – 3000 Å	(72.1)
II	$B^1\Pi \to X^1\Sigma$	Hollow cathode	-	R	16801.2(0,0)		(72.1)
III	$C \to X^1\Sigma$	Hollow cathode	-	R	18891.5(0,0)		(72.1)
IV	$D^1\Pi \to X^1\Sigma$	Hollow cathode	-	R	20034.7(0,0)		(72.1)
V	$E^1\Pi \to X^1\Sigma$	Hollow cathode	-	R	24446.0(0,0)		(72.1)
VI	$F^1\Sigma \to X^1\Sigma$	Hollow cathode	-	R	25813.4(0,0)	Bands observed from 8000 – 3000 Å	(72.1)
VII	$G^1\Sigma \to X^1\Sigma$	Hollow cathode	-	R	33220.1(0,0)		(72.1)

LuF

SPECTROSCOPIC CONSTANTS

State	T_o	ω_e	$x_e\omega_e$	B_e	$\alpha_e \times 10^3$	$D_e \times 10^6$	r_e	Remarks	Bibliography
$G^1\Sigma$	33220.1	599.1	2.6	-	-	-	-		(72.1)
$F^1\Sigma$	25813.4	559.5	2.6	-	-	-	-		(72.1)
$E^1\Pi$	24446.0	542.6	2.3	-	-	-	-		(72.1)
$D^1\Pi$	20034.7	569.7	2.5	-	-	-	-		(72.1)
C	18891.5	605.5	2.5				-		(72.1)
$B^1\Pi$	16801.2	576.3	2.5	-	-	-	-		(72.1)
$A^1\Sigma$	16169.4	586.4	2.5	-	-	-	-		(72.1)
$X^1\Sigma$	0	610.6	2.5						

Dissociation energy is unknown.

BIBLIOGRAPHY

(72.1)   A, B, C, D, E, F, and G→X Systems, Vibrational Analysis,
J. D'Incan, C. Effantin, and R. Bacis,
"Electronic Spectrum of the LuF Molecule,"
J. Phys. B (Atom. Mol. Phys.) 5, L189-90

LuH

Methods of Production and Experimental Technique

Excitation in a hollow cathode of Lu_2O_3 in the presence of hydrogen (72.1).

BAND SYSTEMS

System	Transition	Sources	Wavelength Limits	Degrading	Band Head, $\nu_{0,0}$	Remarks	Bibliography
I	$A^1\Pi - X^1\Sigma$	Hollow cathode	~7800	-	12998.36(0,0)		(73.3, 72.2)
II	$B^1\Sigma - X^1\Sigma$	Hollow cathode	~6500	-	15279.05(0,0)		(73.3, 72.2)
III	$C^1\Pi - X^1\Sigma$	Hollow cathode	~6000	-	16730.49(0,0)		(73.3, 72.2)
IV	$D^3\Pi_1 - X^1\Sigma$	Hollow cathode	~5800	-	17050.8(0,0)		(73.3, 72.2)
V	$E^1\Sigma - X^1\Sigma$	Hollow cathode	~5600	-	17733.0(0,0)		(73.3, 72.2)
VI	$F^1\Pi - X^1\Sigma$	Hollow cathode	~5300	-	18921.4(0,0)		(73.3, 72.2)
VII	$G^1\Sigma - X^1\Sigma$	Hollow cathode	~5100	-	19767.02(0,0)		(72.1, 72.2)
VIII	$H^1\Pi - X^1\Sigma$	Hollow cathode	~4300	-	23524.96(0,0)		(72.1, 72.2)

I. $A^1\Pi \rightarrow X^1\Sigma$ System

Band systems, λ (73.3):

(v', v'')	(0, 0)	(1, 1)	(2, 2)
λ	7699.0	7749.3	7806.4

II. $B^1\Sigma \rightarrow X^1\Sigma$ System

Band systems, λ (73.3):

(v', v'')	(0, 0)	(1, 1)
λ	6548.8	6590.6

III. $C^1\Pi \rightarrow X^1\Sigma$ System

Band systems, λ (73.3):

(v', v'')	(0, 0)	(1, 1)
λ	5980.2	6027.4

IV. $D^3\Pi_1 \rightarrow X^1\Sigma$ System

Band systems, λ (73.3):

(v', v'')	(0, 0)	(1, 1)
λ	5865.1	5910.1

V. $E^1\Sigma \rightarrow X^1\Sigma$ System

Band systems, λ (73.3):

(v', v'')	(0, 0)	(1, 1)
λ	5639.2	5651.1

VI $F^1\Pi \rightarrow X^1\Sigma$ System

Band systems, λ (73.3):

(v', v'')	(0, 0)
λ	5285.0

LuH

SPECTROSCOPIC CONSTANTS

State	T_{oo}	ω_e	$x_e\omega_e$	B_e	α_e	$D_e \times 10^4$	r_e	Remarks	Bibliography
$H^1\Pi$	23525.00	1500	38	4.6139	0.1376	1.69	1.909	$\beta_e = 10 \times 10^{-6}$ cm^{-1}	(73.3, 72.1, 72.2)
$G^1\Sigma$	19767.00	1450	26	4.5458	0.1205	1.78	1.9229	$\beta_e = 11 \times 10^{-6}$ cm^{-1}	(73.3, 72.1, 72.2)
$F^1\Pi$	18921.49	-	-	-	-	-	-		(73.3, 72.2)
$E^1\Sigma$	17732.92	-	-	-	-	-	-		(73.3, 72.2)
$D^1\Pi$	17050.1	1400	29	4.5263	0.1340	2.2	1.9271		(73.3, 72.2)
$C^1\Pi$	16721.9	-	-	-	-	-	-		(73.3, 72.2)
$B^1\Sigma$	15270.00	1400	24	4.5723	0.1171	1.95	1.9173		(73.3, 72.2)
$A^1\Pi$	12988.63	1445	27	-	-	-	-		(73.3, 72.2)
$X^1\Sigma$	0	1520	22	4.6021	0.0990	1.69	1.9111	$\beta_e = 5 \times 10^{-6}$ cm^{-1}	(73.3, 72.2)

Dissociation energy ~ 26200 cm^{-1}.

BIBLIOGRAPHY

(72.1)   H, G→X Systems, Vibrational Rotational Analysis,
J. D'Incan, C. Effantin, and R. Bacis,
"New Electronic Transitions Assigned to the LuH Molecule,"
Can. J. Phys. 50, 1810-4

(72.2)   A, B, C, D, E, F, G, H-X Systems,
J. D'Incan, C. Effantin, and R. Bacis,
"New Electronic Spectra Attributed to the LuH and LuD Molecules,"
J. Phys. B (Atom. Mol. Phys.) 5, L187-8

(73.3)   Rotational and Vibrational Constants,
C. Effantin and J. D'Incan,
"Electronic Spectra of the LuH Molecule,"
Can. J. Phys. 51, 1394-1402

LuO

Methods of Production and Experimental Technique

Emission from an arc or flame with added Lu salts.

BAND SYSTEMS

	System	Transition	Sources	Wavelength Limits	Degrading	Band Head, $\nu_{0,0}$	Remarks	Bibliography
Violet	I	$A \to X^2\Sigma$	Flame	5500-5100	R	19335.9		(57.5, 52.4, 42.3)
	II	$B \to X^2\Sigma$	Flame	4950-4510	R	21445.2		(57.5, 42.3, 38.1)
	III	$C^2\Sigma \to X^2\Sigma$	Arc	4260-3980	R	24419.2		(73.L2, 71.L1, 52.4)

I. $A \rightarrow X^2\Sigma$ System

Most intense band has a head in the region of 5170 Å with complex structure, hence analysis is uncertain.

Characteristic band heads (57.5):

λ	5486.6	5464.0	5448.7	5402.6	5217.7	5195.9	5192.4
	5170.1	5161.0	5119.8	5041.3	5034.6	5019.2	5003.5

II. $B \rightarrow X^2\Sigma$ System

Simple, diffuse bands with $\Delta v = 0$ sequence predominant:

v', v''	λ	Intensity	v', v''	λ	Intensity
9, 9	4780.11	5	2, 2	4684.16	100
8, 8	4764.22	7	1, 1	4672.31	120
7, 7	4749.11	10	0, 0	4661.75	150
6, 6	4735.00	25	7, 6	4575.31	5
5, 5	4720.86	5	6, 5	4560.95	8
4, 4	4708.00	60	4, 3	4533.39	5
3, 3	4695.46	80	2, 1	4508.91	5

III. $C^2\Sigma \rightarrow X^2\Sigma$ System (73.L2, 71.L1)

Band heads:

(v', v'')	(1, 2)	(0, 1)	(1, 1)	(0, 0)	(2, 1)
λ(Intensity)	4252.5(4)	4241.8(6)	4105.8(4)	4094.0(7)	3978.4(2)

LuO

SPECTROSCOPIC CONSTANTS

State	T_e	ω_e	$x_e\omega_e$	B_o	$\alpha_e \times 10^3$	$D_o \times 10^6$	r_e	Remarks	Bibliography
$C^2\Sigma$	24460	733	6	0.34411	-	.297	-	$r_o = -0.4940$	(73.L2, 71.L1, 52.4)
B	21471.8	791.60	3.12	-	-	-	-		(38.1)
A	~19350	-	-	-	-	-	-		(52.4)
$X^2\Sigma$	0	844.5	3.1	0.35806	-	.255	-		(73.L2, 71.L1, 38.1)

Dissociation energy = 7.23 ± 0.12 eV, 166.7 kcal/mole, 58300 cm^{-1} (67.7).

Perturbations and General Information

B-X system has a perturbation in the $\Delta v = 0$ sequence at $v' = 5$ (38.1).

BIBLIOGRAPHY

(38.1) Vibrational Analysis,
W. W. Watson and W. F. Meggers,
J. Res. Nat. Bur. Stand. 20, 125-8

(38.2) Interpretation,
W. W. Watson,
Phys. Rev. 53, 639-42

(42.3) Spectral Reproduction,
A. Gatterer,
Ricerche Spettroscop. Lab. Astrofis. Specola Vaticana 1, 153-79

(52.4) A, C→X Systems,
A. Gatterer and S. G. Krishnamurty,
Proc. Phys. Soc. A, 65, 151-2

(57.5) A, B→X Systems,
A. Gatterer, J. Junkes, E. W. Salpeter, and B. Rosen,
"Molecular Spectra of Metallic Oxides,"
ed. Specola Vaticana

(63.6) Flame Spectra,
R. Herman and C. T. J. Alkemade,
Flame Photometry (Interscience Publ.)

(67.7) Dissociation Energy,
L. L. Ames, P. N. Walsh, and D. White,
J. Phys. Chem. 71, 2707-18

(71.L1) $C^2\Sigma \rightarrow X^2\Sigma$ System,
R. Bacis, A. Bernard, and J. D'Incan,
"Vibrational Structure of the Violet System of the LuO Molecule,"
C. R. Acad. Sci. 273B, 272-4

(73.L2) R. Bacis and A. Bernard,
"Analysis of the $^2\Sigma(b_{\beta J}) \rightarrow {}^2\Sigma(b_{\beta S})$ of the LuO Radical,"
Can. J. Phys. 51, 648-56

LuS

Dissociation energy = 5.21 ± 0.15 eV, 42050 cm^{-1} (69.1).

BIBLIOGRAPHY

(69.1) Dissociation Energy by Mass Spectra,
S. Smoes, P. Coppens, C. Bergman, and J. Drowart,
Trans. Faraday Soc. 65, 682-7

LuSe

Dissociation energy = 4.29 ± 0.15 eV, 34600 cm^{-1} (72.1).

BIBLIOGRAPHY

(72.1) M. Shafi, C. L. Beckel, and R. Engelke,
"Diatomic Molecule Ground State Dissociation Energies,"
J. Mol. Spectrosc. 42, 578-81

LuTe

LuTe

Dissociation energy = 3.33 ± 0.15 eV, 26860 cm^{-1} (72.1).

BIBLIOGRAPHY

(72.1)   M. Shafi, C. L. Beckel, and R. Engelke, "Diatomic Molecule Ground State Dissociation Energies," J. Mol. Spectrosc. 42, 578-81

MgBr

Methods of Production and Experimental Technique

Absorption in a King furnace.

Emission of an arc:flame.

High frequency discharge into a mixture of bromine and argon over heated magnesium metal in a quartz discharge tube (70.6).

Heating of $MgBr_2$ in a furnace (69.5).

BAND SYSTEMS

System	Transition	Sources	Wavelength Limits	Degrading	Band Head, $\nu_{0,0}$	Remarks	Bibliography
I	$A_1{}^2\Pi_{1/2} \rightleftarrows X^2\Sigma^+$	Flame, absorption	3990-3815	V	25775.6		(42.4, 36.3)
	$A_2{}^2\Pi_{3/2} \rightleftarrows X^2\Sigma^+$		3980-3805		25886.2		
II	$B(^2\Sigma) \leftarrow X^2\Sigma^+$	Absorption	~3700	-	~27000		(42.4)
III	$C^2\Sigma \rightarrow X^2\Sigma^+$	Discharge	2540-2720	R	39285.9		(70.6)

MgBr

I. $A^2\Pi \rightleftarrows X^2\Sigma^+$ System

Doublet system with intense Q heads. Predissociates for v' >3 (42.4).

Band heads (P_1):

(v', v'')	(0, 0)	(0, 1)	(1, 0)	(1, 1)	(2, 1)
λ	3880.5	3936.9	3823.4	3877.7	3821.4

II. $B(^2\Sigma) \leftarrow X^2\Sigma^+$ System

State is unstable; responsible for predissociation of $A^2\Pi$ state.

III. $C^2\Sigma \rightarrow X^2\Sigma^+$ System

Band heads, λ(intensity):

(v', v'')	(0, 0)	(0, 1)	(0, 2)	(1, 1)	(2, 2)
λ	2548.8	2573.1	2597.7	2556.0	2563.8

SPECTROSCOPIC CONSTANTS

State	T_e	ω_e	$x_e\omega_e$	B_o	$\alpha_e \times 10^3$	$D_e \times 10^6$	r_e	Remarks	Bibliography
$C^2\Sigma^+$	39285.9	271.9	5.2	–	–	–	–		(70.6)
$A^2\Pi$	25765.2 25876.3	393.8	2.05	0.168	–	–	2.33		(69.5, 36.3)
$X^2\Sigma^+$	0	373.2	1.34	0.164	–	–	2.36		(69.5, 36.3)

Dissociation energy = 3.2 ± 1 eV, 74 kcal/mole, 25810 cm^{-1}.

BIBLIOGRAPHY

(06.1)   A→X System,
C. M. Olmsted,
Z. Wiss. Phot., German 4, 255-333

(28.2)   A←X System,
O. H. Walters and S. Barratt,
Proc. Phys. Soc. A, 118, 120-37

(36.3)   A←X System,
F. Morgan,
Phys. Rev. 50, 603-7

(42.4)   B←X, C←X Systems,
R. E. Harrington,
Thesis, Berkeley

(69.5)   A←X, Rotational Analysis,
M. M. Patel and P. D. Patel,
J. Phys. B, Proc. Phys. Soc. 2, 515-6

(70.6) $C^2\Sigma \rightarrow X^2\Sigma$ System,
B. R. K. Reddy and P. T. Rao,
"The Electronic Emission Spectrum of the MgBr Molecule,"
Current Sci. 39, 509-10

MgF

Methods of Production and Experimental Technique

Absorption in MgF_2 + Mg ($1150 < t < 2000°C$) (64.9, 41.6, 34.3).

Emission from arc and flames.

Heating of MgF_2 + Al in a tantalum Knudsen cell (63.8).

BAND SYSTEMS

System	Transition	Sources	Wavelength Limits	Degrading	Band Head, $\nu_{0,0}$	Remarks	Bibliography
I	$A^2\Pi \rightleftarrows X^2\Sigma^+$	Absorption	3686-3468	V	27829.4(Q_1) 27863.7(Q_2)		(67.11, 34.5, 29.3, 29.4, 21.1)
II	$B^2\Sigma \rightleftarrows X^2\Sigma^+$	Absorption	2742-2630	V	37187.4		(67.11, 34.5, 29.3, 29.4, 21.1)
III	$C^2\Sigma^+ \leftarrow X^2\Sigma^+$	Absorption	2400-2250	V	42580		(41.6)
IV	$D^2\Pi - X^2\Sigma^+$	Absorption	1844-1830	-	54220.5(0,0)(P_1)		(71.19)
V	$E^2\Pi - X^2\Sigma^+$	Absorption	1798-1794	-	55630(0,0)(P_1)		(71.19)
VI	$F^2\Sigma - X^2\Sigma^+$	Absorption	1752-1739	-	57086.4(0,0)		(71.19)
VII	$G^2\Sigma - X^2\Sigma^+$	Absorption	1345-1341	-	74343.9(0,0)		(71.19)

MgF

I. $A^2\Pi \rightleftarrows X^2\Sigma^+$ System

System has three intense doublet sequences. Band heads (41.6, 34.5):

v', v''	P_1	P_2	Q_2
0, 1	3685.8(5)	3682.8	3682.3
1, 2	3680.8(5)	3677.9	3677.4
2, 3	3675.9(4)	3673.1	3672.5
0, 0	3594.2(10)	3592.8	3588.2
1, 1	3590.6(10)	3589.1	3584.6
2, 2	3586.9(9)	3585.4	3581.0
1, 0	3504.4(4)	3502.4	
2, 1	3500.8(4)	3499.9	
3, 2	3498.3(3)	3497.4	
4, 3?	--	3494.8	

II. $B^2\Sigma^+ \rightleftarrows X^2\Sigma^+$ System

Band heads (P) (29.3):

v', v''	λ	Intensity
0, 1	2741.6	2
1, 2	2738.0	2
2, 3	2734.6	2
3, 4	2731.4	2
0, 0	2689.4	6
1, 1	2686.5	5
2, 2	2683.8	5
3, 3	2681.3	4
1, 0	2636.5	3
2, 1	2634.5	3
3, 2	2632.6	3

III. $C^2\Sigma \leftarrow X^2\Sigma^+$ System

Four sequences of simple bands. Band heads (P) (41.6):

v', v''	λ	Intensity
0, 1	2387.1	2
1, 2	2381.4	1
2, 3	(2375.3)	0
0, 0	2347.8	10
1, 1	2342.2	5
1, 0	2303.8	5
2, 1	2298.0	4
3, 2	2294.2	3
4, 3	2289.5	1
2, 0	2262.0	1
3, 1	2257.7	2
4, 2	2253.6	3
5, 3	2249.7	2

IV. $D^2\Pi - X^2\Sigma^+$ System

Band heads (P_1) (71.19):

(v', v'')	(0, 0)	(1, 1)	(2, 2)	(3, 3)
λ	1844.3	1841.8	1839.4	1836.9

V. $E^2\Pi - X^2\Sigma^+$ System

Band heads (P_1) (71.19):

(v', v'')	(0, 0)	(1, 1)	(2, 2)
λ	1797.6	1795.7	1793.9

VI. $F^2\Sigma - X^2\Sigma^+$ System

Band heads (71.19):

(v', v'')	(0, 0)	(1, 1)	(2, 2)	(3, 3)
λ	1751.7	1750.6	1749.4	1748.2

MgF

VII. $\underline{G^2\Sigma - X^2\Sigma^+ \text{ System}}$

Band heads (71.19):

(v', v'')	(0, 0)	(1, 1)	(2, 2)	(3, 3)
λ	1345.1	1343.7	1342.4	1341.1

MgF

SPECTROSCOPIC CONSTANTS

State	T_e	ω_e	$x_e\omega_e$	B_e	$\alpha_e \times 10^3$	$D_e \times 10^6$	r_e	Remarks	Bibliography
$G^2\Sigma$	74304.2	800.0	6.18	–	–	–	–		(71.19)
$F^2\Sigma$	57067.2	756.6	4.06	–	–	–	–		(71.19)
$E^2\Pi$	55694.9	775.9	4.00	–	–	–	–		(71.19)
$D^2\Pi$	54263.6	792.3	4.38	–	–	–	–		(71.19)
$C^2\Sigma^+$	42589.64	823.1	5.04	0.55102	4.19	0.994	1.6988		(67.11)
$B^2\Sigma^+$	37187.45	762.1	5.60	0.53845	5.11	1.084	1.7185		(67.11)
$A^2\Pi$	27829.60	746.0	3.97	0.52004	3.27	0.18	1.7486		(67.11)
$X^2\Sigma^+$	0	721.6	4.94	0.51922	4.70	1.083	1.7500		(67.11)

Dissociation energy = 4.5 ± 0.1 eV, 105.5 kcal/mole, 36296 cm^{-1}.

MgF

Perturbations and General Information

Potential energy curves — RKR potential (70.16):

State	v	U(cm^{-1})	r$_{min}$(Å)	r$_{max}$(Å)	T$_e$+U(cm^{-1})
B$^2\Sigma^+$	0	379.7	1.657	1.787	37531.4
	1	1126.5	1.619	1.846	38278.2
	2	1863.0	1.591	1.885	39014.7
	3	2588.4	1.571	1.921	39740.1
	4	3302.8	1.555	1.954	40454.5
	5	4005.9	1.541	1.985	41157.6
	6	4698.0	1.528	2.014	41849.7
X$^2\Sigma^+$	0	359.6	1.687	1.820	359.6
	1	1067.4	1.645	1.877	1067.4
	2	1766.3	1.618	1.920	1766.3
	3	2457.3	1.597	1.957	2457.3
	4	3140.7	1.580	1.987	3140.7
	5	3815.3	1.567	2.021	3815.3
	6	4481.0	1.554	2.050	4481.0
	7	5140.0	1.543	2.078	5140.0

Franck-Condon factors, r-centroids — Morse potential:

A$^2\Pi$ - X$^2\Sigma^+$ (69.15)

v', v''	0	1	2	3
0	0.965	0.035	0.001	0.000
	1.746	1.523	1.333	1.141
	3588.2	3682.3	-	-
1	0.035	0.901	0.067	0.004
	2.054	1.754	1.533	1.344
	3503.4	3584.6	3677.4	-
2	0.000	0.071	0.833	0.096
	2.594	2.062	1.763	1.542
	-	3500.8	3581.0	3672.5
3	0.000	0.001	0.104	0.768
	-	2.602	2.070	1.772
	-	-	3498.3	-

Top = Franck-Condon factors, middle = r-centroids, and bottom = wavelengths.

MgF

$B^2\Sigma^+ - X^2\Sigma^+$ (68.14)

v', v''	0	1	2	3	4
0	0.890 1.738 2689.4	0.100 1.621 2741.6	0.009 1.528 -	0.002 1.451 -	0.000 1.386 -
1	0.107 1.900 2636.5	0.695 1.745 2686.5	0.173 1.629 2738.0	0.022 1.533 -	0.003 1.459 -
2	0.004 2.135 -	0.197 1.908 2634.5	0.529 1.753 2683.8	0.223 1.636 2734.6	0.043 1.543 -
3	0.000 2.560 -	0.010 2.143 -	0.270 1.916 2632.6	0.392 1.761 2681.3	0.251 1.644 2731.4
4	0.000 - -	0.000 2.568 -	0.020 2.451 -	0.329 1.924 -	0.281 1.769 -

$C^2\Sigma^+ - X^2\Sigma^+$ (68.14)

v', v''	0	1	2	3	4
0	0.723 1.729 2347.8	0.227 1.654 2387.7	0.042 1.580 -	0.006 1.499 -	0.000 1.396 -
1	0.239 1.815 2303.8	0.326 1.739 2342.2	0.310 1.666 2381.4	0.099 1.591 -	0.027 1.515 -
2	0.034 1.916 2262.0	0.349 1.825 2298.9	0.109 1.750 -	0.304 1.678 2375.3	0.167 1.605 -
3	0.003 1.999 -	0.088 1.913 2257.7	0.367 1.873 2294.2	0.016 1.761 -	0.281 1.689 -
4	0.000 2.104 -	0.011 2.009 -	0.146 1.923 2253.6	0.330 1.845 2289.5	0.002 1.771 -

Top = Franck-Condon factors, middle = r-centroids, and bottom = wavelengths.

MgF

Vibrational lasing action has been observed by exploding an Mg wire in F_2 between $12.8 \leq \lambda \leq 13.5$ μm and also by exploding an Mg film in F_2 (73.21).

Quantum yields (n.p. 22):

$$Mg + F_2 \rightarrow MgF (X^2\Sigma^+, A^2\Pi, B^2\Sigma^+) + F$$

$$\frac{A^2\Pi}{MgF_{tot.}} = 2.1 \times 10^{-4}, \quad \frac{B^2\Sigma^+}{MgF_{tot.}} = 9.1 \times 10^{-7}$$

BIBLIOGRAPHY

(21.1) A, B→X Systems,
S. Datta,
Proc. Roy. Soc. A, 99, 436-55

(28.2) Assignment Incorrect,
O. H. Walters and S. Barratt,
Proc. Phys. Soc. A, 118, 120-37

(29.3) A→X, B→X Systems; Analysis Redone, See (34.5, 41.6),
W. Jevons,
Proc. Roy. Soc. A, 122, 211-27

(29.4) A, B→X, Preliminary Analysis,
R. C. Johnson,
Proc. Roy. Soc. A, 122, 189-200

(34.5) A←X, B←X Rotational Analysis,
F. A. Jenkins and R. Grinfeld,
Phys. Rev. 45, 229-33

(41.6) All Systems in Absorption: Vibrational Analysis,
C. A. Fowler, Jr.,
Phys. Rev. 59, 645-52

(50.7) Correction of (34.5),
R. W. B. Pearse and A. G. Gaydon,
Identification of Molecular Spectra
Second Edition, Chapman and Hall, London

(63.8) G. D. Blue, J. W. Green, T. C. Ehlert, and J. L. Margrave,
"Dissociation Energies of the Alkaline Earth Monofluorides,"
Nature 199, 804-5.

(64.9) Dissociation Energy by Mass Spectra,
A. Lagerqvist, H. Neuhaus, and R. Scullman,
Proc. Phys. Soc. 83, 498-9

(66.10) Dissociation Energy by Mass Spectra,
E. Murad, D. L. Hildenbrand, and R. P. Main,
J. Chem. Phys. 45, 263-9

(67.11) C, B, A←X Systems, Rotational Analysis,
R. F. Barrow and J. R. Beale,
Proc. Phys. Soc. 91, 483-8

(68.12) Dissociation Energy by Mass Spectra,
D. L. Hildenbrand,
J. Chem. Phys. 48, 3657-65

(68.13) Theoretical Discussion of the $A^2\Pi$ State,
T. E. H. Walker and W. G. Richards,
J. Phys. B (Proc. Phys. Soc.) 1, 1061-5

(68.14) R. C. Maheshwari, I. D. Singh, and M. M. Shukla,
"Franck-Condon Factors and r-Centroids for the B-X and C-X
Systems of the MgF Molecule,"
J. Phys. B (Proc. Phys. Soc.) 2, 993-6

(69.15) I. D. Singh, M. M. Shukla, and R. C. Maheshwari,
"Vibrational Transition Probabilities and r-Centroids for the A-X
System of the MgF Molecule,"
J. Quant. Spectrosc. Radiative Transfer 1, 533-5

(70.16) T. V. R. Rao and S. V. J. Lakshman,
"The True Potential Energy Curves, r-Centroids and Franck-
Condon Factors of the Bands of the $B^2\Sigma^+ - X^2\Sigma^+$ System of MgF,"
Physica 46, 609-13

(70.17) Hartree-Fock Calculations,
T. E. H. Walker and W. G. Richards,
"The Assignment of Molecular Orbital Configurations on the Basis
of Λ-Type Doubling,"
J. Phys. B (Atom. Mol. Phys.) 3, 271-9

MgF

(70.18) T. V. R. Rao and S. V. J. Lakshman,
"r-Centroids and Franck-Condon Factors for the Bands of the $C^2\Sigma^+ - X^2\Sigma^+$ System of MgF,"
J. Quant. Spectrosc. Radiative Transfer 10, 945-8

(71.19) D, E, F, G←X Systems, Vibrational Analysis,
M. M. Novikov and L. V. Gurvich,
"Study of the Electronic Spectra of MgF and MgF^+,"
Zh. Prikl Spectrosk. 14, 1113-6

(71.20) L. B. Knight, Jr., W. C. Easley, W. Weltner, Jr., and M. Wilson,
"Hyperfine Interaction and Chemical Bonding in MgF, CaF, SrF, and BaF Molecules,"
J. Chem. Phys. 54, 322-9

(73.21) W. W. Rice and W. H. Beattie,
"Metal Atom Oxidation Lasers,"
Chem. Phys. Letters 19, 82-5

(n.p. 22) Quantum Yields,
D. J. Eckstrom, S. A. Edelstein, and S. W. Benson,
(to be published)

MgO

Methods of Production and Experimental Technique

Absorption (45.12, 37.9).

Emission from an arc; arc under vacuum, Mg flame burning in air; exploding wire.

Astrophysics; absorption in stellar atmosphere (45.11).

Hollow cathode lamp; flame burning aqueous magnesium nitrate (69.38, 69.39).

Combustion of aqueous MgO at 4000°K with H_2 and F_2 gases (67.34).

Vaporization of MgO in laser blowoff apparatus.

BAND SYSTEMS

	System	Transition	Sources	Wavelength Limits	Degrading	Band Head, $\nu_{0,0}$	Remarks	Bibliography
Green	I	$B^1\Sigma^+ \to X^1\Sigma^+$	Arc	5205-4830	V	20003.5		(49.14, 45.12, 32.3)
	II	$E^1\Sigma \to X^1\Sigma$	Arc	2687-2652	V	37683.5		(71.42)
Red	III	$B^1\Sigma^+ \to A^1\Pi$	Arc	6870-4717	V	16500.2		(71.42)
	IV	$C^1\Sigma^- \to A^1\Pi$	Arc	3795-3766	-	26500.94		(69.39, 65.32, 64.28, 62.24, 62.25)
	V	$D^1\Delta \to A^1\Pi$	Arc	3830-3798	-	26272.04		(69.39, 65.32)
	VI	$E^1\Sigma \to F$	Arc	3962-3649	-	27073.9(0,0)		(69.39)
	VII	Y - Z	Arc	3688-3672	-	27234.3		(69.39)

MgO

I. $B^1\Sigma^+ \to X^1\Sigma^+$ System (Green)

Band heads (32.3):

v', v''	0, 1	1, 2	0, 0	1, 1	2, 2
λ(intensity)	5205.96(4)	5191.98(4)	5007.32(10)	4996.71(9)	4985.92(8)
v', v''	3, 3	4, 4	5, 5	6, 6	
λ(intensity)	4974.47(7)	4962.10(6)	4949.53(5)	4935.30(4)	

II. $E^1\Sigma^+ \to X^1\Sigma^+$ System

Only (0, 0) band observed (71.42).

III. $B^1\Sigma^+ \to A^1\Pi$ System (Red)

Band heads (32.3):

v', v''	0, 2	0, 1	1, 2	0, 0	1, 0
λ(intensity)	6581.05(3)	6311.75(4)	6246.38(3)	6060.31(6)	5775.50(5)
v', v''	2, 1	2, 0	3, 1	3, 0	
λ(intensity)	5726.34(3)	5518.78(4)	5475.94(3)	5285.74(3)	

IV. $C^1\Sigma^- \to A^1\Pi$ System

Band heads (65.32):

(v', v'')	(1, 1)	(0, 0)	(0, 0)
λ	3777.4(Q)	3772.4(Q)	3766.1(R)

V. $D^1\Delta \to A^1\Pi$ System

Band heads (69.39):

(v', v'')	(0, 0)	(1, 1)	(2, 2)	(0, 1)	(1, 2)
λ	3805.3	3810.3	3815.7	3902.8	3906.8

VI. $\underline{E^1\Sigma \to F \text{ System}}$

Band heads (69.39):

(v', v'')	(0, 0)	(1, 1)	(2, 2)	(1, 0)	(2, 1)
λ	3962.6	3693.6	3694.6	3638.8	3640.7

VII. $\underline{Y \to Z \text{ System}}$

Band heads (69.39):

(v', v'')	(0, 0)	(1, 1)	(2, 2)	(3, 3)
λ	3672.1	3674.6	3677.5	3680.7

MgO

SPECTROSCOPIC CONSTANTS

State	T_e	ω_e	$x_e\omega_e$	B_e	$\alpha_e \times 10^3$	$D_e \times 10^6$	r_e	Remarks	Bibliography
$E^1\Sigma^+$	37683.5	406.3	3.25	0.524[a]	–	1.1[b]	–		(71.42, 69.39)
F	–	412.5	2.80	–	–	–	–		(69.39)
$C^1\Sigma^-$	31080.6	632.4	5.2	0.4984[a]	–	1.27[b]	1.876[c]		(65.32)
$D^1\Delta$	29851.7	632.5	5.3	0.5014	4.8	1.26[b]	1.8716		(65.32)
$B^1\Sigma^+$	19984.0	824.0	4.7	0.5822	4.5	1.14	1.737		(49.14)
$A^1\Pi$	3563.3	664.4	3.9	0.5050	4.0	1.18	1.864		(65.32, 49.14)
$X^1\Sigma^+$	0	785.0	5.1	0.5743	5.0	1.22	1.749		(49.14)

[a] B_o, [b] D_o, [c] r_o.

Dissociation energy = 4.16 ± 0.13 eV, 96 kcal/mole, 33553 cm^{-1} (69.38).

Perturbations and General Information

Franck-Condon factors; r-centroids — Morse potential.

$B^1\Sigma^+ - X^1\Sigma^+$ (64, 31):

v', v''	0	1	2	3	4	5	6	7
0	0.98212	0.01777	0.00012	-	-	-	-	-
1	0.01764	0.94427	0.03767	0.00043	-	-	-	-
2	0.00024	0.03715	0.90100	0.05957	0.00104	-	-	-
3	-	0.00080	0.05844	0.85542	0.08326	0.00207	-	-
4	-	-	0.00176	0.08133	0.80475	0.10845	0.00367	-
5	-	-	-	0.00321	0.10562	0.75032	0.13477	0.00601
6	-	-	-	-	0.00524	0.13100	0.69255	0.16175
7	-	-	-	-	-	0.00798	0.15715	0.63204
8	-	-	-	-	-	-	0.01156	0.18309

$D^1\Delta - A^1\Pi$ (67, 35):

v', v''	0	1	2	3	4
0	0.994	0.006	0.000	0.000	-
	1.881	2.041	2.138	2.223	2.305
1	0.006	0.978	0.016	0.000	-
	-	1.898	2.043	2.138	2.217
2	0.000	0.017	0.952	0.032	0.001
	-	-	1.912	2.046	2.139
3	0.000	0.001	0.032	0.438	0.053
	-	-	-	1.923	2.049
4	-	-	-	-	-
					1.933

Top = Franck-Condon factors, and bottom = r-centroids.

MgO

$C^1\Sigma^-$ - $A^1\Pi$ — RKR potential (70.41):

v', v''	0	1	2	3	4
0	0.989	0.009	-	-	-
	1.874	2.436	-	-	-
1	0.010	0.965	0.023	-	-
	1.347	1.884	2.386	-	-
2	-	0.016	0.935	0.039	-
	-	1.305	1.896	2.376	-
3	-	-	0.035	0.906	0.059
	-	-	1.432	1.907	2.482

Top = Franck-Condon factors, and bottom = r-centroids.

$B^1\Sigma$ - $A^1\Pi$ — Morse potential (62.31):

v', v''	0	1	2	3	4	5
0	1.8306-1	2.8220-1	2.4205-1	1.5297-1	7.9650-2	3.6252-2
1	3.3966-1	8.8133-2	6.1050-3	9.9418-2	1.5503-1	1.3574-1
2	2.8256-1	2.6356-2	1.6410-1	5.5441-2	1.9893-3	6.2800-2
3	1.3872-1	2.0777-1	3.1342-2	6.0528-2	1.1484-1	2.9167-2
4	4.4545-2	2.3041-1	5.5960-2	1.1420-1	4.6824-4	8.1395-2
5	9.8033-3	1.1987-1	2.1144-1	8.4384-6	1.1876-1	2.7056-2
6	1.5068-3	3.6920-2	1.8561-1	1.2679-1	3.3018-2	6.3150-2
7	1.6178-4	7.3153-3	7.9740-2	2.1189-1	4.3965-2	8.5851-2

RKRV potential energy curve for ground state (72.L1):

State	v	U(eV)	r_{min}(Å)	r_{max}(Å)
$X^1\Sigma^+$	0	0.0484	1.685	1.820
	1	0.1429	1.642	1.879
	2	0.2306	1.615	1.922
	3	0.3225	1.594	1.958
	4	0.4131	1.577	1.992
	5	0.5028	1.562	2.023
	6	0.5911	1.549	2.052
	7	0.6787	1.537	2.081
	8	0.7642	1.526	2.108
	9	0.8664	1.517	2.134
	10	0.9509	1.508	2.160

BIBLIOGRAPHY

(29. 1) Preliminary Note to (30. 2),
P. N. Ghosh, B. C. Mookerjee, and P. C. Mahanti,
Nature 124, 303

(30. 2) Green and Red System,
P. N. Ghosh, P. C. Mahanti, and B. C. Mukkerjee,
Phys. Rev. 35, 1491-4

(32. 3) Green and Red System,
P. C. Mahanti,
Phys. Rev. 42, 609-21

(35. 4) Green and Red System,
P. C. Mahanti,
Indian J. Phys. 9, 455-86

(35. 5) Red System. Dissociation Energy,
P. C. Mahanti,
Indian J. Phys. 9, 517-36

(35. 6) K. Korth,
Nachr. Ges. Wiss. Gottingen Math. Phys. Kl. 11 1, 187-94

(35. 7) J. Verhoeghe,
Wis-Natuurkund. Tijdschr. 7, 224-33

(36. 8) H. Lessheim and R. Samuel,
Philos. Mag. 21, 41-64

(37. 9) P. Ledoux,
Thesis, Liege

(43. 10) Green System,
A. Lagerqvist,
Arkiv. Mat. Astron. Fysik A 29A, n° 3, 1-13

(45. 11) H. D. Babcock,
Astrophys. J. 102, 154-67

(45. 12) Green System,
R. F. Barrow and D. V. Crawford,
Proc. Phys. Soc. 57, 12-5

MgO

(48.13) Infrared,
E. Burstein, J. J. Oberly, and E. K. Plyler,
Proc. Indian Acad. Sci. A 28, 388-400

(49.14) Red and Green System,
A. Lagerqvist and U. Uhler,
Arkiv Fysik Sverige 1, 459-75

(49.15) Red System. Preliminary Note to (49.14),
A. Lagerqvist and U. Uhler,
Nature 164, 665-6

(50.16) Dissociation Energy,
L. Hulett and A. Lagerqvist,
Arkiv Fysik 2, 333-6

(54.17) Dissociation Energy,
L. Brewer and R. F. Porter,
J. Chem. Phys. 22, 1867-77

(55.18) Dissociation Energy,
R. F. Porter, W. A. Chupka, and M. G. Inghram,
J. Chem. Phys. 23, 1347-8

(56.19) I. V. Veits and L. V. Gurvich,
Optika Spektrosk. S.S.S.R. 1, 22-33

(57.20) Dissociation Energy,
I. V. Veits and L. V. Burvich,
Zh. Fiz. Khim. S.S.S.R. 31, 2306-11

(59.21) Dissociation Energy,
E. M. Bulewicz and T. M. Sugden,
Trans. Faraday Soc. 55, 720-9

(60.22) Mg Emission in Flame,
V. A. Sokolov and N. A. Nazimova,
Optics Spect. 8, 303-4

(60.23) Isotopic Effect Due to Oxygen,
D. S. Pesic,
Proc. Phys. Soc. 76, 844-8

(62.24) C → A System, Emission,
L. Brewer, S. Trajmar and R. A. Berg,
Astrophys. J. 135, 955-62

(62.25) Ultraviolet Emission of MgO and MgOH,
L. Brewer and S. Trajmar,
J. Chem. Phys. 36, 1585-7

(62.26) Franck-Condon Factors,
S. S. Prasad and K. Prasad,
Proc. Phys. Soc. 80, 311-3

(62.27) Franck-Condon Factors, A-B, B-X Systems,
R. W. Nicholls,
"Transition Probabilities of Molecular Band System. XIX.
Franck-Condon Factors to High Vibrational Quantum Numbers II:,"
Univ. Western Ontario T.R. #AD-262-592

(64.28) Arc Emission in a Vacuum,
S. Trajmar and G. E. Ewing,
J. Chem. Phys. 40, 1170

(64.29) B-X System, Isotope Effect,
D. S. Pesic,
Proc. Phys. Soc. 83, 885-7

(64.30) Dissociation Energy,
J. Drowart, G. Exsteen, and G. Verhaegen,
Trans. Faraday Soc. 60, 1920-33

(64.31) Franck-Condon Factors, B-X,
F. S. Ortenberg, V. B. Glasko, and A. I. Dmitriev,
"Vibrational Transition Probabilities for Band Systems of
Some Diatomic Oxides II,"
Soviet Astronomy-AJ 8, 258-61

(65.32) Rotational Analysis,
S. Trajmar and G. E. Ewing,
Astrophys. J. 142, 77-83

(66.33) Wave Functions for the Valence Levels,
W. G. Richards, G. Verhaegen, and C. M. Moser,
J. Chem. Phys. 45, 3226-30

(67.34) R. P. Main, D. J. Carlson, and R. A. DuPuis,
"Measurement of Oscillator Strengths of the MgO
($B^1\Sigma^+ - X^1\Sigma^+$) and MgH ($A^2\pi - X^2\Sigma^+$) Band Systems,"
J. Quant. Spectrosc. Radiative Transfer 7, 805-11

MgO

(67.35) Franck-Condon Factors, r-Centroids, D-A System of MgO,"
Y. P. Srivastava and R. C. Maheshwari,
"Overlap Integrals and r-Centroids of $D^1\Delta \rightarrow A^1\pi$ System of MgO,"
Proc. Phys. Soc. 90, 1177-8

(67.36) K. Schofield,
"The Bond Dissociation Energies of Group IIA Diatomic Oxides,"
Chem. Rev. 67, 707-15

(68.37) Hartree Fock Calculations,
M. Yoshimine,
"Computed Ground State Properties of BeO, MgO, CaO and SrO in Molecular Orbital Approximation,"
J. Phys. Soc. 25, 1100-19

(69.38) D. H. Cotton and D. R. Jenkins,
"Band-Dissociation Energy of Gaseous Magnesium Oxide,"
Trans. Faraday Soc. 65, 376-9

(69.39) C-A, E-F, Y-Z Systems,
J. Schamps, G. Gandara, and M. Becart,
"Contribution to the Study of the Vibrational Spectra of Magnesium Oxide in the Ultraviolet,"
Can. Spectrosc. 14, 13-6

(69.40) R. P. Main and A. Shadee,
"On the Oscillator Strengths of MgO and F_2,"
J. Quant. Spectrosc. Radiative Transfer 9, 713-4

(70.41) G. Gandara, J. Schamps, and M. Becart,
"Franck-Condon Factors and r-Centroids for the Systems $D^1\Delta - A^1\pi$ and $C^1\Sigma^+ - A^1\pi$ for the Molecule MgO,"
C. R. Acad. Sci. B 270, 121-15

(71.42) E-X System,
M. Singh,
"A New Electronic State of MgO,"
J. Phys. B 4, 565-7

(72.43) Theory,
B. Huron and P. Rancurel,
"Iterative Calculation by Perturbation of the Configuration Interaction for Fundamental State and First Excited States of MgO,"
Chem. Phys. Letters 13, 515-20

(72.44) J. Schamps and H. Lefebvre-Brion,
"SCF Calculations of the Electronic States of Magnesium Monoxide,
J. Chem. Phys. 56, 573-85

(72.L1) Molecular Constants, Potential Energy Curves, Dissociation Energy,
B. Rai and S. N. Rai,
"Dissociation Energies of MgO and BeO Molecules,"
Indian J. Pure Appl. Phys. 10, 401-2

MnCl

Methods of Production and Experimental Technique

Absorption by the heating of $MnCl_2$.

Emission from discharge into a hollow cathode.

BAND SYSTEMS

System	Transition	Sources	Wavelength Limits	Degrading	Band Head, $\nu_{0,0}$	Remarks	Bibliography
I	$(\Pi \rightarrow \Sigma)$	Hollow cathode	9150-8180	R	11414.0(Q)		(55.10)
II	$(\Pi \rightarrow \Sigma)$	Hollow cathode	5055-5010	R	19928.6(Q)		(55.11)
III	$(\Pi \rightarrow \Sigma)$	Hollow cathode	5060-4850	R	20112.0(Q)		(55.11)
IV	$^7\Pi \rightleftarrows X^7\Sigma$	Hollow cathode	3900-3500	V	27017.8(Q_4)		(55.12)
V	$^7\Pi \leftarrow X^7\Sigma$	Absorption	2500-2400	V	40776		(48.5)

I. ($\Pi \to \Sigma$) System

Intense (0, 0) sequence. Analysis uncertain. Band head, λ(intensity):

(Q)8758.8(30) | (R)8752.2(40) | (S)8745.2(0) | (T)8733.4(0)

II. ($\Pi \to \Sigma$) System

Intense (0, 0) sequence. Each band has several heads. Band head, λ(intensity):

(Q")5022.2(2) | (Q', R")5019.7(3) | (R')5017.6(8) | (Q)5016.5(3) | (R)5012.8(10)

III. ($\Pi \to \Sigma$) System

Intense (0, 0) sequence. R and Q heads:

v', v"	λ(Q)(Intensity)	λ(R)(Intensity)
5, 5	4980.5(10)	4975.3(5)
4, 4	4978.3(15)	4973.5(8)
3, 3	4976.2(20)	4971.6(14)
2, 2	4974.2(12)	4969.8(10)
1, 1	4972.4(4)	4968.1(8)
0, 0	4970.8(4)	4966.5(4)

IV. $^7\Pi \rightleftarrows X^7\Sigma$ System

Bands with 7 P heads, 7 Q heads, and 6 O heads. (0, 0) band heads (55.12):

	F_1	F_2	F_3	F_4
λ(O)	3719.2(1)	3715.5(3)	-	3705.2(2)
λ(P)	3716.9(8)	3712.7(8)	3708.8(3)	3702.4(9)
λ(Q)	3715.5(3)	3710.9(8)	3706.4(5)	3700.2(7)

	F_5	F_6	F_7
λ(O)	3698.5(2)	3692.0(5)	-
λ(P)	3695.9(9)	3689.6(10)	3683.2(5)
λ(Q)	3694.1(6)	3688.0(8)	3681.7(7)

MnCl

V. $^7\Pi \rightleftarrows X^7\Sigma$ System

Band heads (48.5):

(v', v'')	(0, 1)	(1, 1)	(0, 0)	(1, 0)
λ	2475.0	2455.7	2451.7	2432.6

SPECTROSCOPIC CONSTANTS

State	T_e	ω_e	$x_e\omega_e$	B_e	$\alpha_e \times 10^3$	$D_e \times 10^6$	r_e	Remarks	Bibliography
$(^7\Sigma)$	40808	320	–						(48.5)
$^7\Pi$	27004.6	407.2	0.3						(48.5)
$X^7\Sigma$	0	384.9	1.4						(48.5)

CONSTANTS FOR UNCERTAIN BANDS

State	T_e	ω'_e	$x'_e\omega'_e$	ω''_e	$x''_e\omega''_e$	Bibliography
I ($\Pi \to \Sigma$)	11420	385.4	-0.56^a	397.9	-0.47^a	(55.10)
II ($\Pi \to \Sigma$)	~19920	~385	–	~410	–	(55.11)
III ($\Pi \to \Sigma$)	20115	378.0	-0.45^b	385.5	-0.50^b	(55.11)

[a] Sign is anomalous and there is an important cross term $0.15(v'+1/2)(v''+1/2)$,

[b] sign of $x_e\omega_e$ is anomalous.

Dissociation energy = 3.7 ± 0.1 eV, 85.3 kcal/mole, 29843 cm^{-1}.

BIBLIOGRAPHY

(39.1) Emission,
P. Mesnage,
Ann. Physique 12, 5-87

(42.2) Interpretation,
E. Miescher and W. Mueller,
Helv. Phys. Acta 15, 319-20

(43.3) Absorption,
W. Mueller,
Helv. Phys. Acta 16, 3-32

(47.4) Complex Spectra,
J. Bacher and E. Miescher,
Helv. Phys. Acta 20, 245-7

(48.5) Precise Measurements, Detailed Interpretation,
J. Bacher,
Helv. Phys. Acta 21, 379-402

(48.6) Complex Spectra,
E. Miescher,
J. Phys. Rad. 9, 153-5

(49.7) Erroneous Interpretation,
P. T. Rao,
Indian J. Phys. 23, 301-8

(49.8) Supplementary Data to (48.5),
P. T. Rao,
Indian J. Phys. 23, 517-24

(52.9) P. T. Rao,
"The Complex Band Spectrum of Diatomic Manganese Chloride in the Visible Region,"
Proc. Nat. Inst. Sci., India 18, 481-6

(55.10) System I. Vibration Analysis,
W. Hayes and T. E. Nevin,
Nuovo Cim. Suppl., Italian 2, 734-41

(55.11) Systems II and III. Vibrational Analysis,
W. Hayes,
Proc. Phys Soc. A, 68, 1097-1106

(55.12)   System IV. Vibrational Analysis,
          W. Hayes and T. E. Nevin,
          Proc. Roy. Irish Acad. A, 57, 15-30

(61.13)   Dissociation Energy,
          E. M. Bulewicz, L. F. Phillips, and T. M. Sugden,
          Trans. Faraday Soc. 57, 921-31

MnF

Methods of Production and Experimental Technique

Absorption.

Emission from a high current discharge.

Hollow cathode discharge in MnF_2 (55.5).

BAND SYSTEMS

System	Transition	Sources	Wavelength Limits	Degrading	Band Head, $\nu_{0,0}$	Remarks	Bibliography
I	$(\Pi \rightarrow \Sigma)$	Hollow cathode	8495-8180	R and V	12179.6		(55.5)
II	$(O^\pm \rightarrow {}^5\Sigma)$	Hollow cathode	7300-6230	R	14500.1		(55.7)
III	$(O^\pm \rightarrow \Sigma)$	Hollow cathode	5100-4990	R ?	19869.3 19823.7 ?		(55.6)
IV	$({}^5\Pi \rightarrow {}^5\Sigma)$	Hollow cathode	4990-4920	R	20293.5		(55.6)
V	${}^7\Pi \rightleftarrows {}^7\Sigma$?	Absorption discharge	3700-3300	R and V	28478.1(Q_4)		(62.8)
VI	${}^7\Sigma \leftarrow {}^7\Sigma$	Absorption	2500-2300	V	40776		(48.3, 39.1)

I. $(\Pi \rightarrow \Sigma)$ System

   Only (0, 0) sequence; band heads:

v', v"	0, 0	1, 1	2, 2	3, 3	11, 11
λ(Q)(Intensity)	8208.2(1)	8219.5(1)	8229.9(1)	8241.1(1)	8364.3(6)

II. $(O^{\pm} \rightarrow {}^{5}\Sigma)$ System

   Five sequences with (0, 0) most intense, λ(intensity):

   (T)6890.71(10) | (S)6892.94(10) | (R)6894.27(20) | (Q)6894.60(15)

III. $(O^{\pm} \rightarrow \Sigma)$ System

   Only (0, 0) sequence R and Q heads. Band heads:

   (Q)5043.19(9) | (R)5031.5(10)

IV. $({}^{5}\Pi \rightarrow {}^{5}\Sigma)$ System

   Complex band structure with (0, 0) most intense:

v', v"	0, 0	1, 1	2, 2	3, 3	4, 4
λ(R)(Intensity)	4922.6(1)	4924.9(2)		4931.1(5)	4934.3(6)
λ(Q)(Intensity)	4926.3(3)	4929.3(3)	4932.9(1)	4936.3(10)	4940.0(2)

V. ${}^{7}\Pi \rightleftarrows {}^{7}\Sigma?$ System

   Band heads:

(v', v")	(0, 2)	(0, 1)	(0, 0)	(1, 0)	(2, 0)
λ	3673.8	3594.5	3519.7	3439.1	3361.5

VI. ${}^{7}\Sigma \leftarrow {}^{7}\Sigma$ System

   Simple bands arranged in sequences:

(v', v")	(0, 0)	(0, 1)	(1, 0)
λ	2424.1	2460.5	2387.7

MnF

SPECTROSCOPIC CONSTANTS

State	T_e	ω_e	$x_e\omega_e$	B_e	$\alpha_e \times 10^3$	$D_e \times 10^6$	r_e	Remarks	Bibliography
$^7\Sigma$	41231.5	637.2	4.46						(39.1)
$^7\Pi$	28461.5	669.5	2.45					$A \sim 24$ cm^{-1}	(62.8)
$X^7\Sigma$	0	618.1	2.25						(62.8)

CONSTANTS FOR UNCERTAIN BANDS

	State	T_e	ω_e'	$x_e'\omega_e'$	ω_e''	$x_e''\omega_e''$	Bibliography
I	$(\Pi \rightarrow \Sigma)$	~12150	–	–	–	–	(55.5)
II	$(0^{\pm} \rightarrow {}^5\Sigma)$	14527.7	595.4	3.15	645.4	3.2	(55.7)
III	$(0^{\pm} \rightarrow \Sigma)$	~19800	–	–	–	–	(55.6)
IV	$({}^5\Pi \rightarrow {}^5\Sigma)$	20298.2	637.1	1.9	649.1	1.5	(55.6)

Dissociation energy = 5.2 ± 1.5 eV, 120 kcal/mole, 41942 cm^{-1}.

BIBLIOGRAPHY

(39.1) System VI,
G. D. Rochester and E. Olsson,
Z. Physik 114, 495-9

(47.2) Systems IV and V,
J. Bacher and E. Miecher,
Helv. Phys. Acta 20, 245-7

(48.3) Systems IV and V,
J. Bacher,
Helv. Phys. Acta 21, 379-402

(53.4) P. T. Rao,
"The $^5\Pi$ - $^7\Sigma$ Electronic Transition in MnBr and MnF,"
Proc. Nat. Inst. Sci., India 19, 149-51

(55.5) System I,
W. Hayes and T. E. Nevin,
Nuovo Am. Suppl., Italian 2, 734-41

(55.6) Systems III and IV. Vibrational Analysis,
W. Hayes,
Proc. Phys. Soc. A, 68, 1097-1106

(55.7) System II. Vibrational Analysis,
W. Hayes and T. E. Nevin,
Proc. Phys. Soc. A, 68, 665-9

(62.8) $^7\Pi$ - $^7\Sigma$ Systems. Vibrational Analysis,
S. V. K. Rao, S. P. Reddy, and P. T. Rao,
Proc. Phys. Soc. 79, 741-4

(64.9) Dissociation Energy by Mass Spectra,
R. A. Kent, T. C. Ehlert, and J. L. Margrave,
J. Am. Chem. Soc. 86, 5090-3

MnH

Methods of Production and Experimental Technique

Absorption in a King furnace.

Emission from a discharge tube, King furnace, high pressure arc in H_2.

BAND SYSTEMS

	System	Transition	Sources	Wavelength Limits	Degrading	Characteristic Bands, λ	Remarks	Bibliography
5677 Å	I	$A^7\Pi \rightarrow X^7\Sigma$	Discharge	6900-5000	V	5677.5(0,0)	Perturbed	(53.6, 48.5, 45.4, 42.3)
4800 Å	II	-	Discharge	~4800	R	4794	Complex quintet structure	(53.6, 38.2, 36.1)
4500 Å	III	-	Discharge	4700-4400	R	4475	Complex quintet structure	(53.6, 38.2)
Infrared	IV	-	Discharge	9000-6900	-	-	Complex quintet structure	(53.6)

I. $A^7\Pi \rightarrow X^7\Sigma$ System

Band heads:

(v', v'')	(0, 2)	(0, 1)	(0, 0)	(1, 0)
λ	6849	6244	5724	5212

MnH

SPECTROSCOPIC CONSTANTS

State	T_e	ω_e	$x_e\omega_e$	B_e	α_e	$D_o \times 10^4$	r_e	Remarks	Bibliography
$A^7\Pi$	~17700	(1708)	–	6.425	0.187	3.62	(1.703)		(73.L1, 53.6)
$X^7\Sigma$	0	1548.0	28.8	5.6841	0.1570	3.024	1.722		(57.7)

Dissociation energy = 2.4 ± 0.3 eV, 55 kcal/mole, 19358 cm^{-1}.

Perturbations and General Information

A$^7\Pi$ state is perturbed in the two known vibrational states.

BIBLIOGRAPHY

(36.1) Note,
T. Heimer,
Naturwissenschaften 24, 521-2

(38.2) Analysis,
R. W. B. Pearse and A. G. Gaydon,
Proc. Phys. Soc. 50, 201-6

(42.3) Analysis,
T. E. Nevin,
Proc. Roy. Irish Acad. A 48, 1-42

(45.4) Analysis,
T. E. Nevin,
Proc. Roy. Irish Acad. A 50, 123-37

(48.5) Analysis,
T. E. Nevin and P. J. Doyle,
Proc. Roy. Irish Acad. A 52, 35-50

(53.6) Analysis,
T. E. Nevin and D. V. Stephens,
Proc. Roy. Irish Acad. A 55, 109-16

(57.7) Analysis,
W. Hayes, P. O. MacCarvill, and T. E. Nevin,
Proc. Phys. Soc. A 70, 904-5

(63.8) I. Kovacs,
"On the Anomalous Splitting of the Multiplet States of Diatomic Molecules,"
Bull. Sci. Fac. Chim. Ind., Bulgaria 21, 44-50

(73.L1) Experimental and Theoretical Spectroscopic Constants,
P. S. Bagus and H. F. Schaefer III,
"$^7\Sigma^+$ and $^7\Pi$ States of Manganese Hydride,"
J. Chem. Phys. 58, 1844-8

MnO

Methods of Production and Experimental Technique

Absorption after flash photolysis.

Emission from arc, flame, explosions (71.11).

Arc of manganese electrodes.

BAND SYSTEMS

System	Transition	Sources	Wavelength Limits	Degrading	Band Head, $\nu_{0,0}$	Remarks	Bibliography
I	A ⇌ X	Arc	6700-4500	R	17910		(59.7, 57.5)
II	B ← X	Absorption	2580-2550	-	39000		(60.8)

MnO

I. **A ⇌ X System**

Most intense band heads, λ(intensity):

v', v''	0	1	2	3	4	5
0	5582	5853(6)				
1			5879.5(9)	6174.75(7)		
2	5158.9(4)	5390.0(7)		5910.1(5)	6203.6(7)	
3		5192.0(5)	5423.2(4)			6236.5(6)

II. **B ← X System**

Bands appear like a continuum with a maximum near 2556-2578 Å; faint heads (60.8).

MnO

SPECTROSCOPIC CONSTANTS

State	T_e	ω_e	$x_e\omega_e$	B_e	$\alpha_e \times 10^3$	$D_e \times 10^6$	r_e	Remarks	Bibliography
B	39000	–	–						(60.8)
A	17949	762.75	9.6					$y_e\omega_e = 0.06$ cm^{-1}	(59.7)
X	0	839.55	4.79						(59.7)

Dissociation energy = 3.71 ± 0.17 eV, 85.4 kcal/mole, 29900 cm^{-1} (67.10).

Perturbations and General Information

A state is perturbed so several values can be assigned for the vibration constants (57.5).

Franck-Condon factors and r-centroids — Morse potential (71.11):

v'', v'	0	1	2
0	0.7930 1.820	0.3697 1.736	
1	0.3173 1.868	0.2160 1.826	
2	0.7987 1.900	0.2032 1.874	0.8368 1.832
3	0.0029 --		

Top = Franck-Condon factors, and bottom = r-centroids.

BIBLIOGRAPHY

(27.1) R. Mecke,
Z. Physik 42, 390-425

(34.2) A. K. SenGupta,
Z. Physik 91, 471-4

(49.3) Dissociation Energy,
L. Brewer and D. F. Mastick,
UCRL 571

(51.4) Flame Photometry. Dissociation Energy,
L. Huldt and A. Lagerqvist,
Arkiv Fysik 3, 525-31

(57.5) A. Gatterer, J. Junkes, E. W. Salpeter, and B. Rosen,
"Molecular Spectra of Metallic Oxides,"
ed. Specola Vaticana

MnO

(59.6) Equilibrium in Flames. Dissociation Energy,
P. J. Padley and T. M. Sugden,
Trans. Faraday Soc. 55, 2054-61

(59.7) Emission Vibrational Analysis,
J. M. Das Sarma,
Z. Physik 157, 98-105

(60.8) Ultraviolet Band in Flash Photolysis. A-X System in Absorption,
A. B. Callear and R. G. W. Norrish,
Proc. Roy. Soc. A, 259, 304-24

(67.10) Dissociation Energy,
C. J. Cheetham and R. F. Barrow,
Advances High Temp. Chem. 1, 7-41

(71.11) Franck-Condon Factors, Morse Potential, A-X System,
P. S. Dube, A. K. Chaudhry, G. O. Basuah, and D. K. Rai,
"Variation of the Electronic Transition Moment and the Effective Vibrational Temperature in 4800-6700 Å System of MnO Molecule,"
Applied Spectrosc. 25, 554-6

MnS

Methods of Production and Experimental Technique

Thermal emission from a King furnace at 2200°C.

BAND SYSTEMS

System	Transition	Sources	Wavelength Limits	Degrading	Band Head, $\nu_{0,0}$	Remarks	Bibliography
I	A(?) → X	Emission	4900-5890	R	5300.9 (0, 0)		(73. L1)
II	B(?) → X	Emission	4200-4750	R	4479 (0, 0)		(73. L1)

MnS

I. <u>A(?) → X System</u>

v', v''	0	1	2	3
0	5300.8	5440.9	5586.5	5739.9
1	5199.8			
2	5102.9			
3	5009.8			
4	4919.8			

II. <u>B(?) → X System</u>

v', v''	0	1	2
0	4479	4576.6	4679.9
1	4388.4		
2	4301.5		
3	4217.5		

SPECTROSCOPIC CONSTANTS

State	T_e	ω_e	$x_e\omega_e$	B_e	$\alpha_e \times 10^3$	$D_e \times 10^6$	r_e	Remarks	Bibliography
B(?)	22332	466.0	2.0						(73.L1)
A(?)	18920.3	369.2	1.16						(73.L1)
X	0	490.4	2.6						(73.L1)

Dissociation energy = 2.85 ± 0.11 eV, 65.7 kcal/mole, 23000 cm^{-1} (67.2, 65.1).

BIBLIOGRAPHY

(65.1) H. Wiedemeier and P. W. Gilles,
J. Chem. Phys. 42, 2765-9

(67.2) J. Drowart, A. Pattoret, and S. Smoes,
Proc. Brit. Ceram. Soc. 8, 67-89

(73.L1) A. Monjazeb and H. Mohan,
"Thermal Emission Spectra of Two New Diatomic Emitters — CrS and MnS,"
Spectrosc. Letters 6, 143-6

NBr

Methods of Production and Experimental Technique

Infrared absorption of the radical formed by ultraviolet flash-photolysis of NBr$_3$.

Emission (Br$_2$ + active nitrogen).

BAND SYSTEMS

	System	Transition	Sources	Wavelength Limits	Degrading	Band Head, $\nu_{0,0}$	Remarks	Bibliography
	I	b$^1\Sigma^+ \to$ X$^3\Sigma^-$	Active N$_2$	6500-5500	V	14837		(61.2, 39.1)

NBr

I. $b^1\Sigma^+ \to X^3\Sigma^-$ System

Band heads (P):

v', v''	λ	Intensity	v', v''	λ	Intensity
6, 6	6487.5	3	4, 2	6047.9	6
7, 7	6445.7	4	5, 3	6019.5	7
1, 0	6404.8	5	6, 4	5990.7	8
2, 1	6370.1	5	7, 5	5962.4	9
3, 2	6335.4	6	8, 6	5933.8	9
4, 3	6300.2	6	9, 7	5905.0	8
5, 4	6265.9	6	10, 8	5876.2	5
6, 5	6231.1	6	7, 4	5741.9	6
7, 6	6196.0	5	8, 5	5718.6	7
8, 7	6163.5	4	9, 6	5695.1	7
3, 1	6075.9	3	10, 7	5671.5	5

NBr

SPECTROSCOPIC CONSTANTS

State	T_e	ω_e	$x_e\omega_e$	B_e	$\alpha_e \times 10^3$	$D_e \times 10^6$	r_e	Remarks	Bibliography
$b^1\Sigma^+$	14787.3	785.5[a]	4.363[a]	0.4733	15.2	–	1.729		(61.2)
$X^3\Sigma^-$	0	691.75[a]	4.720[a]	0.444[b]	4.0[b]	–	1.79		(61.2)

[a] Measured from heads, [b] possibly in error.

Dissociation energy = 2.8 ± 0.2 eV, 65 kcal/mole, 22584 cm^{-1}.

NBr

Perturbations and General Information

Potential energy curves — RKR potential (66.4):

State	v	U(cm^{-1})	r_{min}(Å)	r_{max}(Å)
$X^3\Sigma$	0	345.3	1.725	1.853
	1	1028.3	1.685	1.908
	2	1702.3	1.659	1.949
	3	2364.3	1.640	1.985
	4	3017.3	1.623	2.017
	5	3661.3	1.609	2.048
	6	4295.3	1.597	2.076
	7	4920.3	1.586	2.104
	8	5535.3	1.576	2.131
$b^1\Sigma^+$	0	391.4	1.672	1.792
	1	1164.4	1.634	1.844
	2	1931.4	1.610	1.882
	3	2691.4	1.591	1.914
	4	3441.4	1.576	1.944
	5	4181.4	1.562	1.971
	6	4916.4	1.550	1.997
	7	5638.4	1.540	2.022
	8	6354.4	1.530	2.046

BIBLIOGRAPHY

(39.1) Spectral Production,
A. Elliott,
Proc. Roy. Soc. A, 169, 469-75

(61.2) Rotational Analysis,
E. R. V. Milton, H. B. Dunford, and A. E. Douglas,
J. Chem. Phys. 35, 1202-11

(64.3) Infrared Absorption in Matrix,
D. E. Milligan and M. E. Jacox,
J. Chem. Phys. 40, 2461-6

(66.4) R. B. Singh and D. K. Rai,
"Potential Curves for Some Diatomic Molecules: P_2, PN, SiN, NBr, BaO, BeF, SiF, and SnF,"
Indian J. Pure Appl. Phys. 4, 102-5

NCl

Methods of Production and Experimental Technique

Absorption after flash-photolysis of NCl_3. Infrared absorption in a solid matrix.

Emission of $Cl_2 + N_2$ in afterglow of a microwave discharge.

BAND SYSTEMS

System	Transition	Sources	Wavelength Limits	Degrading	Band Head, $\nu_{0,0}$	Remarks	Bibliography
I	$b^1\Sigma^+ \to X^3\Sigma^-$	Afterglow	~6648	V	15038.94		(67.3)
II	$A^3\Pi \leftarrow X^3\Sigma^-$	Flash photolysis	2401-2392	R	41666		(64.2)

NCl

I. $b^1\Sigma^+ \to X^3\Sigma^-$ System

Band heads ($^Q P$), λ:

v', v''	0	1	2	3
0	6646.7	6261.7		
1	7002.9	6599.3	6223.8	
2		6970.1	6554.6	6186.8
3			6913.7	6507.3

II. $A^3\Pi \leftarrow X^3\Sigma^-$ System

Bands of greatest intensity:

λ | 2401.3 | 2397.8 | 2392.5

SPECTROSCOPIC CONSTANTS

State	T_e	ω_e	$x_e\omega_e$	B_o	$\alpha_e \times 10^3$	$D_o \times 10^6$	r_e	Remarks	Bibliography
$A^3\Pi$	~41600	550	–	–	–	–	–		(64.2)
$b^1\Sigma^+$	14984.6	935.6[a]	5.4[a]	0.6828	–	1.65	1.5653	$\lambda = 1.77$ cm^{-1}	(67.3)
$X^3\Sigma^-$	0	827.0[a]	5.1[a]	0.6468	–	1.78	1.6083	$\gamma = 0.0071$ cm^{-1}	(67.3)

[a] Measured from heads.

Dissociation energy = 2.82 ± 0.2 eV, 65 kcal/mole, 22734 cm^{-1} (73.L1)

BIBLIOGRAPHY

(64.1) Infrared Absorption in a Matrix,
D. E. Milligan and D. E. Jacox,
J. Chem. Phys. 40, 2461-6

(64.2) A ← X System,
A. G. Briggs and R. G. W. Norrish,
Proc. Roy. Soc. A, 278, 27-34

(67.3) b → X System,
R. Colin and W. E. Jones,
Can. J. Phys. 45, 301-9

(73.L1) Dissociation Energy,
D. L. Hildenbrand (unpublished)

NF

Methods of Production and Experimental Technique

Absorption in a low temperature matrix.

Emission of NF_3 + Ar in afterglow of microwave discharge.

Microwave discharge in N_2F_4 + reaction with atom (70.6).

BAND SYSTEMS

System	Transition	Sources	Wavelength Limits	Degrading	Band Head, $\nu_{0,0}$	Remarks	Bibliography
I	$a^1\Delta \rightarrow X^3\Sigma^-$	Microwave discharge	8742	-	10866.7	One band	(70.6, 67.3, 64.1)
II	$b^1\Sigma^+ \rightarrow X^3\Sigma^-$	Microwave discharge	5600-5255	V	18905.20		(70.6, 66.2)

NF

I. $a^1\Delta \rightarrow X^3\Sigma^-$ System

One intense band at 8742 Å.

II. $B^1\Sigma^+ \rightarrow X^3\Sigma^-$ System

Bands of greatest intensity:

(v', v'')	(0, 1)	(1, 2)	(0, 0)	(1, 1)	(2, 2)
λ(Intensity)	5622(10)	5598(1)	5288(100)	5272(10)	5255(1)

NF

SPECTROSCOPIC CONSTANTS

State	T_e	ω_e	$x_e\omega_e$	B_e	$\alpha_e \times 10^3$	$D_e \times 10^6$	r_e	Remarks	Bibliography
$b\,^1\Sigma^+$	18877.05	1197.49	8.64	1.2377	14.48	5.28	1.3001		(66.2)
$a\,^1\Delta$	11435.16[a]	-	-	1.222[b]	-	4.5[c]	1.3082[d]		(67.3)
$X\,^3\Sigma^-$	0	1141.37	8.99	1.2056	14.92	5.39	1.3173	$\lambda = 1.215$ cm^{-1} $\gamma = 0.0048$ cm^{-1}	(66.2)

[a] T_o, [b] B_o, [c] D_o, [d] r_o.

Dissociation energy = 3.25 ± 0.22 eV, 75 kcal/mole, 26232 cm^{-1} (73.L1)

Perturbations and General Information

Radiative lifetime $b^1\Sigma^+ \rightarrow X^3\Sigma^-$ (70.6): $\tau = 0.2$ sec.

Electronic quenching: $NF(^1\Sigma^+) + N_2 \rightarrow NF(^3\Sigma^-) + N_2$ (70.6)
$k = 2 \times 10^{-16}$ cm^3/sec.

BIBLIOGRAPHY

(64.1) Infrared Spectrum in a Matrix,
D. E. Milligan and M. E. Jacox,
J. Chem. Phys. 40, 2461-6

(66.2) b-X System, Rotational Analysis,
A. E. Douglas and W. E. Jones,
Can. J. Phys. 44, 2251-8

(67.3) a-X System, Rotational Analysis,
W. E. Jones,
Can. J. Phys. 45, 21-6

(67.4) Theory, LCAO-MO Calculations,
R. C. Sahni,
"Quantum Mechanical Treatment of Molecules: Part 3, Predictions of Electronic States of NF by a Comparative Study of the Electronic States of N_2, CO, O_2, and NF Molecules,"
Trans. Faraday Soc. 63, 801-5

(69.5) S. P. Ionov and G. V. Ionova,
"Dissociation Energy and Electronic Structure of Certain Diatomic Molecules,"
Russ. J. Phys. Chem. 43, 1234-6

(70.6) Production by Chemical Reaction,
M. A. A. Clyne and I. F. White,
"Electronic Energy Transfer Processes in Fluorine-Containing Radicals: Singlet NF,"
Chem. Phys. Letters 6, 465-7

(71.7) Electron Paramagnetic Resonance Spectrum,
A. H. Curran, R. G. MacDonald, A. J. Stone, and B. A. Thrush,
"The Electron Paramagnetic Resonance Spectrum of NF($^1\Delta$) in the Gas Phase,"
Chem. Phys. Letters **8**, 451-3

(73.L1) Dissociation Energy,
D. L. Hildenbrand (unpublished)

NH

Methods of Production and Experimental Technique

Absorption after flash-photolysis of HNCO in flames containing NH_3.

Emission from discharge tubes, low pressure arcs, hollow cathodes containing NH_3 in $H_2 + N_2$ chemiluminescence.

Discharge into mixture of Helium, Nitrogen, and Hydrogen.

Shock tube methods.

Shock wave through Xenon carrying ammonia.

Combustion in Hydrogen flames.

Flash photolysis of HN_3.

BAND SYSTEMS

System	Transition	Sources	Wavelength Limits	Degrading	Characteristic Bands, λ	Remarks	Bibliography
I	$A^3\Pi_i \rightleftarrows X^3\Sigma^-$	Discharge	3680-3020	-	3360		(70.50, 66.31, 66.33, 60.15, 60.16, 59.13)
II	$c^1\Pi \rightarrow a^1\Delta$	Discharge	3650-3030	R	3353.4(0,0)		(66.33, 64.21, 34.3, 33.2)
III	$c^1\Pi \rightarrow b^1\Sigma^+$	Discharge	~4502	R	4523.2(0,0)		(68.42, 66.33, 35.4)
IV	$d^1\Sigma^+ \rightarrow c^1\Pi$	Discharge	~2530	V	2530.2(0,0)		(67.35, 66.30, 66.33, 36.6)
V	$d^1\Sigma^+ \leftarrow a^1\Delta$	Atmospheric absorption	~2540	-	2530.2(0,0)		(72.L2)

NH

I. $A^3\Pi_i \rightleftarrows X^3\Sigma^-$ System

Band heads (Q), λ:

v', v''	0	1	2
0	3360	3637.7	
1	3050.3	3370	3635.0
2			3383

II. $c^1\Pi \rightarrow a^1\Delta$ System

Band heads (Q):

(v', v'')	(0, 0)	(1, 0)	(0, 1)
λ	3353.4	3042.6	3627.2

III. $c^1\Pi \rightarrow b^1\Sigma^+$ System

Band heads (Q):

(v', v'')	(0, 0)
λ	4523.2

IV. $d^1\Sigma^+ \rightarrow c^1\Pi$ System

Band heads (Q):

(v', v'')	(0, 0)
λ	2530.2

V. $d^1\Sigma^+ \leftarrow a^1\Delta$ System

Band heads (Q and R):

(v', v'')	(0, 0)
λ	2557(R)
λ	2530.2(Q)

NH

SPECTROSCOPIC CONSTANTS

State	T_e	ω_e	$x_e\omega_e$	B_e	α_e	$D_o \times 10^3$	r_o	Remarks	Bibliography
$d^1\Sigma^+$	83413.0[a]	-	-	14.409	0.634	1.666	1.125		(67.35, 36.6)
$c^1\Pi$	43902.6[a]	2503	194	14.697	1.015	2.19	1.112		(67.35, 64.21)
$A^3\Pi_i$	29776.8	-	-	16.3181	0.7440	1.778	1.046	$H_o = 7.80$ $\times 10^{-8}$ cm^{-1}	(59.13)
$b^1\Sigma^+$	21797[a]	3354.7	74.4	16.7326	0.6049	1.654	1.043		(68.42, 35.4)
$a^1\Delta$	13147[a]	3231.0	98.5	16.439[b]	-	1.62	1.041		(70.50, 64.21)
$X^3\Sigma^-$	0	3203.2	78.3	16.666	0.648	1.699	1.045	$H_o = 1.048$ $\times 10^{-7}$ cm^{-1}	

[a] Theoretical prediction (68.44), [b] B_o.

Dissociation energy = 3.2 ± 0.16 eV, 74 kcal/mole, 25810 cm^{-1}.

Perturbations and General Information

(0, 0) band of $d^1\Sigma^+ \to c^1\Pi$ system predissociates for $J'' \geq 24$ and (1, 1) band for $J'' \geq 16$, and is caused by interaction of $d^1\Sigma^+$ with repulsive $^1\Pi$ state (69.45).

Electronic quenching cross sections (64.2):

Species	Partner	Cross section ($\times 10^{-16}$ cm^2)
$c^1\Pi$	Ar	≤ 0.01
	N_2	0.5
	H_2	3
	NH_3	~ 16
$A^3\Pi$	N_2H_4	~ 40

Formation of $NH(A^3\Pi)$ by shock heating NH_3: $2900°K \leq T \leq 9600°K$

$$k = 4 \times 10^{-14} \exp\left[\frac{-54000}{T}\right] cm^3/\text{molecule-sec}.$$

Rate is second order in initial NH_3 concentration.

Radiative lifetimes (69.19):

$$A^3\Pi - X^3\Sigma^- \qquad \tau = 455 \pm 90 \text{ nsec}$$
$$c^1\Pi - a^1\Delta \qquad \tau = 480 \pm 90 \text{ nsec}$$
$$c^1\Pi - b^1\Sigma \qquad \tau = 485 \pm 90 \text{ nsec}$$
$$d^1\Sigma - c^1\Pi \qquad \tau = 18 \pm 3 \text{ nsec}$$

Franck-Condon factors, r-centroids, A factors — RKR potential (71.54):

$A^3\Pi - X^3\Sigma^-$

v', v''	0	1	2
0	0.9998-0 1.078 0.220	0.204-3	0.202-4
1	0.198-3	0.997-0 1.148 0.220	0.258-2
2	0.284-4	0.250-2	0.987-0 1.218 0.220

Top = Franck-Condon factors, middle = r-centroids, and bottom = $A_{v',v''}(\times 10^7 \text{ sec}^{-1})$.

$c^1\Pi - a^1\Delta$

v'', v'	0	1	2	3
0	0.756-0 1.082 0.140	0.198-0 0.958 0.037	0.381-1 0.826 0.0074	0.700-2 0.672 0.001
1	0.199-1 1.265 --	0.296-0 1.111 --	0.304-0 1.023 --	0.133-0 0.931 --
2	0.390-1 1.367 --	0.312-0 1.300 --	0.190-1 1.038 --	0.185-0 1.076 --
3	0.500-2 1.513 --	0.150-0 1.387 --	0.209-0 1.354 --	0.420-1 1.375 --
4	0.100-2 1.495 --	0.360-1 1.536 --	0.267-0 1.427 --	0.280-1 1.471 --

Top = Franck-Condon factors, middle = r-centroids, and bottom = $A_{v',v''}(\times 10^7 \text{ sec}^{-1})$.

$c^1\Pi - b^1\Sigma$

v'', v'	0	1	2	3
0	0.788-0 1.087 1.58	0.178-0 0.957 0.47	0.285-1 0.810 0.051	0.387-2 0.612 --
1	0.178-0 1.288 --	0.377-0 1.127 --	0.293-0 1.034 --	0.105-0 0.937 --
2	0.304-1 1.375 --	0.302-0 1.333 --	0.617-1 1.126 --	0.219-0 1.107 --
3	0.238-2 1.597 --	0.119-0 1.410 --	0.262-0 1.391 --	0.172-1 1.475 --

Top = Franck-Condon factors, middle = r-centroids, and bottom = $A_{v', v''}(\times 10^7 \text{ sec}^{-1})$.

$d^1\Sigma - c^1\Pi$

v'', v'	0	1	2	3	4
0	0.995-0 1.135 5.54	0.339-3 -- --	0.404-2 0.993 0.021	0.241-3 -- --	0.144-3 0.759 --
1	0.245-3 -- --	0.974-0 1.191 --	0.686-2 2.884 --	0.174-1 1.210 --	0.254-3 0.600 --
2	0.367-2 1.434 --	0.111-1 0.217 --	0.862-0 1.251 --	0.616-1 2.006 --	0.578-1 1.421 --
3	0.820-3 1.087 --	0.569-2 1.817 --	0.103-0 0.949 --	0.541-0 1.300 --	0.169-0 1.849 --

Top = Franck-Condon factors, middle = r-centroids, and bottom = $A_{v', v''}(\times 10^7 \text{ sec}^{-1})$.

Potential energy curves (69.45):

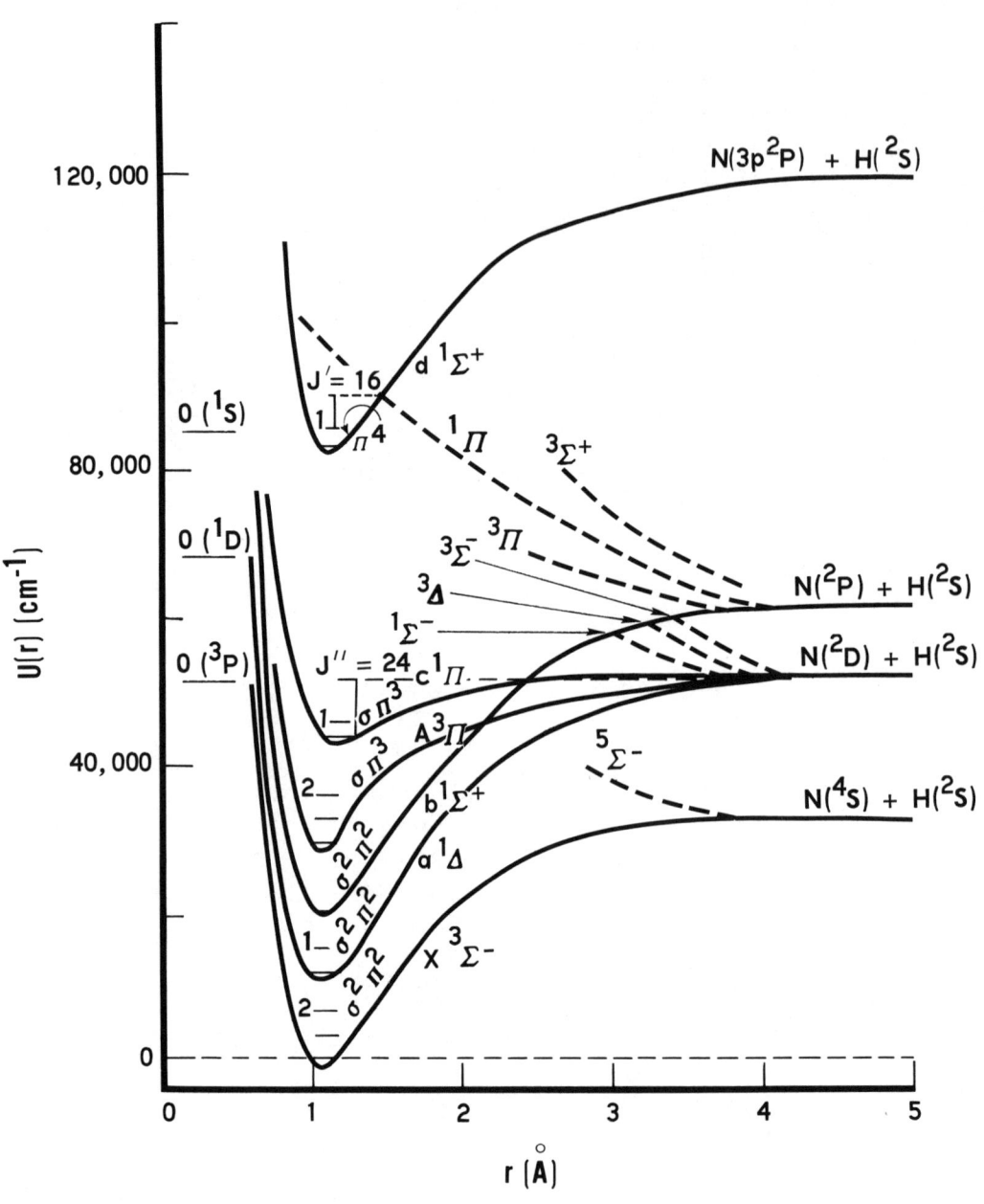

BIBLIOGRAPHY

(893. 1) First Observations,
J. M. Edler,
Denkschr. Kais. Akad. Wiss. Wien 60, 1

(33. 2) c-a System,
R. W. B. Pearse,
Proc. Roy. Soc. A 143, 112-23

(34. 3) c-a System,
G. Nakamura and T. Shiedi,
Japan J. Phys. 10, 5-10

(35. 4) c-b System,
R. W. Lunt, R. W. B. Pearse, and E. C. W. Smith,
Proc. Roy. Soc. A 151, 602-9

(35. 5) A-X System,
G. W. Funke,
Z. Physik 96, 787-98

(36. 6) d-c System,
R. W. Lunt, R. W. B. Pearse, and E. C. W. Smith,
Proc. Roy. Soc. A 155, 173-82

(36. 7) A-X System,
G. W. Funke,
Z. Physik 101, 104-12

(55. 8) A-X System,
G. Pannetier, H. Guenebout, and A. G. Gaydon,
C. R. Acad. Sci. 246, 958-60

(58. 9) LCAO-MO-SCF Calculations of the Different Valence Levels,
M. E. Boyd,
J. Chem. Phys. 29, 108-15

(58. 10) Ground State Studied by LCAO-MO-SCF Method,
M. Krauss,
J. Chem. Phys. 28, 1021-6

(58. 11) Study of Configuration Interaction Using the Wave Function
of (58. 10),
M. Krauss and J. F. Wehner,
J. Chem. Phys. 29, 1287-97

(58. 12) Study of the Ground and Excited States by LCAO-MO-SCF-VB Method,
A. C. Hurley,
Proc. Roy. Soc. A 248, 118-35

(59. 13) Analysis of the A-X System,
R. N. Dixon,
Can. J. Phys. 37, 1171-86

(59. 14) Study of the Ground and Excited States by LCAO-MO-SCF-VB Method,
A. C. Hurley,
Proc. Roy. Soc. A 249, 402-13

(60. 15) A-X System,
H. Guenebaut and G. Pannetier,
C. R. Acad. Sci. 250, 3613-5

(60. 16) A-X System,
H. Guenebaut, G. Pannetier, and P. Goudmand,
C. R. Acad. Sci. 251, 1166-8

(61. 17) Study of the Ground and Excited States by the LCAO-MO-SCF-VB Method,
A. C. Hurley,
Proc. Roy. Soc. A 261, 237-45

(62. 18) G. Pannetier, L. Marsigny, and M. Ben-Caid,
"Spectroscopic Study of the Reaction of Atomic Oxygen in Nitrogen with Hydrocarbons Containing Chlorine and Bromine: Observation of the Transition ($B^2\Sigma_u^+ \rightarrow X^2\Sigma_g^+$) of the N_2^+ Molecule in Certain Reactions,"
C. R. Acad. Sci. 12, 1270-1

(63. 19) LCAO-SCF Calculations of $X^3\Sigma^-$ Over a Gaussian Base,
C. M. Reeves,
J. Chem. Phys. 39, 1-10

(64. 20) Valence-Band Calculation of a Simple Configuration of the Ground State,
D. M. Bishop and H. R. Hoyland,
Molecular Phys. 7, 161-4

(64. 21) c-a System,
M. Shimauchi,
Science Light 13, 53-62

(64.22) K. H. Becker and K. H. Welge,
"Fluorescent Study and Photochemical Primary Processes in the Vacuum Ultraviolet in NH_3, N_2H_4, PH_3 and Reactions of the Electronically Excited Radiacals $NH^*(^1\pi)$, $NH^*(^3\pi)$, $PH^*(^3\pi)$,"
Z. Naturforsch. A 19, 1006-15

(64.23) E. Fink and K. H. Welge,
"Lifetimes of the Electronic States of $N_2(C^3\pi_u)$, $N_2^+(B^2\Sigma^+)$, $NH(A^3\pi)$, $NH(c^1\pi)$, $PH(^3\pi)$,"
Z. Naturforsch. A 19, 1193-1201

(64.24) D. E. Milligan and M. E. Jacox,
"Infrared Studies of the Photolysis of HN_3 in Inert and Reactive Matricies: The Infrared Spectrum of NH,"
J. Chem. Phys. 41, 2838-41

(64.25) M. W. P. Cann and S. W. Kash,
"Formation of NH from Shock-Heated Ammonia,"
J. Chem. Phys. 41, 3055-60

(65.26) LCAO-MO-SCF Calculations of the $X^3\Sigma^-$ Level,
J. B. Lounsbury,
J. Chem. Phys. 42, 1549-54

(65.27) LCAO-SCF Calculations of $X^3\Sigma^-$ over a Gaussian Base,
C. M. Reeves and R. Fletcher,
J. Chem. Phys. 42, 4073-81

(65.28) LCAO-MO-SCF Calculations of the $X^3\Sigma^-$ Level,
B. D. Joshi,
J. Chem. Phys. Suppl. n°10 (part 2) 43, 40-58

(66.29) Spin-Orbital Interaction of $X^3\Sigma^-$ Using the Wave Function of (58.9),
J. W. MacIver Jr., and H. F. Hameka,
J. Chem. Phys. 45, 767-73

(66.30) d-c System,
F. Grimaldi, A. Lecourt, H. Lefebvre-Brion and C. M. Moser,
J. Mol. Spectrosc. 20, 341-6

(66.31) The $^3\pi$ State,
T. Murai and M. Shimauchi
Science Light 15, 48-67

NH

(66.32) M. Shimauchi,
"Rotational Isotope Effect in the ($A^3\pi_i - X^3\Sigma^-$) System of NH and NO,
Science Light 15, 161-5

(66.33) K. H. Welge,
"Formation of NH($c^1\pi$) and NH($A^3\pi$) in the Vacuum-UV Photolysis of HN_3,"
J. Chem. Phys. 45, 4373-4

(67.34) Study of the Bonding Transitions Using the Wave Function Characteristics, (67.40),
R. F. W. Bader, I. Keaverny, and P. E. Cade,
J. Chem. Phys. 47, 3381-402

(67.35) F. L. Whittaker,
"The $\Delta v=0$ Sequence in the $d^1\Sigma^+ - c^1\pi$ System of NH and ND,"
Proc. Phys. Soc. 90, 535-541

(67.36) P. E. Cade,
"The Electron Affinities of the Diatomic Hybrids CH, NH, SiH and PH,"
Proc. Phys. Soc. 91, 842-54

(67.37) M. Horani, J. Rostas, and H. Lefebvre-Brion,
"Fine Structure of $^3\Sigma^-$ and $^3\pi$ States of NH, OH^+, PH and SH^+,"
Can. J. Phys. 45, 3319-31

(67.38) LCAO-MO-SCF Calculations of the Ground State. Hartree Fock Limit Reached,
P. E. Cade and W. M. Huo,
J. Chem. Phys. 47, 614-48

(67.39) Calculation Using a Base at a Centered Null Field,
J. B. Lounsbury,
J. Chem. Phys. 46, 2193-200; Erratum (67.39a)

(67.39a) J. B. Lounsbury,
J. Chem. Phys. 47, 1566

(67.40) J. B. Lounsbury,
J. Chem. Phys. 47, 1566

(67.41) J. W. MacIver Jr., and H. F. Hameka,
J. Chem. Phys. 46, 825-6

(68.42) F. L. Whittaker,
"The $c^1\pi - b^1\Sigma^+$ Band System of NH and ND,"
J. Phys. B: (Proc. Phys. Soc.) 2, 977-82

(68.43) W. M. Huo,
"Valence Excited States of NH and CH and Theoretical Transition Probabilities,"
J. Chem. Phys. 49, 1482-92

(68.44) Hartree Fock Calculations, a-X System,
P. E. Cade,
"Theoretical Prediction of the Singlet Triplet Intercombination Separations for NH, OH^+, PH, SH,"
Can. J. Phys. 46, 1989-91

(69.45) d-c System, Potential Energy Curves,
G. Krishnamurty and N. A. Narasimham,
"Predissociations in the $d^1\Sigma^+-c^1\pi$ Bands of NH,"
J. Mol. Spectrosc. 29, 410-4

(69.46) d-c, c-b, c-a, A-X Lifetimes,
W. H. Smith,
"Lifetimes and Total Transition Probabilities of NH, SiH and SiD,"
J. Chem. Phys. 51, 520-4

(69.47) R. M. Dagnall, D. J. Smith, K. C. Thompson, and T. S. West,
"Emission Spectra Obtained from the Combustion of Organic Compounds in Hydrogen Flames,"
Analyst 94, 871-8

(69.48) I. V. Sushavin and S. M. Kishko,
"Excitation of the Ultraviolet Bands of NH,"
Ukranian Phys. J. 14, 1892-4

(70.49) W. H. Smith,
"Franck-Condon Factor,"
Princeton University T.R. DAHC04 69 C 0081

(70.50) J. Malicet, J. Brion, and H. Guenebaut,
"Contribution to the Spectroscopic Study of the Transition $A^3\pi_i-X^3\Sigma^-$ of the NH Radical,"
J. Chim. Phys. 67, 25-30

(70.51) J. Kouba and Y. Öhm,
"Natural-Orbital Valence-Shell CI Studies of Diatomic Molecules. I. Potential Energy Curves and Spectra of Imidogen,"
J. Chem. Phys. 52, 5387-94

NH

(71.52)   LCAO-MO-SCF Calculations, Potential Energy Curves,
H. P. D. Liu and G. Verhaegen,
"Electronic States of NH and OH^+,"
Indian J. Quant. Chem. 5, 103-17

(71.53)   E. C. Shane and W. Brennen,
"Chemiluminescence of the Atomic Oxygen-Hydrogen Reaction,"
J. Chem. Phys. 55, 1479-80

(71.54)   W. H. Smith and H. S. Liszt,
"Franck-Condon Factors and Absolute Oscillator Strengths for NH, SiH, S_2 and SO,"
J. Quant. Spectrosc. Radiative Transfer 11, 45-54

(72.55)   L. Veseth,
"Fine Structure of $^3\pi$ and $^3\Sigma^-$ States in Diatomic Molecules,"
J. Phys. B: Atom. Mol. Phys. 5, 229-41

(72.56)   G. Schotz and M. J. Kaufman,
"Chemiluminescent Reactions of Atomic Fluorine,"
Princeton University T.R. NR 092-531, February 1972

(73.57)   W. J. Stevens,
"Abinitio Calculations of the Dissociation Energy of the $X^3\Sigma^-$ State of Imidogen,"
J. Chem. Phys. 58, 1264-6

(73.58)   J. M. Lents,
"An Evaluation of Molecular Constants and Transition Probabilities for the NH Free Radical,"
J. Quant. Spectrosc. Radiative Transfer 13, 297

(71.L1)   S. V. O'Neil and H. F. Schaefer III,
"Configuration Interaction Study of the $X^3\Sigma^-$, $a^1\Delta$, $b^1\Sigma^+$ States of NH,"
J. Chem. Phys. 55, 394-401

(72.L2)   R. W. Nicholls,
"Identification of Stratospheric NH,"
Nature 240, 142-3

NI

Dissociation energy = 1.64 ± 0.14 eV, 37.8 kcal/mole, 13200 cm^{-1} (68.1).

BIBLIOGRAPHY

(68.1) L. F. Phillips,
        Can. J. Chem. <u>46</u>, 1429-34

NO

Methods of Production and Experimental Technique

Absorption (gas at liquid oxygen temperature).

Emission of a discharge into a stream of NO mixed with a rare gas.

Microwave discharge afterglow (73.141).

Election impact excitation. Chemiluminescence.

BAND SYSTEMS
IN EMISSION

A - Non-Rydberg Systems

	System	Transition	Sources	Wavelength Limits	Degrading	Band Head, $v_{0,0}$	Remarks	Bibliography
β	I	$B^2\Pi \rightarrow X^2\Pi$	Discharge	6500-2000	R	2201.7(0,0)		(54.77, 50.68, 44.52, 26.3)
β'	II	$B'^2\Delta \rightarrow X^2\Pi$	Discharge	2000-1400	R	1719.4(0,1) 1685.1(1,1)		(59.87, 53.71, 53.72)
	III	$B'^2\Delta \rightarrow B^2\Pi$	Discharge	8000-5000	R	6373.89(1,0) 5937.47(2,0)		(64.104, 56.82, 53.74)

BAND SYSTEMS IN EMISSION

B - Rydberg Systems

	System	Transition	Sources	Wavelength Limits	Degrading	Band Head, $\nu_{0,0}$	Remarks	Bibliography
γ	I	$A^2\Sigma^+ \to X^2\Pi$	Discharge	3400-1950	V	2269.4(0,0)		(63.100, 58.86, 55.79, 48.66)
δ	II	$C^2\Pi \to X^2\Pi$	Discharge	2100-1840	V	1981.43(0,1) 1910.26(0,0)		(69.133, 65.113, 64.103, 54.75)
ε	III	$D^2\Sigma^+ \to X^2\Pi$	Discharge	1900-1700	V	1912.41(0,0) 1828.7(1,0)		(68.128, 55.79, 53.72)
	IV	$F^2\Delta \to X^2\Pi$	Discharge	1620-1450	V	1611.92(0,0) 1496.76(2,0)		(66.116, 65.110)
	V	$N^2\Delta \to X^2\Pi$	Discharge	1600-1350	R, V	1565.53(0,2) 1429.31(1,0)		(66.116, 65.110)
s-s	VI	$E^2\Sigma^+ \to A^2\Sigma^+$	High pressure discharge	6060-5950		6000.99(0,0) 6000.81(2,2)		(68.126, 64.104, 60.91, 59.89)
p-s	VII	$C^2\Pi \to A^2\Sigma^+$	High pressure discharge	12350-12050		12237(0,0)(Q)		(69.133, 69.134, 64.104, 63.100)
p-s	VIII	$D^2\Sigma^+ \to A^2\Sigma^+$	High pressure discharge	11300-10850		10998.5(0,0) 11074.6(1,1)	No head	(69.134, 68.124, 68.128, 64.104)
s-p	IX	$E^2\Sigma^+ \to D^2\Sigma^+$	High pressure discharge	13300-13100		13215(0,0)	No head	(64.104, 60.91)

	System	Transition	Sources	Wavelength Limits	Degrading	Band Head, $\nu_{0,0}$	Remarks	Bibliography
d-s	X	$H^2\Sigma^+$, $H'^2\Pi \to A^2\Sigma^+$	High pressure discharge	5410-5380	V	5400.4, 5402.5 (0,0)	ℓ-decoupling $Q(\Pi^- - \Sigma^+)$ $P(\Pi^+ - \Sigma^+)$ $R(\Pi^+ - \Sigma^+)$	(69.132, 64.104)
d-p	XI	$H^2\Sigma^+$, $H'^2\Pi \to D^2\Sigma^+$	High pressure discharge	10660-10500	V	10606.2, 10619 (0,0)	ℓ-decoupling $Q(\Pi^- - \Sigma^+)$ $P(\Pi^+ - \Sigma^+)$ $R(\Sigma^+ - \Sigma^+)$	(69.134, 64.104)
d-p	XII	$H^2\Sigma^+$, $H'^2\Pi \to C^2\Pi$	High pressure discharge	9750-9550	V	9691.8, 9661.8 (0,0)	ℓ-decoupling $R(\Sigma^+ - \Pi^+)$ $R(\Pi^- - \Pi^-)$ $P(\Pi^- - \Pi^-)$	(69.134, 64.104)
d-p	XIII	$F^2\Delta \to C^2\Pi$	High pressure discharge	10450-10100	R and V	10341.3(0,0)(Q) 10256.4(1,1)(Q)		(69.134, 64.104)
d-p	XIV	$O'^2\Pi \to D^2\Sigma^+$	High pressure discharge	6830-6730	V	6800.9(0,0) 6772.8(1,1)	ℓ-decoupling $Q(\Pi^- - \Sigma^+)$	(69.132, 68.126)

BAND SYSTEMS IN EMISSION

B - Rydberg Systems (cont.)

	System	Transition	Sources	Wavelength Limits	Degrading	Band Head, $\nu_{0,0}$	Remarks	Bibliography
d-p	XV	$O'^2\Pi - C^2\Pi$	High pressure discharge	6450-6350	V	6400.5(0,0)	l-decoupling $R(\Pi^- - \Pi^-)$ $P(\Pi^- - \Pi^-)$	(69.132, 68.126)
d-p	XVI	$N^2\Delta - C^2\Pi$	High pressure discharge	6630-6520	R	6568.8(R) (0,0) 6537.5(R)		(69.132, 68.126)

C - Quartet-Quartet Systems in Emission

$\underline{b^4\Sigma^- \to a^4\Pi \text{ System}}$

Three groups of bands with multiple heads, $9800 > \lambda > 7500$ $(v'-v'' = 0)$ $\lambda = 9700$, $(v'-v'' = 1)$ $\lambda = 8650$, and $(v'-v'' = 2)$ $\lambda = 7800$.

SPECTROSCOPIC CONSTANTS

State	T_e	ω_e	$x_e\omega_e$	B_e	$\alpha_e \times 10^3$	$D_e \times 10^6$	r_e	Remarks	Bibliography
$G^2\Sigma^-$	62911.7	1085.5	11.08	1.2523	20	-	1.3426	$y_e\omega_e = 0.144$ cm^{-1}	(64, 102)
$B'^2\Delta$	60364.2	1217.4	15.61	1.332	21	-	1.302	$A = -2.4$ cm^{-1}	(66, 116)
$B^2\Pi_{3/2}$	45947.3	1038.3	7.455	1.178	18.9	-	1.383		(30, 22, 27.8)
$B^2\Pi_{1/2}$	45919.5	1036.9	7.460	1.076	11.6	-	1.449		(30, 22, 27.8)
Rydberg	-	2377	16.4	2.002	20	-	1.1062		(64, 104)
$X^2\Pi_{3/2}$	121	1903.68	13.97	1.7046	17.8	-	1.1508	$y_e\omega_e = -0.001\frac{2}{1}$ cm^{-1} $A = 124.2$ cm^{-1}	(39, 35)
$X^2\Pi_{1/2}$	0	1904.03							

Dissociation energy = 6.50 ± 0.01 eV, 149.9 kcal/mole, 52427 cm^{-1}.

ν_{oo} of the Rydberg levels is given by $\nu_{oo} = 74770 - \dfrac{R}{(n-a)^2}$ (65, 111).

NO

Perturbations and General Information

Electronic quenching cross sections: $NO(^2\Sigma) + M \rightarrow NO(^2\Pi) + M$

M	Cross section ($\overset{\circ}{A}{}^2$)
NO	14
N_2	≤ 0.02
CO	0.6
H_2O	30
CO_2	5

Rate of radiative recombination (73.141): $N + O \rightleftarrows (C^2\Pi) \rightarrow NO(X^2\Pi) + h\nu$

$k = (2.9 \pm 0.5) \times 10^{-18}$ cm^3/sec at 300°K

Radiative lifetimes:

$A^2\Sigma^+ - X^2\Pi$ $\tau = 178 \pm 19$ nsec (72.L12)
                     $\tau = 108 \pm 6$ nsec (72.L13)
                     $\tau = 410$ nsec (68.130)

$B^2\Pi - X^2\Pi$ $\tau = 3.16$ μsec (64.106)

$D^2\Sigma - X^2\Pi$ $\tau = 18.4$ nsec

Electronic transition laser action has been observed at 1.02 μm from the transition $F^2\Delta \rightarrow C^2\Pi$, (1,1) (66.120, n.p. 145).

Vibrational laser action has been observed between $5.8 \leq \lambda \leq 6.4$ μm (71.138a, 66.120a, 66.120b).

NO

Franck-Condon factors (followed by factor of 10) — Morse potential (64.107):

$B^2\Pi - X^2\Pi$ (β bands)

v', v''	0	1	2	3	4	5	6
0	2.2641-5	3.1095-4	2.0460-3	8.5693-3	2.5605-2	5.7994-2	1.0327-1
1	1.8507-4	2.0886-3	1.0933-2	3.4872-2	7.4613-2	1.1026-1	1.0966-1
2	7.9208-4	7.2777-3	2.9747-2	6.9589-2	9.8353-2	7.7835-2	2.2280-2
3	2.3640-3	1.7488-2	5.4452-2	8.8170-2	7.0117-2	1.4830-2	3.9847-3
4	5.5296-3	3.2490-2	7.4499-2	7.5692-2	2.1921-2	3.2797-3	4.7031-2
5	1.0802-2	4.9586-2	7.9752-2	4.1836-2	3.7157-5	3.3852-2	4.8235-2
6	1.8342-2	6.4437-2	6.7470-2	1.0719-2	1.3818-2	4.9225-2	1.3380-2
7	2.7819-2	7.2876-2	4.3780-2	2.3794-5	3.7263-2	3.0752-2	6.8564-4
8	3.8444-2	7.2602-2	1.9422-2	9.9449-3	4.3958-2	5.0880-3	1.9595-2
9	4.9137-2	6.3933-2	3.6587-3	2.7687-2	3.0196-2	1.2400-3	3.5473-2
10	5.8772-2	4.9435-2	2.9977-4	3.9271-2	1.0517-2	1.5594-2	2.7802-2

$A^2\Sigma^+ - X^2\Pi$ (γ bands)

v', v''	0	1	2	3	4	5	6
0	1.6725-1	2.6456-1	2.3741-1	1.5978-1	9.0065-2	4.5066-2	2.0735-2
1	3.3431-1	1.0242-1	1.1580-3	7.4263-2	1.3583-1	1.3330-1	9.7259-2
2	2.9262-1	1.7957-2	1.5684-1	7.1223-2	1.7613-4	3.7296-2	9.0778-2
3	1.4727-1	2.0259-1	4.2298-2	4.3396-2	1.1315-1	4.6214-2	9.8246-5
4	4.7062-2	2.4039-1	4.6886-2	1.2242-1	6.9213-4	6.1636-2	8.5428-2
5	9.9525-3	1.2656-1	2.1521-1	5.6888-4	1.1140-1	4.1408-2	8.1102-3
6	1.4083-3	3.7778-2	1.9572-1	1.2393-1	4.2608-2	4.8568-2	8.3044-2
7	1.3177-4	6.9230-3	8.2277-2	2.2299-1	3.9266-2	9.5223-2	4.7082-3
8	7.8606-6	7.9343-4	1.9317-2	1.3398-1	2.0249-1	1.3818-3	1.1050-1
9	2.7802-7	5.5665-5	2.6807-3	4.0138-2	1.8008-1	1.4909-1	1.1266-2
10	5.0323-9	2.2372-6	2.1883-4	6.6919-3	6.8965-2	2.0966-1	8.6078-2

$C^2\Pi - X^2\Pi$ (δ bands)

v', v''	0	1	2	3	4	5	6
0	1.5303-1	2.5430-1	2.3802-1	1.6619-1	9.6817-2	4.9932-2	2.3636-2
1	3.1826-1	1.1664-1	3.4184-8	6.0911-2	1.2789-1	1.3451-1	1.0312-1
2	2.9469-1	7.9811-3	1.4637-1	8.5132-2	2.7400-3	2.5826-2	8.0610-2
3	1.6035-1	1.7343-1	6.0846-2	2.6051-2	1.0916-1	6.0224-2	2.7988-3
4	5.7021-2	2.3879-1	2.3254-2	1.2922-1	7.2589-3	4.3165-2	8.7758-2

NO

$D^2\Sigma^+ - X^2\Pi$ (ϵ bands)

v', v''	0	1	2	3	4	5	6
0	1.6588-1	2.5830-1	2.3329-1	1.6030-1	9.3089-2	4.8237-2	2.3038-
1	3.5011-1	1.0080-1	9.9080-4	6.8939-2	1.2782-1	1.2813-1	9.6030-
2	3.0414-1	2.7624-2	1.5930-1	6.5280-2	6.1723-5	3.5537-2	8.5003-
3	1.3914-1	2.4466-1	2.6133-2	5.7407-2	1.0916-1	3.7944-2	1.7524-
4	3.5527-2	2.4918-1	9.4737-2	1.0409-1	1.2430-3	7.3266-2	7.5852-
5	4.9046-3	9.9762-2	2.7168-1	1.1081-2	1.2446-1	1.7093-2	2.2750-
6	3.1618-4	1.8287-2	1.7000-1	2.2885-1	4.2543-3	9.1441-2	5.4692-
7	5.7018-6	1.3859-3	4.0353-2	2.2847-1	1.6074-1	3.5644-2	4.5282-
8	9.6899-9	2.3825-5	3.4892-3	6.8911-2	2.6769-1	9.6204-2	7.0993-
9	5.6809-9	1.8790-7	5.2947-5	6.6130-3	1.0096-1	2.8811-1	4.8315-
10	7.3032-12	3.8313-8	1.3231-6	7.9711-5	1.0461-2	1.3381-1	2.9422-

$B'^2\Delta - X^2\Pi$ (β' bands)

v', v''	0	1	2	3	4	5	6
0	1.8358-2	8.3318-2	1.7687-1	2.3362-1	2.1542-1	1.4749-1	7.7849-
1	5.8594-2	1.5426-1	1.4162-1	3.4431-2	6.8166-3	9.3789-2	1.7119-
2	1.0320-1	1.3858-1	2.4259-2	2.4264-2	1.0688-1	6.5626-2	3.5856-
3	1.3331-1	7.1848-2	5.6975-3	8.8161-2	3.8636-2	8.5167-3	8.6451-
4	1.4163-1	1.6810-2	5.0973-2	5.7444-2	2.7502-3	7.3045-2	3.2065-
5	1.3163-1	2.4605-5	7.5678-2	7.0741-3	4.7790-2	3.9753-2	7.4184-
6	1.1117-1	1.3118-2	5.9638-2	5.5592-3	5.9402-2	1.8304-4	5.4105-

$E^2\Sigma^+ - X^2\Pi$ (γ' bands)

v', v''	0	1	2	3	4	5	6
0	1.8292-1	2.7657-1	2.3718-1	1.5256-1	8.2230-2	3.9369-2	1.7346-
1	3.4419-1	8.8265-2	4.8132-3	9.0257-2	1.4532-1	1.3244-1	9.1120-
2	2.8488-1	2.8268-2	1.6337-1	5.8596-2	6.1843-4	5.0870-2	1.0194-
3	1.3638-1	2.1667-1	3.0458-2	5.8301-2	1.1413-1	3.4941-2	8.4098-
4	4.1780-2	2.3390-1	6.1865-2	1.1475-1	1.8000-4	7.5107-2	8.1670-
5	8.5623-3	1.1620-1	2.2212-1	1.5384-4	1.1958-1	3.0105-2	1.6595-

$a^4\Pi - X^2\Pi$ (M bands)

v', v''	0	1	2	3	4	5	6
0	3.6173-5	4.7769-4	2.9981-3	1.1897-2	3.3501-2	7.1222-2	1.1875-
1	2.5281-4	2.7560-3	1.3784-2	4.1550-2	8.2978-2	1.1254-1	9.9641-
2	9.4105-4	8.4217-3	3.3078-2	7.3176-2	9.5485-2	6.6157-2	1.2844-
3	2.4832-3	1.8123-2	5.4838-2	8.4501-2	6.1186-2	9.2341-3	1.0044-

$D^2\Sigma^+ - A^2\Sigma^+$ (Feast 1 bands)

v', v''	0	1	2	3	4	5	6
0	9.9969-1	1.8711-4	1.2792-4	4.2490-7	1.9841-7	5.0558-8	6.3344-9
1	1.8680-4	9.9943-1	2.3871-7	3.8609-4	2.6853-6	7.9047-7	2.4905-7
2	1.2809-4	7.3738-7	9.9843-1	6.3266-6	8.1095-4	1.0154-5	1.8354-6
3	3.5713-7	3.8558-4	7.2150-4	9.9407-1	3.3205-3	1.4801-3	1.4801-3
4	5.8411-7	1.1705-7	7.2576-4	3.8448-3	9.8354-1	9.3034-3	2.5250-3
5	5.0012-8	2.6448-6	6.7206-7	1.0513-3	1.1049-2	9.6385-1	1.9746-2
6	2.7548-9	3.5170-7	6.7176-6	9.1700-6	1.2300-3	2.4096-2	9.3205-1
7	2.2567-10	2.5801-8	1.3823-6	1.2174-5	3.9537-5	1.1380-3	4.4657-2
8	3.7741-11	2.2890-9	1.3468-7	3.9507-6	1.6793-5	1.0863-4	7.3893-4
9	8.7194-12	3.7557-10	1.3324-8	5.1088-7	9.0455-6	1.7403-5	2.2596-4
10	2.1009-12	9.1238-11	2.1272-9	5.8012-8	1.5584-6	1.7336-5	1.2044-5

$E^2\Sigma^+ - A^2\Sigma^+$ (Feast 2 bands)

v', v''	0	1	2	3	4	5	6
0	9.9870-1	1.2883-3	1.0997-5	4.9375-9	1.0915-9	1.0213-10	5.6952-12
1	1.2730-3	9.9633-1	2.3741-3	3.4489-5	3.2079-8	4.7978-9	6.0333-10
2	2.5361-5	2.3180-3	9.9433-1	3.2624-3	7.1876-5	1.1942-7	1.2403-8
3	3.2103-7	7.3484-5	3.1471-3	9.9271-1	3.9593-3	1.2441-4	3.3486-7
4	2.7010-9	1.2658-6	1.4188-4	3.7731-3	9.9144-1	4.4718-3	1.9316-4
5	9.3350-13	1.4230-8	3.1152-6	2.2815-4	4.2094-3	9.9050-1	4.8079-3

$B'^2\Delta - B^2\Pi$ (Ogawa 1 bands)

v', v''	0	1	2	3	4	5	6
0	2.2557-1	2.8877-1	2.2351-1	1.3509-1	7.0194-2	3.2915-2	1.4329-2
1	4.0580-1	4.6505-2	2.3550-2	1.1453-1	1.4223-1	1.1537-1	7.3699-2
2	2.7386-1	1.1578-1	1.5641-1	1.7228-2	1.6543-2	7.7701-2	1.0553-1
3	8.3679-2	3.3228-1	3.4276-3	1.2714-1	7.4062-2	2.3923-3	2.1451-2
4	1.0751-2	1.8393-1	2.6978-1	1.7606-2	5.9222-2	9.5322-2	2.7899-2
5	3.5658-4	3.1720-2	2.6413-1	1.8477-1	5.8488-2	1.4888-2	8.1219-2
6	1.2371-6	1.0260-3	5.7520-2	3.1892-1	1.1736-1	8.5432-2	3.5003-4

$b^4\Sigma - a^4\Pi$ (Ogawa 2 band)

v', v''	0	1	2	3
0	2.5072-1	2.8316-1	2.0542-1	1.2337-1
1	4.1515-1	3.3835-2	2.6702-2	1.0337-1
2	2.5352-1	1.3774-1	1.5569-1	1.9506-2
3	7.0990-2	3.3382-1	3.7309-3	1.2598-1

Potential energy curves (71. L2):

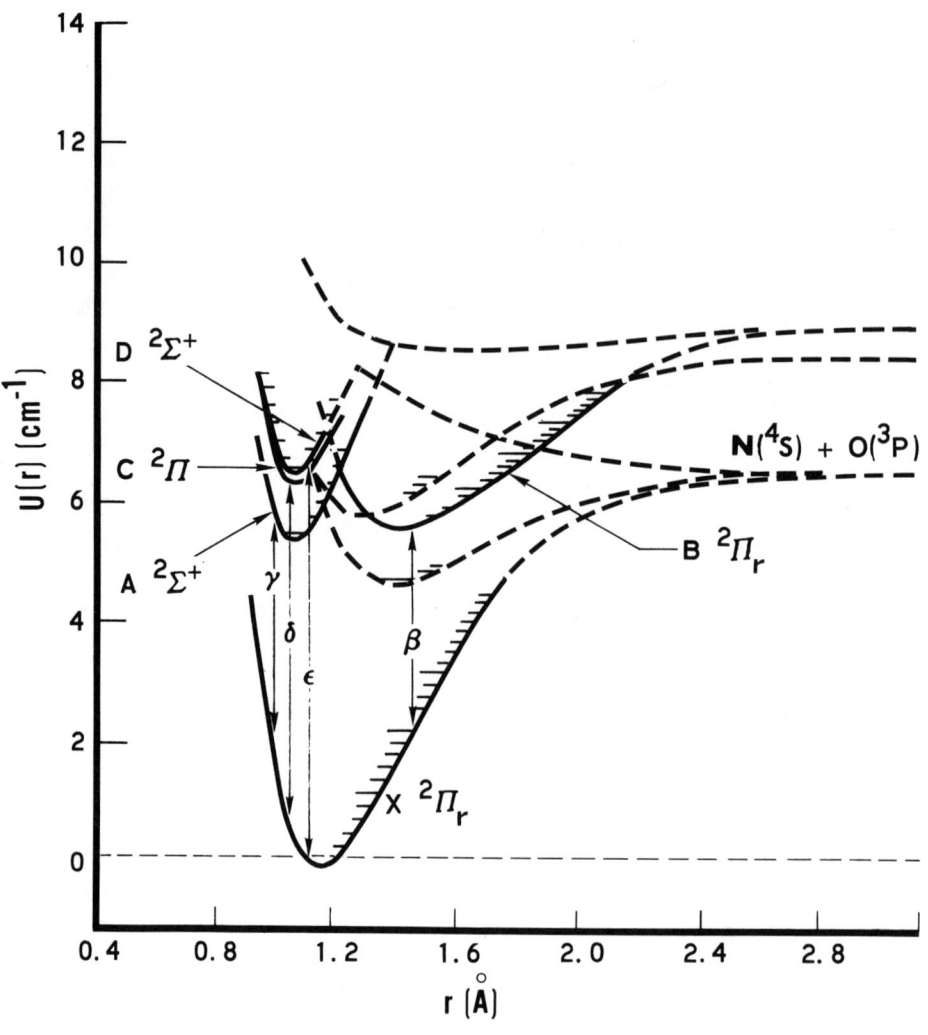

BIBLIOGRAPHY

(20.1) γ System,
W. H. Bair,
Astrophys. J. 52, 301-16

(26.2) 2000-1000 Å, Absorption,
S. W. Leifson,
Astrophys. J. 63, 73-89

(26.3) β System: Excited Active Nitrogen,
R. C. Johnson and H. G. Jenkins,
Philos. Mag. 2, 621-32

(26.4) Generalities,
R. Mecke,
Z. Physik 36, 795-802

(27.5) Preliminary Note to (30.8),
M. Lambrey,
C. R. Acad. Sci. 185, 382-4

(27.6) Preliminary Notes to (27.7) and (27.8),
F. A. Jenkins, H. A. Barton, and R. S. Mulliken,
Nature 119, 118-9

(27.7) β System,
H. A. Barton, F. A. Jenkins, and R. S. Mulliken,
Phys. Rev. 30, 175-88

(27.8) β System: Intensity Distribution,
F. A. Jenkins, H. A. Barton, and R. S. Mulliken,
Phys. Rev. 30, 150-74

(27.9) γ System: Emission,
M. Guillery,
Z. Physik 42, 121-45

(27.10) Generalities,
R. Mecke,
Z. Physik 42, 390-425

(28.11) Preliminary Note to (30.8),
M. Lambrey,
C. R. Acad. Sci. 186, 1112-4

(28.12) Preliminary Note to (30.18),
M. Lambrey,
C. R. Acad. Sci. 187, 210-2

(28.13) δ System: Excited Active Nitrogen,
H. P. Knauss,
Phys. Rev. 32, 417-26

(29.14) Preliminary Note to (30.18),
M. Lambrey,
C. R. Acad. Sci. 189, 574-6

(29.15) γ System: Photometric Intensities,
B. Pogany and R. Schmid,
Z. Physik 54, 779-87

(29.16) β and γ Systems,
R. Schmid,
Z. Physik 59, 42-7

(29.17) Intensity Distribution,
R. Schmid,
Z. Physik 59, 850-6

(30.18) β and γ Systems,
M. Lambrey,
Ann. Physique 14, 95-183

(30.19) δ Systems,
R. Schmid,
Matem. Term. Ertesito, Hungary 47, 534-42

(30.20) See (30.22),
R. Schmid, T. König, and D. von Farkas,
Matem. Term. Ertesito, Hungary 47, 485-532

(30.21) Isotope Effect,
R. Ruedy,
Phys. Rev. 35, 125

(30.22) β and γ Systems,
R. Schmid, T. König and D. von Farkas,
Z. Physik 64, 84-120

(31.23) β System: Intensity Distribution,
J. Kaplan,
Phys. Rev. 37, 1406-11

(33.24) Predissociation Inferred,
V. Kondratjew and L. Polak,
Phys. Z. Soviet Union, Ger. 4, 764-86

(34.25) Predissociation Inferred,
O. R. Wulf,
Phys. Rev. 46, 316

(35.26) Generalities,
E. Kondratjewa and V. Kondratjew,
Acta Physicochim. U.R.S.S. 3, 1-10

(35.27) Photochemical Dissociation,
P. J. Flory and H. L. Johnston,
J. Am. Chem. Soc. 57, 2641-51

(36.28) E-A Bands,
M. Duffieux and L. Grillet,
C.R. Acad. Sci. 202, 937-9

(37.29) Generalities,
W. M. Tschulanowsky,
Acta Physicochim. U.R.S.S. 7, 27-48

(37.30) E-A Band,
G. Jausseran, L. Grillet, and M. Duffieux,
C.R. Acad. Sci. 205, 39-41

(37.31) Generalities,
R. S. Mulliken,
J. Phys. Chem. 41, 5-45

(37.32) Bands Attributed to NO,
V. Kondratjew,
J. Exper. Theo. Phys. U.S.S.R. 7, 477-82

(37.33) ϵ System,
M. Hellermann,
Z. Physik 104, 417-29

(38.34) Absorption Coefficient,
P. H. Brodersen,
Verhandl. Dtsch. Phys. Ges. 19, 32

(39.35) Molecular Constants of X^2: I.R.,
R. H. Gillette and E. H. Eyster,
Phys. Rev. 56, 1113-9

(40.36) Predissociation,
G. Herzberg and L. G. Mundie,
J. Chem. Phys. 8, 263-73

(40.37) Mass Spectra,
H. D. Hagstrum and J. T. Tate,
Phys. Rev. 57, 561

(40.38) Dissociation Energy,
C. H. D. Clark,
Trans. Faraday Soc. 36, 370-6

(40.39) Potential Energy,
G. W. Linnett,
Trans. Faraday Soc. 36, 1123-34

(41.40) Night Sky,
C. T. Elvey, P. Swings, and W. Linke,
Astrophys. J. 93, 337-48

(41.41) Mass Spectra,
H. D. Hagstrum and J. T. Tate,
Phys. Rev. 59, 354-70

(41.42) Empirical Formula,
C. H. D. Clark,
Trans. Faraday Soc. 37, 299-302

(42.43) Theory,
K. Fajans,
J. Chem. Phys. 10, 759-60

(42.44) Absorption <1000Å,
Y. Tanaka,
Sci. Papers Inst. Phys. Chem. Res. 39, 456-64

(42.45) Thermochemical Dissociation Energy,
J. W. Linnett,
Trans. Faraday Soc. 38, 1-9

(42.46) Potential Energy Curves,
H. Zeise,
Z. Elektrochem. 48, 476-509

(43.47) B. M. Anand,
Indian J. Phys. 17, 246-51

(43.48) ε System,
L. Gerő, R. F. Schmid, and K. F. von Szily,
Naturwissenschaften 31, 203

NO

(44.49) δ and ε Systems,
L. Gerö, R. F. Schmid, and F. K. von Szily,
Physica, Pays-Bas 11, 144-50

(44.50) R. F. Barrow,
Proc. Phys. Soc. 56, 204-10

(44.51) γ and δ Systems,
A. G. Gaydon,
Prec. Phys. Soc. 56, 95-103

(44.52) ε and β Systems,
A. G. Gaydon,
Proc. Phys. Soc. 56, 160-74

(45.53) Emission 2000-1800Å. Bands ∼2700Å,
P. Migeotte,
Bull. Soc. Roy. Sci. Liege 14, 40-8

(45.54) Perturbations, Predissociations,
P. Migeotte and B. Rosen,
Bull. Soc. Roy. Sci. Liege 14, 49-57

(45.55) Generalities,
D. Sharma,
Proc. Nat. Acad. Sci. 14, 37 44

(45.56) Dissociation Energy,
A. G. Gaydon and W. G. Penney,
Proc. Roy. Soc. A 183, 374-87

(46.57) Photochemical Dissociation,
P. J. Flory and H. L. Johnston,
J. Chem. Phys. 14, 212-3

(46.58) Perturbations, Predissociations,
B. Rosen,
Physica, Pays-Bas 12, 184-8

(47.59) U.V. Absorption of Liquid NO,
H. J. Bernstein and G. Herzberg,
J. Chem. Phys. 15, 77

(47.60) Perturbations,
I. Kovacs and A. Budo,
J. Chem. Phys. 15, 166-73

NO

(47.61) Dissociation Energy,
A. G. Gaydon,
"Dissociation Energies and Spectra of Diatomic Molecules,"
Landen, Chapman, and Hall

(48.62) Photochemical Dissociation,
G. Glockler,
J. Chem. Phys. 16, 604-8

(48.63) Dissociation Energy by Mass Spectra,
H. D. Hagstrum,
J. Chem. Phys. 16, 848-9

(48.64) γ System. Predissociation,
L. Gerö and R. Schmid,
Proc. Phys. Soc. 60, 533-40

(48.65) Potential Energy Curves and Predissociation,
B. Rosen,
"Contribution to the Study of Molecular Structure,"
Vol Comm Victor Henri 1947-1948 pp 181-90

(48.66) Absorption and Emission, 2000-1800 A,
P. Migeotte,
Thesis, Liege

(49.67) Absorption, 2000-1000 Å,
Y. Tanaka,
J. Sci. Res. Inst. 43, 1198-1199, 28-35

(50.68) Perturbations, Predissociations. β System,
P. Migeotte and B. Rosen,
Bull. Soc. Roy. Sci. Liege 19, 343-8

(50.69) $E^2\Sigma^+ - A^2\Sigma^+$, $D^2\Sigma^+ - A^2\Sigma^+$ Systems,
M. W. Feast,
Can. J. Res. 28, 488-97

(50.70) G. Herzberg,
"Molecular Spectra and Molecular Structure I,"
2nd Ed. New York, D. van Nostrand Co.

(53.71) B'-X System. U.V. Emission,
P. Baer and E. Miescher,
Helv. Phys. Acta 26, 91-110

(53.72) U. V. Emission,
Y. Tanaka,
J. Chem. Phys. 21, 788-93

(53.73) Intensity in Absorption,
F. F. Marmo,
J. Opt. Soc. Am. 43, 1186-90

(53.74) Visible Emission,
M. Ogawa,
Science Light 2, 87-94

(54.75) "Afterglow" Emission,
Y. Tanaka,
J. Chem. Phys. 22, 2045-8

(54.76) Photoionization,
K. Watanabe,
J. Chem. Phys. 22, 1564-70

(54.77) β Bands. Quartet Bands,
M. Brook and J. Kaplan,
Phys. Rev. 96, 1540-2

(54.78) Quartet Bands,
M. Ogawa,
Science Light 3, 39-46

(55.79) β and Other Bands,
M. Ogawa,
Science Light 3, 90-128

(56.80) Rotational Analysis,
E. Miescher,
Helv. Phys. Acta 29, 135-44

(56.81) β' Bands,
E. Miescher,
Helv. Phys. Acta 29, 401-9

(56.82) β-β' Bands,
M. Ogawa and M. Shimouchi,
Science Light 5, 147-61

(56.83) R. R. Kadesch,
Univ. Microfilms 16176, 79pp

(57.84) A, B, D-X System. Rydberg States,
I. Deezsi and T. Matrai,
"Further Bands in the γ, ϵ, and β Band Systems of the Molecular Structure of Nitric Oxide,"
Acta Phys. 7, 111-6

(58.85) Franck-Condon Factors, Transition Probabilities, A, B → X,
R. W. Nicholls, P. A. Fraser, and W. R. Jarmain,
"Transition Probability Parameters of Molecular Spectra Arising from Combustion Processes,"
Combustion and Flame 26, 13-38

(58.86) γ Bands,
I. Deezsi,
Acta Phys. Acad. Sci. 9, 125-50

(58.87) A. Lagerqvist and E. Miescher,
Helv. Phys. Acta 31, 221-62

(59.88) F Values,
G. W. Bethke,
J. Chem. Phys. 31, 662-8

(59.89) $E^2\Sigma^+ - A^2\Sigma^+$ System,
M. L. Csaszar,
Magyar Fiz. Folyoirat 7, 489-95

(60-90) Quartet-Doublet,
H. P. Broida and M. Peyron
J. Chem. Phys. 32, 1068-71

(60.91) Los Alamos Scientific Lab,
Report - U.S. At. Energy Comm.
LA-2335
D. F. Heath, 79 pp

(61.92) 950-600 Å Absorption,
K. P. Huber,
Helv. Phys. Acta 34, 929-53

(61.93) Electronic Quenching Cross-Sections for A-X,
N. Basco, A. B. Callear, and R. G. W. Norrish,
"Fluorescence and Vibrational Relaxation of Nitric Oxide Studied by Kinetic Spectroscopy,"
Proc. Roy. Soc. A 260, 459-74

(62.94) Table of Bands,
L. Wallace,
Astrophys. J. Supplement 7, 165-290

(62.95) H, H' Perturbations. Rydberg Levels,
E. Miescher and K. P. Huber,
"Absorption Spectrum of the NO Molecule III,"
AFCRL-62-1118

(62.96) Upper Rydberg States,
E. Miescher,
"Spectrum and Energy Levels of the NO Molecule,"
J. Quant. Spectrosc. Radiative Transfer 2, 421-5

(62.97) Spectroscopic Constants of F, H, H', K, M and S States,
A. Lagerqvist and E. Miescher,
"Absorption Spectrum of the NO Molecule II,"
Can. J. Phys. 40, 352-7

(63.98) K. P. Huber and E. Miescher,
Helv. Phys. Acta 36, 257-68

(63.99) Chemiluminescence,
R. A. Young and R. L. Sharpless,
J. Chem. Phys. 39, 1071-102

(63.100) Fluorescence γ, Lifetimes of $A^2\Sigma^+$
A. B. Callear and I. W. M. Smith,
Trans. Faraday Soc. 59, 1720-34

(63.101) K. P. Huber, M. Huber, and E. Miescher,
"Rydberg-Series of the NO Molecule in the Visible and Infrared Emission Spectra,"
Phys. Letters 3, 315-6

(64.102) A. Lofthus and E. Miescher,
Can. J. Phys. 42, 848-59

(64.103) Fluorescence,
A. B. Callear and I. W. M. Smith,
Discuss. Faraday Soc. 37, 96-111

(64.104) 3d-3p, 3d-3s, Rotational Analysis,
M. Huber,
Helv. Phys. Acta 37, 329-47

NO

(64.105) Quartet–Doublet,
R. P. Frosch and G. W. Robinson,
J. Chem. Phys. 41, 367-74

(64.106) Radiative Lifetimes A, B → X,
M. Jeunehonume and A. B. F. Duncan,
"Lifetime Measurements of Some Excited States of Nitrogen, Nitric Oxide and Formaldehyde,"
J. Chem. Phys. 41, 1692-9

(64.107) Franck-Condon Factors, Many Systems,
R. W. Nicholls,
"Franck-Condon Factors to High Vibrational Quantum Numbers. IV. NO Band Systems,"
J. Res. Nat. Bur. Stand. A 68, 535-40

(64.108) D. J. Flinn, R. J. Spindler, S. Fifer, and M. Kelly,
"Franck-Conden Factors for the β and γ Bands of the NO Molecule Based on Realistic Potential Functions,"
J. Quant. Spectrosc. Radiative Transfer 4, 271-82

(65.109) K. Dressler and E. Miescher,
Astrophys. J. 141, 1266-83

(65.110) C. Jungen and E. Miescher,
Astrophys. J. 142, 1660-1

(65.111) LCAO-MO-SCF Calculations of the Rydberg Levels,
H. Lefebvre-Brion and C. M. Moser,
J. Mol. Spectrosc. 15, 211-9

(65.112) $\beta(0,0)$ Bands,
A. B. Callear and I. W. M. Smith,
Trans. Faraday Soc. 61, 1303-7

(65.113) Fluorescence, δ,
A. B. Callear and I. W. M. Smith,
Trans. Faraday Soc. 61, 2383-94

(65.114) Radiative Lifetime, D-X,
J. E. Hesser and K. Dressler,
Astrophys. J. 142, 389

(66.115) Theory,
C. Jungen,
Can. J. Phys. 44, 3197-216

(66.116) $^2\Delta$ State Interactions,
P. Felenbok and H. Lefebvre-Brion,
Can. J. Phys. 44, 1677-83

(66.117) A. Lagerqvist and E. Miescher,
Can. J. Phys. 44, 1525-39

(66.118) E. Miescher,
J. Mol. Spectrosc. 20, 130-40

(66.119) LCAO-MO-SCF Calculations of the Valence Levels of NO and NO^+,
"Semi-Empirical Calculations of Spin Orbital Constants,"
H. Lefebvre-Brion and C. M. Moser,
J. Chem. Phys. 44, 2951-4

(66.120) C. Jungen, E. Miescher, and R. Suter,
"Level Crossings $C^2\pi-B^2\pi$, $F^2\Delta-B'^2\Delta$, $B'^2\Delta-C^2\pi$ and the Lasing Combinations Transition B'^2-C^2 of the NO Molecule,"
Phys. Letters 21, 36-7

(66.120a) T. F. Deutsch,
"NO Molecular Laser,"
Appl. Phys. Letters 9, 295-7

(66.120b) M. A. Pollack,
"Molecular Laser Action in Nitric Oxide by Photodissociation of NOCl,"
Appl. Phys. Letters 9, 94-6

(67.121) 900-600 Å Absorption,
P. H. Metzger, G. R. Cook, and M. Ogawa
Can. J. Phys. 45, 203-17

(67.122) 900-600 Å Absorption,
P. H. Metzger, G. R. Cook, and M. Ogawa,
Can. J. Phys. 45, 203-17

(68.123) NO^+
K. P. Huber,
Can. J. Phys. 46, 1691-6

(68.124) C. Jungen and E. Miescher,
Can. J. Phys. 46, 987-1003

(68.125) F. Ackermann and E. Miescher,
Chem. Phys. Letters 2, 351-2

NO

(68.126) Visible Fluorescence,
R. A. Young, G. Black and T. G. Slanger,
J. Chem. Phys. 48, 2067-70

(68.127) Photoelectrons, NO^+,
J. A. R. Samson,
Phys. Letters A 28, 391-2

(68.128) ϵ Fluorescence, Lifetime of $D^2\Sigma^+$ State,
A. B. Callear, M. J. Pilling, and I. W. M. Smith,
Trans. Faraday Soc. 64, 2296-303

(68.129) B-X Emission by Electron Impact,
D. E. Rothe and D. J. McCaa,
"Emission Spectra of Molecular Gases Excited by 10 keV Electrons,"
Cornell Aero. Lab TR #165

(68.130) Radiative Lifetime, A-X,
E. H. Fink and K. H. Welge,
Z Naturforsch A 23, 358

(69.131) I. C. Jungen and E. Miescher,
Can. J. Phys. 47, 1769-87

(69.132) 4d-3p,
R. Suter,
Can. J. Phys. 47, 881-91

(69.133) F. Ackermann and E. Miescher,
J. Mol. Spectrosc. 31, 400-5

(69.134) Shocktube f Values,
K. L. Wray,
J. Quant. Spectrosc. Radiative Transfer 9, 255-76

(69.135) Production of B-X System in Flames,
R. M. Dagnall, D. J. Smith, K. C. Thompson, and T. S. West,
"Emission Spectra Obtained from the Combustion of Organic Compounds in Hydrogen Flames,"
Analyst 94, 871-8

(70.136) J. I. Generosa and R. A. Harris,
"Effects of High Rotational Quantum Numbers on Rydberg-Klein-Reese Franck-Condon Factors: The Nitric Oxide (NO) Beta Band System,"
J. Chem. Phys. 53, 3147-52

(70.137) Franck-Condon Factors, B-X System,
B. Petropoules, O. Dessaux, and P. Goudmand,
"Quantitative Study of the Spectra Obtained by the Action of Active Nitrogen on the Oxygen Compounds of Sulfur and Selenium,"
C. R. Acad. Sci. 270, 1223-6

(71.138) Reaction Producing B-X System,
E. C. Shane, and W. Brennen,
"Chemiluminescence of the Atomic Oxygen-Hydrogen Reaction,"
J. Chem. Phys. 55, 1479-80

(71.138a) CRC Handbook of Lasers,
Ed. R. J. Pressley,
Chemical Rubber Company, Ohio 1971

(72.139) A. J. D. Farmes, V. Hasson, and R. W. Nicholls,
"Absolute Oscillator Strength Estimates for Some Bands of the β-System of Nitric Oxide,"
J. Quant. Spectrosc. Radiative Transfer 12, 635-8

(72.140) M. M. Porch, V. V. Shubenich, and T. P. Zapesochmji,
"Excitation of Nitric Oxide Molecule by Slow Electrons,"
Opt. Spectrosc. 32, 565

(73.141) M. Mandelman, T. Carrington, and R. A. Young,
"Predissociation and its Inverse, Using Absorption $NO(C^2\pi) \rightleftarrows N+O$,
J. Chem. Phys. 58, 84-90

(np 142) $N^2\Delta - C^2\pi$ bands,
F. Ackermann

(np 143) $I^2\Sigma^+ - X^2\pi$ Bands,
E. Miescher

(np 144) Dissociation Energy 2000-1600 Å Absorption,
E. Miescher and B. Rosen

(np 145) $F^2\Delta - C^2\pi$ Laser System,
H. P. Broida and E. Miescher

(71.L1) H. H. Michels,
"Calculation of the Integrated Band Intensities of NO,"
J. Quant. Spectrosc. Radiative Transfer 11, 1735-9

NO

(71. L2) V. Hasson and R. W. Nicholls,
"Absolute Spectral Absorption Measurements of the NO - ($B^2\pi - X^2\pi$) Band System of Nitric Oxide,"
J. Phys. B(Atom. Mol. Phys.) 4, 1769-77

(71. L3) D. Spence and G. J. Schulz,
"Vibrational Excitation and Compound States in NO,"
Phys. Rev. A 3, 1968-76

(71. L4) H. Lefebvre-Brion,
"Intensity Anomaly in the Photoelectron Spectrum of NO,"
Chem. Phys. Letters 9, 463-4

(71. L5) K. R. German, R. N. Zare and D. R. Crosley,
"Reinvestigation of the Hanle Effect for the NO $A^2\Sigma^+$ State,"
J. Chem. Phys. 54, 4039-44

(71. L6) E. Miescher,
"Absorption Spectrum of the NO Molecule X. The 3d Rydberg Complex, its Vibrational Structure, Spin-Orbit Coupling, and Interactions with Non-Rydberg States,"
Can. J. Phys. 49, 2350-65

(72. L7) J. L. Bahr, A. J. Blake, J. H. Carver, J. L. Gardner, and V. Kumar,
"Photoelectron Spectra and Partial Photoionization Cross Sections for NO, N_2O, CO, CO_2 and NH_3,"
J. Quant. Spectrosc. Radiative Transfer 12, 59-73

(72. L8) R. K. Hinkley, J. A. Hall, T. E. H. Walker, and W. G. Richards,
"Λ-Doubling in $^2\pi$ States of Diatomic Molecules,"
J. Phys. B (Atom. Mol. Phys.) 5, 204-12

(72. L9) A. J. D. Farmer, V. Hasson, and R. W. Nicholls,
"Absolute Oscillator Strength Measurements of the ($V'' = 0$, $V' = 0-3$) Bands of the ($A^2\Sigma - X^2\pi$) γ-System of Nitric Oxide,"
J. Quant. Spectrosc. Radiative Transfer 12, 627-33

(72. L10) W. T. King and B. Crawford, Jr.,
"The Integrated Intensity of the Nitric Oxide Fundamental Band,"
J. Quant. Spectrosc. Radiative Transfer 12, 443-7

(72. L11) E. J. Stone and E. C. Zipt,
"Electron-Impact Excitation of Nitric Oxide,"
J. Chem. Phys. 56, 2870-4

(72.L12) E. M. Weinstock, R. N. Zare and L. A. Melton,
"Lifetime Determination of the NO $A^2\Sigma^+$ State,"
J. Chem. Phys. 56, 3456-62

(72.L13) G. E. Copeland,
"Lifetimes of the v' = 0, 1, 2 Levels of the $A^2\Sigma^+$ Electronic State of NO,"
J. Chem. Phys. 56, 689-91

(72.L14) J. E. Mentald and H. D. Morgan,
"Vacuum-Ultraviolet Excitation Cross Sections by Electron Impact on NO,"
J. Chem. Phys. 56, 2271-7

(72.L15) J. M. Brown, A. R. H. Cole, and F. R. Honey,
"Magnetic Dipole Transitions in the Far Infrared Spectrum of Nitric Oxide,"
Mol. Phys. 23, 287-95

(72.L16) S. Green,
"Calculated Properties for NO $X^2\pi$ and $A^2\Sigma^+$,"
Chem. Phys. Letters 13, 552-6

(72.L17) G. Gouedard,
"Optical Resonance and Hanle Effect on the NO Molecule,"
Ann. Phys. Fr. 7, 159-98

(73.L18) P. A. Bonczyk,
"Pressure Broadening of Magnetically--Turned Infrared Absorption Spectrum of NO Using a CO Laser,"
Chem. Phys. Letters 18, 147-9

(73.L19) E. A. Andreev, S. Y. Umansky, and A. A. Zembekov,
"Mechanism of Vibrational Relaxation of NO Upon Collisions with Ar,"
Chem. Phys. Letters 18, 567-9

(73.L20) G. Chandraiah and C. W. Cho,
"A Study of the Fundamental and First Overtone Bands of NO in NO-Rare Gas Mixtures at Pressures up to 10,000 psi,"
J. Mol. Spectrosc. 47, 134-147

(73.L21) G. L. Zarur and Y. N. Chiu,
"Cooperative Optical Phenomena II. Spin Forbidden Lifetime of the $a^4\pi$ State of Nitric Oxide,"
J. Chem. Phys. 59, 82-8

NS

Methods of Production and Experimental Technique

Absorption (52.4, 50.2).

Emission from a microwave discharge (2450 MHz) (69.25, 64.13, 63.9, 62.8)

Discharge tube (66.18, 63.10, 54.5, 51.3, 32.1).

Afterglow from a discharge into $N_2 + SF_6$ (67.21). Active N_2 flame + SCl_2, S_2Cl_2, H_2S (67.20, 64.12, 62.6).

Ground state studied by microwave techniques (69.24, 69.26, 68.23).

BAND SYSTEMS

	System	Transition	Sources	Wavelength Limits	Degrading	Characteristic Bands, λ	Remarks	Bibliography
β	I	$A^2\Delta \to X^2\Pi$	Discharge	2700-2390	R	2518.35 2506.65 (0,0)		(69.25, 66.18, 62.7)
	II	$B^2\Pi \to X^2\Pi$	Discharge	5100-2650	R	Depends on method of excitation		(69.25, 67.20, 67.21)
γ	III	$C^2\Sigma \rightleftarrows X^2\Pi$	Discharge	2520-2160	V	2317.32 2316.06 2305.37 (0,0) 2304.24		(69.25, 66.18, 62.20)
	IV	$G^2\Pi_i \to X^2\Pi$	Microwave discharge	2665-2550	R	2556.89(0,4)		(71.31)
	V	$B'^2\Sigma \to X^2\Pi$	Discharge	2399-2268	R	2327.6(0,0)		(66.18)
	VI	$D^2\Sigma \to X^2\Pi$	Discharge	2342-2207	R	(45061)(0,0)		(66.18)
	VII	$E(^2\Delta \text{ or } ^2\Pi) \to X^2\Pi$	Discharge	2045-1947	No head	2210.2(0,0)		(66.18)
	VIII	$F(^2\Sigma) \to X^2\Pi$	Discharge	1867-1779	V	1787.2(0,0)		(66.18)
	IX	$D^2\Sigma - C^2\Sigma$	Discharge	4900-3900	R	-		(64.15)

I. $A^2\Delta \rightarrow X^2\Pi$ System (β)

Double headed bands (69.25):

(v', v'')	(0, 1)	(0, 0)	(1, 0)
λ	2597.09	2518.35	2460.56
	2584.69	2506.65	2448.62

II. $B^2\Pi \rightarrow X^2\Pi$ System

Double headed bands (66.18):

(v', v'')	(0, 1)	(0, 0)	(1, 0)
λ	2384.6(5)	2318.2(8)	2246.4(4)
	2372.1(5)	2306.2(8)	2233.6(5)

III. $C^2\Sigma \rightleftarrows X^2\Pi$ System (γ)

Double bands with double heads (69.25):

(v', v'')	(0, 1)	(0, 0)	(1, 0)
λ	2383.75	2317.32	--
	2382.55	2316.06	2244.04
	2371.10	2305.37	--
	2370.04	2304.24	2232.84

IV. $G^2\Pi_i \rightarrow X^2\Pi$ System

Double banded with double heads (71.31).

Band origins $^2\Pi_{1/2} - {}^2\Pi_{1/2}$ sub-band:

$$\nu_o(0-4) = 39095.41 \text{ cm}^{-1}$$
$$\nu_o(0-5) = 37949.30 \text{ cm}^{-1}$$

V. $\underline{B'\,^2\Sigma \to X\,^2\Pi \text{ System}}$

Double headed bands similar to II (66.18):

(v', v'')	(0, 1)	(0, 0)	(1, 0)
λ	2269.5(5)	2327.6(5)	2394.8(5)
	2268.9(5)	2326.9(5)	2393.8(5)

VI. $\underline{D\,^2\Sigma \to X\,^2\Pi \text{ System}}$

Three bands with double heads (66.8):

(v', v'')	(0, 0)	(0, 1)	(0, 2)
λ	2219.2(3)	2280.4(3)	2343.6(2)
	2208.3(3)	2268.9(5)	--

VII. $\underline{E(^2\Delta \text{ or } ^2\Pi) \to X\,^2\Pi \text{ System}}$

Three bands with no heads (66.18):

(v', v'')	(0, 0)	(0, 1)	(0, 2)
λ	1949.4(3)	1996.2(2)	2044.6(1)
	1948.0(3)	1994.6(2)	2042.9(1)

VIII. $\underline{F(^2\Sigma) \to X\,^2\Pi \text{ System}}$

Three bands with double heads (66.18):

(v', v'')	(0, 0)	(0, 1)	(0, 2)
λ	1787.2(5)	1826.6(4)	1867.0(1)
	1780.2(5)	1819.2(4)	1859.3(1)

SPECTROSCOPIC CONSTANTS

State	T_e	ω_e	$x_e\omega_e$	B_e	$\alpha_e \times 10^3$	$D_e \times 10^6$	r_e	Remarks	Bibliography
$C^2\Sigma$	43387.3[a]	1403	8	0.6267[b]	—	0.77[c]	1.447[d]		(52.4, 51.3)
$A^2\Delta_{5/2}$	39911.08[a]	940.4	4.8	0.6960	6.9	1.4	1.576		(54.5)
$A^2\Delta_{3/2}$	39875.70[a]	970.6	8.6	0.595	4.0	2	1.706		
$B^2\Pi_{3/2}$	30363.3	803.3	3.82	0.596	6.1	1.2	1.495		(69.25)
$B^2\Pi_{1/2}$	30297.9	796.3	3.63					$A = 223$ cm^{-1}	(51.3)
X^2	0	1219.1	7.5	0.7736					

[a] T_o, [b] B_o, [c] D_o, [d] r_o.

Dissociation energy = 4.8 ± 0.25 eV, 110.7 kcal/mole, 38715 cm^{-1} (70.30).

Perturbations and General Information

Formulas for the ultraviolet progressions (66.18).

Vibrational perturbations of $C^2\Sigma$ at v' = 2-3 (66.18, 62.7). Vibrational perturbations and abnormal doubling of $A^2\Delta$ (66.18, 54.5).

Franck-Condon factors — RKR potential (70.28):

$\underline{B^2\Pi - X^2\Pi}$

v', v''	0	1	2	3	4
0	.008	.035	.081	.129	.16
1	.043	.118	.150	.107	.036
2	.112	.159	.069	.0009	.031
3	.185	.094	.000	.062	.077
4	.217	.093	.065	.074	.002
5	.191	.021	.100	.002	.060

Potential energy curves — (67.20):

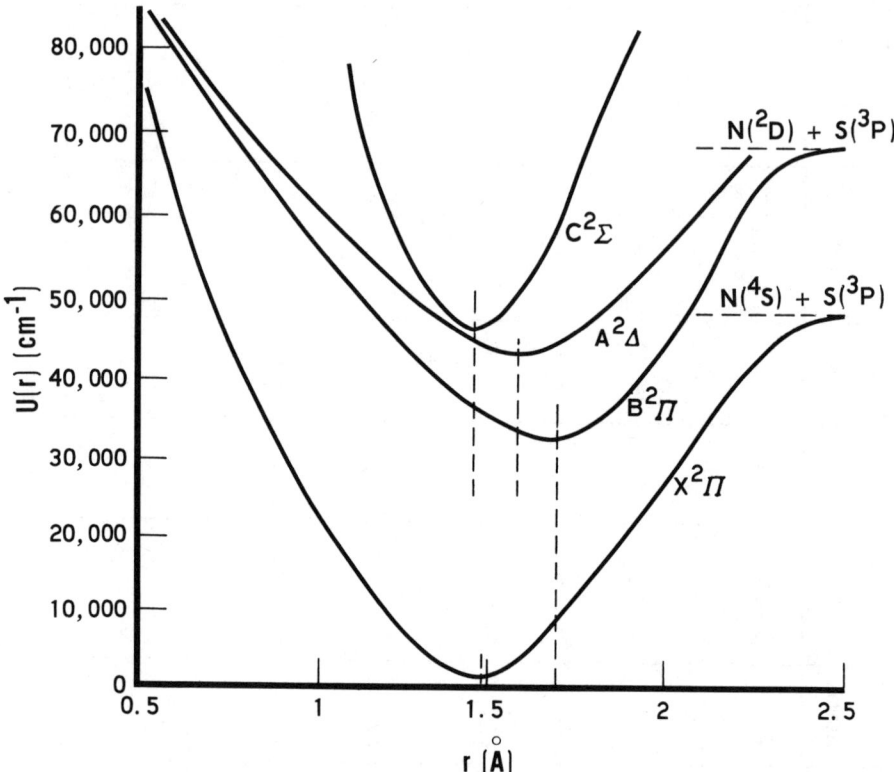

BIBLIOGRAPHY

(32.1) A-X, C-X Systems. Vibrational Analysis,
R. C. Johnson and R. C. Turner,
Proc. Roy. Soc. A, 142, 574-87

(50.2) C-X System in Absorption,
R. F. Barrow,
Trans. Faraday Soc. 9, 81

(51.3) C-X System, Rotational and Vibrational Analysis,
P. B. Zeeman,
Can. J. Phys. 29, 174-85

(52.4) A-X, C-X Systems. C-X System in Absorption,
R. F. Barrow, A. R. Downie, and R. K. Laird,
Proc. Phys Soc. A, 65, 70-1

(54.5) A-X System. Rotational and Vibrational Analysis,
R. F. Barrow, G. Drummond, and P. B. Zeeman,
Proc. Phys. Soc. A, 67, 365-77

(62.6) Flames,
G. Pannetier, P. Goodmand, O. Dessaux, and N. Tavernier,
C. R. Acad. Sci. 255, 91-3

(62.7) A-X, C-X Systems, ^{15}N Study,
N. A. Narasimham and K. Srikameswaran,
Proc. Indian Acad. Sci. A, 56, 316-24

(62.8) B-X System, ^{15}N Study,
N. A. Narasimham and K. Srikameswaran,
Proc. Indian Acad. Sci. A, 56, 325-8

(63.9) A, B, C-X Systems,
N. A. Narasimham and K. Srikameswaran,
Nature 197, 370

(63.10) Bands in Emission,
M. M. Patel,
Z. Physik 173, 347-51

(64.11) Flames,
G. Pannetier, O. Dessaux, I. Arditi, and P. Goudmand,
C. R. Acad. Sci. 259, 2198-9

(64.12) Flames. B-X System, Study of ^{34}S and ^{15}N,
J. J. Smith and B. Meyer,
J. Mol. Spectrosc. 14, 160-72

(64.13) B-X System, Study of ^{15}N,
N. A. Narasimham and K. Srikameswaran,
Proc. Indian Acad. Sci. A, 59, 227-40

(64.14) Preliminary Note to (64.15),
J. A. S. Bett and C. A. Winkler,
"The Reaction of Active Nitrogen With Sulfur,
J. Phys. Chem. 68, 2501-8

(64.15) J. A. S. Bett and C. A. Winkler,
"Emission of the $D(^2\Sigma) \to C(^2\Pi)$ System of Nitric Sulfide in the Reaction of Sulfur Vapor With Active Nitrogen,"
J. Phys. Chem. 68, 2735-6

(64.16) D-C System,
G. Pannetier, P. Goudmand, O. Dessaux, and N. Tavernier,
"Reactions of Some Atoms or Free Radicals on H_2S and D_2S,"
J. Chim. Phys. 61, 395-406

(65.17) B-X System,
G. Pannetier, P. Goudmand, O. Dessaux, and I. Arditi,
"Reactions of Atomic Nitrogen on Sulfur and Selenium Chlorides,"
C. R. Acad. Sci. 1, 117-27

(66.18) A-X, C-X Systems. Three Ultraviolet Systems,
K. C. Joshi,
Z. Physik 191, 126-36

(67.19) Franck-Condon Factors (B-X),
B. Petropoulos, O. Dessaux, D. Chaffiol, and P. Goudmand,
C. R. Acad. Sci. B, 265, 355-8

(67.20) Flames, B-X System,
P. Goodmand and O. Dessaux,
J. Chim. Phys. 64, 135-40

(67.21) Postluminescence. A, B, C-X Systems,
M. Peyron and Lan Thanh My,
J. Chim. Phys. 64, 129-34

(67.22) C, A-X Systems,
K. C. Joshi,
"The γ and β-Systems of NS,"
Z. Phys. Chem. Neue Folge 55, 173-8

(68.23) Electron Resonance in the Gas Phase,
A. Carrington, B. J. Howard, D. H. Levy, and J. C. Robertson,
Mol. Phys. 15, 187-200

(69.24) Microwave Absorption $^{14}N^{32}S$, $^{14}N^{34}S$,
T. Amano, S. Saito, E. Hirota, and Y. Morino,
J. Mol. Spectrosc. 32, 97-107

(69.25) A, B, C-X, $^2\Sigma - X^2\Pi$ Systems. Study of ^{34}S,
N. A. Narasimham and T. K. B. Subramanion,
J. Mol. Spectrosc. 29, 294-304

(69.26) Electron Paramagnetic Resonance,
H. Ueharo and Y. Morino,
Mol. Phys. 17, 239-48

(69.27) B. Vidal, O. Dessaux, J. P. Marteel, and P. Goudmand,
"On the Emission of New Bands of the $B^2\Pi - X^2\Pi$ System of NS in Chemiluminescent Reactions,"
C. R. Acad. Sci. 268, 2140-1

(70.28) Franck-Condon Factors, r-Centroids, B-X,
B. Petropoulos, O. Dessaux, and P. Goudmand,
"Quantitative Study of the Spectra Obtained by the Action of Active Nitrogen on the Oxygen Compounds of Sulfur and Selenium,"
C. R. Acad. Sci. 270, 1223-6

(70.29) B. Petropoulos, O. Dessaux, D. Chaffiol, and P. Goudmand,
"Franck-Condon Factors for the $B^2\Pi - X^2\Pi$ Transition of the NS Radical,"
C. R. Acad. Sci. B, 271, 296-7

(70.30) P. A. G. O'Hare,
"Dissociation Energies, Enthalpies of Formation, Ionization Potentials, and Dipole Moments of NS and NS^+,"
J. Chem. Phys. 52, 2992-6

(71.31) N. A. Narasimham and T. K. Balasubramanian,
"A New $^2\Pi_i - X^2\Pi_r$ Band System of NS,"
J. Mol. Spectrosc. 40, 511-8

NSe

Methods of Production and Experimental Technique

Flame of atomic N + $SeCl_4$.

Microwave discharge in nitrogen gas over selenium metal powder (71.17).

BAND SYSTEMS

System	Transition	Sources	Wavelength Limits	Degrading	Band Head, $\nu_{0,0}$	Remarks	Bibliography
I	$A^2\Pi \to X^2\Pi$	Microwave discharge	5675-3950	R	22825(0, 1) (3/2 - 3/2) 22326.8(0, 2) (1/2 - 1/2)		(70.13)
II	$B^2\Sigma \to X^2\Pi$	Microwave discharge	3173-2991	R	33431.2 | (0, 0) 34322.0		(71.15)
III	$C^2\Delta \to X^2\Pi$	Microwave discharge	3072-2843	-	34431.5(0, 0)		(71.15)

I. $A^2\Pi \to X^2\Pi$ System

Band origins (70.13):

$A^2\Pi_{3/2} \to X^2\Pi_{3/2}$

(v', v'')	(0, 1)	(0, 2)	(1, 1)	(1, 2)
λ	4381.1	4568.5	4268.4	3999.8

$A^2\Pi_{1/2} \to X^2\Pi_{1/2}$

(v', v'')	(0, 2)	(0, 3)	(1, 2)	(1, 3)
λ	4478.9	4672.7	4350.8	4533.1

II. $B^2\Sigma \to X^2\Pi$ System

Band origins (71.15):

$B^2\Sigma \to X^2\Pi_{1/2}$

(v', v'')	(0, 0)	(0, 1)	(0, 2)
λ	2913.6	2996.1	3082.4

$B^2\Sigma \to X^2\Pi_{3/2}$

(v', v'')	(0, 0)	(0, 1)	(0, 2)
λ	2991.2	3078.1	3169.1

III. $C^2\Delta \to X^2\Pi$ System

Only one subsystem observed (71.15)

$C^2\Delta_{5/2} \to X^2\Pi_{3/2}$

(v', v'')	(1, 0)	(1, 1)	(1, 2)
λ	2904.3	2986.1	3071.7

NSe

SPECTROSCOPIC CONSTANTS

State	T_e	ω_e	$x_e\omega_e$	B_e	$\alpha_e \times 10^3$	$D_e \times 10^6$	r_e	Remarks	Bibliography
$C^2\Delta_{5/2}$	34431.5[a]	–	–	0.4568[f]	–	–	–		(71.15)
$B^2\Sigma$	34322.0[a]	–	–	0.4503[c]	–	2.8[d]	–	$\gamma = -0.035$ cm^{-1}	(71.15)
$A^2\Pi_{1/2}$	24357.50	657.20[b]	–	0.4117[c]	–	0.583[d]	1.8537[e]	$A = 453$ cm^{-1}	(71.15, 70.13)
$A^2\Pi_{3/2}$	23765.47	612.60[b]	–	0.4141	-7.4	–	1.8483		(70.13)
$X^2\Pi_{3/2}$	890.84	949.7	5.73	0.5192	3.9	0.615	1.6506		(70.13)
$X^2\Pi_{1/2}$	0	957.50	5.69	0.5182	4.0	0.6	1.6522		

[a] T_o, [b] $\Delta G_{1/2}$, [c] B_o, [d] D_o, [e] r_o, [f] B_1.

Dissociation energy ~ 3.97 eV, 91.5 kcal/mole, 32000 cm^{-1} (71.15).

BIBLIOGRAPHY

(65.1) Vibrational Analysis,
G. Pannetier, P. Goudmand, O. Dessaux, and I. Arditi,
C. R. Acad. Sci. 260, 2155-8

(65.2) G. Pannetier, P. Goudmand, O. Dessaux, and I. Arditi,
"Reactions of Atomic Nitrogen Upon the Chlorides of Sulfur and Selenium,"
C. R. Acad. Sci. 1, 117-27

(67.3) Vibrational Analysis,
P. Goudmand and O. Dessaux,
J. Chim. Phys. 64, 135-40

(68.4) O. Dessaux and P. Goudmand,
"New Systems of NSe Emitted in Chemiluminescent Reactions of Active Nitrogen on Selenium Chloride,"
C. R. Acad. Sci. C, 267, 1198-1201

(69.5) A-X, Rotational Analysis,
B. Pascat, D. Daumont, A. Jenouvrier, and H. Guenebaut,
C. R. Acad. Sci. 269, 1309-12

(70.6) B-X, Rotational Analysis,
B. Pascat, D. Daumont, A. Jenouvrier, and H. Guenebaut,
C. R. Acad. Sci. C, 270, 20-3

(70.7) A-X, Vibrational Analysis,
D. Daumont, A. Jenouvrier, and B. Pascat,
"On the Vibrational Analysis of the $A(^2\Pi_r) - X(^2\Pi_r)$ System of NSe. Study of the Two Subsystems and Evidence for the Strong Vibrational Perturbations in the 3/2 Subsystem,"
C. R. Acad. Sci. C, 271, 712-5

(70.8) A. Jenouvrier, D. Daumont, and B. Pascat,
"Vibrational Analysis of Three New Electronic Systems of NSe,"
C. R. Acad. Sci. C, 271, 1358-61

(70.9) B. Petropoulos, O. Dessaux, and P. Goudmand,
"Quantitative Study of the Spectra Obtained by the Action of Active Nitrogen on Oxygn Compounds of Sulfur and Selenium,"
C. R. Acad. Sci. C, 270, 1223-6

NSe

(70.10) C→X System,
M. P. Bassez, B. Vidal, O. Dessaux, and P. Goudmand,
"Rotational Analysis of the (0,5) and (0,6) Bands of the $5/2 - X^2\Pi_{3/2}$ System of NSe,"
C. R. Acad. Sci. C, 270, 377-80

(70.11) A-X System,
R. V. Subbaram and D. R. Rao,
"Electronic Spectrum of NSe,"
Chem. Phys. Letters 4, 653-55

(70.12) D. Daumont, A. Jenouvrier, B. Pascat, and H. Guenebaut,
"Study of the Three New v' Progressions of NSe. Vibrational Analysis and Rotational Study of the Two Bands,"
C. R. Acad. Sci. B, 271, 120-3

(70.13) A-X System,
K. V. Subbaram and D. R. Rao,
$^2\Pi - X^2\Pi$ System of $^{14}N^{80}Se$ and $^{14}N^{78}Se$,"
J. Mol. Spectrosc. 36, 163-182

(71.14) A. Jenouvrier, D. Daumont, B. Pascat, and H. Guenebaut,
"Rotational Analysis of the Two Systems A-X, and A-X of NSe Between 3500 and 5500 Å,"
C. R. Acad. Sci. C, 271, 1627-30

(71.15) C, B→X System, Vibrational Analysis,
L. Harding, W. E. Jones, K. K. Yee, A. Jenouvrier, D. Daumont, B. Pascat, and H. Guenebaut,
"Electronic Spectrum of the $N^{80}Se$ Molecule in the Region 2840-3200
Can. J. Phys. 49, 2033-51

(71.16) A-X System,
D. Daumont, A. Jenouvrier, B. Pascat, and H. Guenebaut,
"Vibrational Analysis of a New System of Bands of NSe Located in 3500 to 5500 Å,"
C. R. Acad. Sci. C, 272, 1545-7

(71.17) A-X System, Vibrational-Rotational Analysis,
K. K. Yee and W. E. Jones,
"New Electronic Emission Spectra of the NSe Radical,"
J. Mol. Spectrosc. 37, 304-313

NbO

Methods of Production and Experimental Technique

Absorption in a graphite furnace.

Emission from an arc.

Discharge in NbCl$_5$.

BAND SYSTEMS

	System	Transition	Sources	Wavelength Limits	Degrading	Band Head, $\nu_{0,0}$	Remarks	Bibliography
Red	I	$B^4\Pi - X^4\Sigma^-$	Arc, Ar matrix	7700-5800	R	~15000		(73. L1, 57. 4, 53. 2)
Blue	II	$A^4\Sigma^- - X^4\Sigma^-$	Arc, Ar matrix	5300-4300	R	21316		(73. L1, 69. 6, 54. 3)

NbO

I. $^4\Pi - X\,^4\Sigma^-$ System

Preliminary analysis has been conducted. Bands of greatest intensity are in the region $6750 > \lambda > 6450$ Å.

Band heads, λ:

λ	7027.3	6969.4	6950.9	6737.3	6633.3	6591.0	6576.0	6499.6
	6495.9	6484.4	6473.9	6228.3	6154.9	6124.5	6120.0	

II. $A\,^4\Sigma^- - X\,^4\Sigma^-$ System

Band heads of greatest intensity, λ:

v', v''	0	1	2
0	4688.6	4914.6	
1	4510.4		4946.7
2	4346.1	4540.2	
3		4375.2	

SPECTROSCOPIC CONSTANTS

State	T_e	ω_e	$x_e\omega_e$	B_o	$\alpha_e \times 10^3$	$D_e \times 10^6$	r_e	Remarks	Bibliography
$A\,^4\Sigma^-$	21316.2	840.48	3.3	0.3992	1.9	-	1.75		(69.6, 54.3)
$B\,^4\Pi$	~15000	~900	0	-	-	-	-		(57.4, 53.2)
$X\,^4\Sigma^-$	0	989.0	3.8	0.4310	2.1	-	1.691		(69.6, 54.3)

Dissociation energy = 7.81 ± 0.11 eV, 180.0 kcal/mole, 63000 cm^{-1} (66.5).

NbO

Perturbations and General Information

Franck-Condon factors, r-centroids — Morse potential (72.7):

v', v''	0	1	2	3	4	5
0	1.725 0.439 4689.0	1.769 0.373 4915.1	1.815 0.145 5161.8	1.861 0.034 -	1.910 0.006 -	1.960 0.005 -
1	1.688 0.345 4510.7	1.732 0.013 -	1.776 0.273 4946.4	1.820 0.233 -	1.867 0.179 -	1.915 - -
2	1.652 0.150 4346.5	1.695 0.226 4540.3	1.738 0.047 -	1.782 0.128 4976.8	1.827 - -	1.873 - -
3	1.616 0.045 -	1.658 0.223 -	1.701 0.097 -	1.744 - -	1.788 - -	1.832 - -
4	1.581 1.016 -	1.623 0.148 -	1.665 - -	1.707 - -	1.750 - -	1.793 - -
5	1.546 0.000 -	1.588 - -	1.630 - -	1.671 - -	1.712 - -	1.756 - -

Top = r-centroids, middle = Franck-Condon factors, and bottom = wavelengths.

BIBLIOGRAPHY

(50.1) V. R. Rao,
Indian J. Phys. 24, 35-49

(53.2) 7680-6030 Å System,
V. R. Rao and D. Premaswap,
Indian J. Phys. 27, 399-405

(54.3) A-X System, Rotational Analysis,
U. Uhler,
Arkiv Fysik 8, 265-79

NbO

(57.4) Vibrational Analysis, B-X,
A. Gatterer, J. Junkes, E. W. Salpeter, and B. Rosen,
"Molecular Spectra of Metallic Oxides,"
ed. Specola Vaticana

(66.5) Dissociation Energy,
S. A. Shchukarev, G. A. Semenov, and K. E. Frantseva,
Zh. Neorg. Khim., S.S.S.R. 11, 233-6

(69.6) A-X, Vibrational Analysis,
D. Richards,
Thesis, Oxford

(69.7) T. M. Dunn and K. M. Rao,
"Hyperfine Splittings in the Optical System of NbO^+,"
Nature 222, 266-7

(72.8) Franck-Condon Factors, r-Centroids, A-X,
P. D. Singh and M. M. Shukla,
"Vibrational Transition Probabilities and r-Centroids for Some Diatomic Molecular Band Systems,"
J. Quant. Spectrosc. Radiative Transfer 12, 1249-52

(73.L1) D. W. Green, W. Korfmacher, and D. M. Gruen,
"infrared Absorption Spectra of Isotopic NbN and NbO Isolated in an Ar Matrix,"
J. Chem. Phys. 58, 404-5

NdF

Dissociation energy = 5.88 ± 0.13 eV, 47426 cm^{-1} (72.1).

BIBLIOGRAPHY

(72.1) M. Shafi, C. L. Beckel, and R. Engelke,
"Diatomic Molecule Ground State Dissociation Energies,"
J. Mol. Spectrosc. 42, 578-581

NdO

Methods of Production and Experimental Technique

Emission from an oxyhydrogen or carbon flame.

Band Systems

Bands observed $8850 > \lambda > 6000$ Å, degrading red. No analysis.

Characteristic bands (63.4, 57.3):

λ	λ	λ
8718.0	7497.6	6773.8
8663.4	7466.4	6747.8
8659.5	7420.2	6630.3(b)
8386.4	7404.6	6598.4(b)
8383.2	7157.5(b)	6501.5(a, b)
8379.8	7033.1(b)	6498.9(a, b)
8004.3(a)	7010.6(b)	6349.4
7792.3	6897.1	
7591.9	6857.2	

(a) - degrades V, and (b) most intense bands.

Spectroscopic Constants

Dissociation energy = 7.48 ± 0.12 eV, 172.5 kcal/mole, 60300 cm^{-1} (67.5).

BIBLIOGRAPHY

(35.1) Excitation in an $H_2 - O_2$ Flame,
G. Piccardi,
Att. Accad. Lincei, Rend. Cl. Sci. Fis. Mat. Nat. 21, 584-8

(38.2) E. N. Eremin, K. S. Bogomolow, N. I. Kobosew, and S. S. Wassiliew,
J. Phys. Chem. 11, 33-44

NdO

(57.3) Wavelengths,
A. Gatterer, J. Junkes, E. W. Salpeter, and B. Rosen,
"Molecular Spectra of Metallic Oxides,"
ed. Specola Vaticana

(63.4) Wavelengths in Flames,
R. Hermann and C. T. J. Alkemade,
"Flame Photometry,"
Interscience Publication

(67.5) Dissociation Energy,
L. L. Ames, P. N. Walsh, and D. White,
J. Chem. Phys. 71, 2707-18

NdS

Dissociation energy = 4.87 ± 0.15 eV, 112.2 kcal/mole, 39250 ± 1200 cm^{-1} (69.1).

BIBLIOGRAPHY

(69.1) Dissociation Energy,
S. Smoes, P. Coppens, C. Bergman, and J. Drowart,
Trans. Faraday Soc. 65, 682-7

NdSe

Dissociation energy = 3.93 ± 0.15 eV, 31700 cm^{-1} (72.1).

BIBLIOGRAPHY

(72.1) M. Shafi, C. L. Beckel, and R. Engelke,
"Diatomic Molecule Ground State Dissociation Energies,
J. Mol. Spectrosc. 42, 578-581

NdTe

Dissociation energy = 3.12 ± 0.15 eV, 25165 cm^{-1} (72.1).

BIBLIOGRAPHY

(72.1) M. Shafi, C. L. Beckel, and R. Engelke,
"Diatomic Molecule Ground State Dissociation Energies,"
J. Mol. Spectrosc. 42, 578-581

NiBr

Methods of Production and Experimental Technique

Emission from high frequency discharge in $NiBr_2$.

Band Systems

Numerous bands $4800 > \lambda > 3930$ Å degrading R (62.5, 60.3, 52.2, 39.1).

Spectroscopic Constants

Analysis is uncertain (60.3)

System	Transitions	T_{00}	ω'_e	$x'_e\omega'_e$	ω''_e	$x''_e\omega''_e$
α_1	$\Delta\Lambda = \pm 1$	24311.9	297.0	3.35	311.6	1.7
α_2	$\Delta\Lambda = 0$	23904.7	~262	-	293.2	-
α_3	$\Delta\Lambda = 0$	23786.4	305.4	0.2	315.0	0.5
β_1	$\Delta\Lambda = 0$	22960.4	293.8	2.5	322.8	1.0
β_2	$\Delta\Lambda = 1$	22427.4	257	-	274	-
γ	$\Delta\Lambda = 1$	21779.8	292.5	1.1	323.0	1.2

Dissociation energy = 3.7 ± 0.1 eV, 85 kcal/mole, 29843 cm^{-1}.

BIBLIOGRAPHY

(39.1) Observation of a Complex Spectrum,
P. Mesnage,
Ann. Physique 12, 5-87

(52.2) Attempted Analysis,
V. G. Krishnamurty,
Indian J. Phys. 26, 429-41

(60.3) Attempted Analysis,
S. P. Reddy and P. T. Rao,
Proc. Phys. Soc. 75, 275-9

(61.4) Dissociation Energy,
E. M. Bulewicz, L. F. Phillips, and T. M. Sugden,
Trans. Faraday Soc. 57, 921-31

(62.5) Attempted Analysis,
N. Sundarachary,
Proc. Nat. Acad. Sci. India A, 32, 311-9

NiCl

Methods of Production and Experimental Technique

Emission from a high frequency discharge in $NiCl_2$.

Band Systems

Large number of bands $5300 > \lambda > 3700$ Å degrading R. Also, many bands $8800 > \lambda > 6800$ Å.

Band heads of greatest intensity (39.2, 38.1):

λ	Intensity	λ	Intensity
5013.2	2	4399.2	9
5006.9	1	4396.6	8
4942.9	2	4375.4	4
4941.4	2	4317.8	7
4935.5	1	4313.2	8
4824.1	2	4308.9	9
4708.9	5	4304.9	10
4705.4	4	4284.4	7
4630.1	5	4216.1	6
4625.5	7	4215.7	4
4622.5	3	4146.1	9
4603.0	3	4143.1	7
4596.9	7	4109.1	9
4566.8	7	4095.3	9
4563.9	8	4081.3	7
4489.8	6	4061.4	5
4403.0	7		

In the region $8750 > \lambda > 6350$ Å five systems are observed; F, G, and I are single-headed, H and J are double-headed (69.7).

SPECTROSCOPIC CONSTANTS

State	T_e	ω_e	$x_e\omega_e$	B_e	$\alpha_e \times 10^3$	$D_e \times 10^6$	r_e	Remarks	Bibliography
$F(^2\Delta_{5/2})$				0.145	-3.7	0.1	2.290		(69.7)
$G(^2\Delta_{5/2})$				0.153	1.8		2.226		(69.7)
$X(^2\Delta_{5/2})$				0.167		2.0	2.137		(69.7)

Dissociation energy = 3.8 ± 0.2 eV, 88 kcal/mole, 30650 cm^{-1}.

NiCl

CONSTANTS FOR UNCERTAIN SYSTEMS

System	T_{oo}	Heads	ω'_e	$x'_e\omega'_e$	ω''_e	$x''_e\omega''_e$	Bibliography
A_1	24613.4	Q	394.5	0.35	415.5	1.3	(60.4)
A_2	2441.6	Q	-	-	-	-	(60.4)
A_3	24129.8	R	380	-	397	-	(60.4)
B_1	23333.6	R	375	-	402	-	(60.4)
B_2	23223.0	R	406.6	2.75	426.3	1.9	(60.4)
C	22738.8	Q	397.8	0.75	418.2	0.7	(60.4)
D	21905.2	Q	398	-	417	-	(60.4)
E_1	21747.3	R	405.2	1.3	435.3	1.85	(60.4)
E_2	21639.4	Q	383.1	-	404.0	1.65	(60.4)
I	20534.4	Q	409.6	0.80	431.1	0.20	(69.7)
II	20264.0	Q	382.8	-	423.9	1.35	(69.7)
III	19967.0	Q	403.6	0.05	430.0	0.30	(69.7)
IV	18679.1	-	412.6	1.60	433.2	1.65	(69.7)
F	14434.2	R	407.2	2.5	420	-	(62.6)
G	12961.6	R	395.0	1.55	424	-	(62.6)
H	12270.8	Q	416.0	-	428	-	(62.6)
I	11902.9	R	395.0	1.40	402	-	(62.6)
J	11508.0	Q	398.7	1.45	405.2	-	(62.6)

BIBLIOGRAPHY

(38. 1) Partial Analysis,
K. R. More,
Phys. Rev. 54, 122-5

(39. 2) Measurements,
P. Mesnage,
Ann. Physique 12, 5-87

(52. 3) Vibrational Analysis,
V. G. Krishnamurty,
Indian J. Phys. 26, 207-25

(60. 4) Vibrational Analysis,
S. P. Reddy and P. T. Rao,
Proc. Phys. Soc. 75, 275-9

(61. 5) Dissociation Energy,
E. M. Bulewicz, L. F. Phillips, and T. M. Sugden,
Trans. Faraday Soc. 57, 921-31

(62. 6) Vibrational Analysis,
S. V. K. Rao, S. P. Reddy, and P. T. Rao,
Z. Physik 166, 261-4

(69. 7) I, II, III, IV Systems, Rotational Analysis, F, G, X Systems,
N. V. K. Rao and P. T. Rao,
"The Emission Spectrum of NiCl,"
Current Sci. 38, 589-90

NiF

Methods of Production and Experimental Technique

Emission from a high current discharge.

Band Systems

Intense (0, 0) sequence of complex bands degrading R (53.1).

Band heads (R_1):

(v', v'')	(0, 0)	(0, 1)	(0, 2)
λ(Intensity)	4535.8(6)	4695.1(6)	4859.7(5)

Spectroscopic Constants

Dissociation energy = 3.8 ± 1.5 eV, 88 kcal/mole, 30650 cm^{-1}.

Perturbations and General Information

Vibrational lasing action has been seen by exploding an Ni wire in F_2 (73.4) between $10.5 \leq \lambda \leq 24$ μm.

BIBLIOGRAPHY

(53.1) λ Spectral Reproduction,
V. G. Krishnamurty,
Indian J. Phys. 27, 354-8

(64.2) Gas Composition,
T. C. Ehlert, R. A. Kent, and J. L. Margrave,
J. Am. Chem. Soc. 86, 5093-5

(64.3) Balkaski, Moche, and Parisot,
"Infrared Spectra of Nickel Fluoride Vibrations in Matrix Isolation,"
C. R. Acad. Sci. 258, 2785

(73.4) W. W. Rice and W. H. Beattie,
"Metal Atom Oxidation Lasers,"
Chem. Phys. Letters 19, 82-5

NiH

Methods of Production and Experimental Technique

Absorption in a King furnace.

Emission from flames containing $Ni(CO)_4$, arcs, and King furnace.

BAND SYSTEMS

System	Transition	Sources	Wavelength Limits	Degrading	Characteristic Bands, λ	Remarks	Bibliography
I	$A^2\Delta_{5/2} - X^2\Delta_{5/2}$	King furnace	6650-6425	R	6425(0, 0)		(37.4, 34.1)
II	$B^2\Delta_{5/2} - X^2\Delta_{5/2}$	King furnace	6260-4900	R	6246(0, 0)		(37.4, 34.1)
III	$C^2\Delta - X^2\Delta$	King furnace	4660-4150	R	4201(0, 0)		(65.6, 37.3)

NiH

I. $A\,^2\Delta_{5/2} - X\,^2\Delta_{5/2}$ System

Band head of (0, 0), $\lambda = 6425.1$.

II. $B\,^2\Delta_{5/2} - X\,^2\Delta_{5/2}$ System

Band heads:

(v', v'')	(0, 0)	(1, 0)	(2, 0)	(3, 0)
λ	6246.0	5712.5	5290.0	4952.1

III. $C\,^2\Delta - X\,^2\Delta$ System

$^2\Delta_{5/2} - {}^2\Delta_{5/2}$

Band heads:

(v', v'')	(0, 1)	(0, 0)
λ	4570.5	4200.6

$^2\Delta_{3/2} - {}^2\Delta_{3/2}$

Band head of (0, 0), $\lambda = 4328$

$^2\Delta_{3/2} - {}^2\Delta_{5/2}$

Band head of (0, 0), $\lambda = 4151$

SPECTROSCOPIC CONSTANTS

State	T_o	ω_e	$x_e\omega_e$	B_o	$\alpha_e \times 10^3$	$D_o \times 10^4$	r_e	Remarks	Bibliography
$C^2\Delta_{3/2}$	24081.2	-	-	6.311	-	7.6	-		(65.6)
$C^2\Delta_{5/2}$	23760.7	~1780	~340	6.156	-	6.15	-		(65.6)
$B^2\Delta_{5/2}$	15977.3	1587.4	47.4	5.113	-	-	-		(37.4)
$A^2\Delta_{5/2}$	15520.1	-	-	6.283	-	5.0 / 4.5	-		(37.4)
$X^2\Delta_{3/2}$	980.4	-	-	7.781	-	5.9	-		(65.6)
$X^2\Delta_{5/2}$	0	~2000	~40	7.700	-	4.81	1.4754		(65.6)

Dissociation energy = 2.6 ± 0.3 eV, 60 kcal/mole, 20971 cm^{-1}.

BIBLIOGRAPHY

(34. 1) Note,
A. G. Gaydon and R. W. B. Pearse,
Nature 134, 287

(35. 2) Analysis,
A. G. Gaydon and R. W. B. Pearse,
Proc. Roy. Soc. A, 148, 312-35

(37. 3) Analysis,
A. Heimer,
Z. Physik 105, 56-72

(37. 4) Analysis,
A. Heimer,
Thesis, Stockholm

(63. 5) Note,
E. Andersen, A. Lagerqvist, H. Neuhaus, and N. Aslund,
Proc. Phys. Soc. 82, 637-8

(65. 6) General Analysis,
N. Aslund, H. Neuhaus, A. Lagerqvist, and E. Andersen,
Arkiv Fysik 28, 271-83

NiI

Dissociation energy = 2.99 ± 0.21 eV, 69 kcal/mole, 24100 cm^{-1} (61.1).

BIBLIOGRAPHY

(61.1)   E. M. Bulewicz, L. F. Phillips, and T. M. Sugden, Trans. Faraday Soc. 57, 921-31

NiO

Methods of Production and Experimental Technique

Emission from a flame containing Ni compounds, glow from a discharge between Ni electrodes, exploding wires.

BAND SYSTEMS

System	Transition	Sources	Wavelength Limits	Degrading	Characteristic Bands, λ	Remarks	Bibliography
I	a → (X)	Exploding wires	9195-7900	R	7900(0, 0) |8300(0, 1)[a]		(45, 2)
II	b → (X)[b]	Exploding wires	6605-5525	R	6153(3, 3) |6134(2, 2)[a]		(45, 2)
III	f → c[b]	Exploding wires	4850-4145	R	4662(2, 1) |4547(3, 1)[a]		(45, 2)

[a] Classification uncertain, [b] several other unclassified bands in this region which can be represented by:

$\nu = 13638 + 490\ v' - 590\ v''$ — weak bands

$= 19314 + 590\ v' - 825\ v''$ — (v', v'') (0,1) (0,0)
λ 5407 5176

$= 19602 + 590\ v' - 820\ v''$ — (v', v'') (0,1) (0,0)
λ 5323 5100

NiO

I. **a → (X) System**

Bands of greatest intensity (45.2):

v', v''	0	1	2	3	4	5	6
0	7900	8300	8739				
1		7983	8392	8843			
2			8094	8502	8950		
3				8187	8600	9070	
4							9195

II. **b - (X) System**

Band heads, λ(intensity), classification uncertain (45.2):

v', v''	0	1	2	3	4
0		6325(1)	6581(1)		
1		6111(1)	6350(3)	6605(1)	
2	5703(3)	5910(3)	6134(3)	6368(1)	
3	5525(2)		5930(2)	6153(3)	6386(2)

III. **f - c System**

Band heads, λ(intensity), classification uncertain (45.2):

v', v''	0	1	2
0	4730(2)		
1	4608(2)	4790(1)	
2	4490(1)	4662(2)	4848(1)
3	4382(2)	4547(2)	4720(1)
4		4436(2)	
5		4334(1)	

NiO

SPECTROSCOPIC CONSTANTS

State	T_e	ω_e	$x_e\omega_e$	B_e	$\alpha_e \times 10^3$	$D_e \times 10^6$	r_e	Remarks	Bibliography
f	21262 + x	(570)							(45. 2)
e	19719 + x	(590)							(45. 2)
d	19431 + x	(590)							(45. 2)
c	x	(825)							(45. 2)
b	(16447)	(560)							(45. 2)
a	(12725)	(475)							(45. 2)
(X)	0	(615)							(45. 2)

Dissociation energy ≤ 4.2 eV, ≤ 97 kcal/mole, 33876 cm^{-1}.

BIBLIOGRAPHY

(35.1) H. G. Howell and G. D. Rochester,
Proc. Univ. Durham Phil. Soc. 9, 126-34

(45.2) Vibrational Analysis,
L. Malet and B. Rosen,
Bull. Soc. Roy. Sci., Liege 14, 382-9

(45.3) Preliminary Note to (45.2),
B. Rosen,
Nature 156, 570

(48.4) General,
B. Rosen,
J. Phys. Radium 9, 155-8

(61.5) Dissociation Energy,
R. T. Grimley, R. P. Burns, and M. G. Inghram,
J. Chem. Phys. 35, 551-4

NiS

Dissociation energy = 3.53 ± 0.15 eV, 81.5 kcal/mole, 28500 cm^{-1} (67.1).

BIBLIOGRAPHY

(67.1) J. Drowart, A. Pattoret, and S. Smoes,
Proc. Brit. Ceram. Soc. <u>8</u>, 67-89

NiSi

Dissociation energy = 3.26 ± 0.17 eV, 75.2 kcal/mole, 26300 cm^{-1} (69.1).

BIBLIOGRAPHY

(69.1) A. Vander Auwera-Mahieu, N. S. MacIntyre, and J. Drowart, Chem. Phys. Letters 4, 198-200

OH

OH

Methods of Production and Experimental Technique

Absorption from a stream of H_2O vapor over an oven in a shock tube. Emission from flames, hollow cathode, continuous discharge.

BAND SYSTEMS

System	Transition	Sources	Wavelength Limits	Degrading	Characteristic Bands, λ	Remarks	Bibliography
I	$A^2\Sigma^+ \rightleftarrows X^2\Pi_i$	Microwave discharge	4107-2608	R	3064(0,0)		(67.171, 62.145)
II	$C^2\Sigma^+ \to X^2\Pi_i$	Microwave discharge	1900-1700	R	1854(0,16) 1839(0,15)		(64.154)
III	$B^2\Sigma^+ - A^2\Sigma^+$	Microwave discharge	5660-4216	R	5480(0,8) 5125(0,7)		(69.186, 68.182, 63.149, 56.110)
IV	$C^2\Sigma^+ - A^2\Sigma^+$	Microwave discharge	2600-2249	R	2545(0,8) 2465(0,7)		(69.186, 63.149, 57.118)

OH

I. $A^2\Sigma^+ \rightarrow X^2\Pi_i$ System

Band heads (R), λ:

v', v''	0	1	2	3	4
0	3064	3428	3843		
1	2811	3122	3484	3898	
2	2609	2875	3185		3959
3	2444	2677	2945	3254	
4		2517	2753	3022	3331

II. $C^2\Sigma^+ - X^2\Pi_i$ System

Band heads, λ(approximate values):

v', v''	0	...	10	11	12	13	14	15	16
0			1635	1686	1734	1777	1813	1839	1854
1			1578	1624	1668	1708	1741	1765	1779

III. $B^2\Sigma^+ \rightarrow A^2\Sigma^+$ System

Band heads (R), λ:

v', v''	0	...	5	6	7	8	9
0			4337	4730	5124	5480	
1			4216	4587	4957		5534

IV. $C^2\Sigma^+ \rightarrow A^2\Sigma^+$ System

Band heads (R), λ:

v', v''	0	...	6	7	8	9
0				2465	2545	2600
1			2249	2334		2455

OH

SPECTROSCOPIC CONSTANTS

State	T_e	ω_e	$x_e\omega_e$	B_e	$\alpha_e \times 10^2$	$D_e \times 10^3$	r_e	Remarks	Bibliography
$C^2\Sigma^+$	89500	1232.9	19.1	4.247	7.80	–	2.16	γ (63.149, 57.118)	(69.186, 68.182, 62.145, 56.110)
$B^2\Sigma^+$	69775	940	105	5.54	65	1.5	1.80	γ (68.182, 63.149)	(68.182, 62.145, 56.110)
$A^2\Sigma^+$	32682.5	3184.28	97.84	17.355	80.7	2.04	1.0121	γ (68.182, 63.149, 62.145)	(68.182, 62.145, 56.110)
$X^2\Pi_i$	0	3735.21	82.81	18.871	71.4	1.88	0.9706	A, γ (62.145)	(71.194, 68.182, 62.145, 56.110)

Dissociation energy = 4.40 ± 0.01 eV, 101.5 kcal/mole, 35489 cm^{-1}.

Perturbations and General Information

$A^2\Sigma^+$ state predissociates at k = 30 of v = 0 level and k = 21 of v = 1 level caused by interaction with a dissociative $^4\Sigma^-$ state (73. 202).

$A^2\Sigma^+$ state predissociates at v = 5 caused by interaction with dissociative $^4\Pi$ and $^2\Sigma^-$ states (71. 194).

Potential energy curves (73. 202):

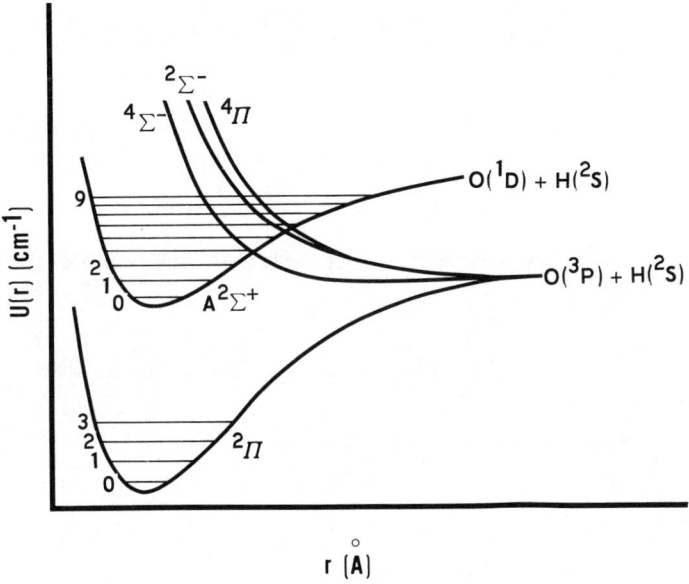

Electronic quenching cross sections: $OH(A^2\Sigma^+) + M \rightarrow OH(X^2\Pi_i) + M$

M	Cross Section (Å^2)	Reference
O_2	7	(59. 128)
CO_2	16	(59. 128)
H_2O	35	(59. 128)

Dipole moment $A^2\Sigma^+$(v=0): $\mu = 1.98 \pm 0.08$ D.

Radiative lifetime (72. 200):

$$A^2\Sigma^+ \rightarrow X^2\Pi_i \qquad \tau(v'=0, k'=1) = 803 \pm 80 \text{ nsec.}$$

OH

Franck-Condon factors — Morse potential (58.126):

$\underline{A^2\Sigma - X^2}$

v', v''	0	1	2	3	4
0	3064	3428	--	--	--
	0.90	0.08	0.00	--	--
1	2811	3122	3484	--	--
	0.08	0.71	0.18	0.01	--
2	--	2875	3185	--	--
	0.00	0.16	0.51	0.28	0.03
3	--	--	2945	3254	--
	0	0.02	0.23	0.32	0.32
4	--	--	--	--	--
	--	0.00	0.05	0.25	--

Top = wavelengths, and bottom = Franck-Condon factors.

Laser action in the infrared (71.197) laser action observed v = 3-2, 2-1, 1-0 (3410-3050 cm^{-1}).

BIBLIOGRAPHY

(27.1) Theory of Rotational Distortion,
        E. C. Kemble,
        Phys. Rev. 10, 387-99

(27.2) Historical View,
        R. Mecke,
        Z. Physik 42, 390-425

(28.3) Historical View,
        E. D. Wilson
        J. Opt. Soc. Am. 17, 37-46

(28.4) Sensitized Fluorescence,
        E. Gaviola and R. W. Wood,
        Philos. Mag. 6, 1191-210

(29.5) Astrophysics,
        R. Willett,
        Z. Physik 54, 856-79

(30.6) Zeeman Effect,
        G. M. Almy,
        Phys. Rev. 35, 1495-512

(31.7) Molecular Constants,
        G. M. Almy and G. D. Rahrer,
        Phys. Rev. 38, 1816-7

(31.8) Selective Excitation,
        M. Kaczynska,
        Z. Physik 67, 601-4

(32.9) Astrophysics,
        R. W. Shaw,
        Astrophys. J. 76, 202-9

(32.10) Electron Affinity,
        E. Lederle,
        Z. Phys. Chem. B17, 362-8

(33.11) Proton Spin and Hyperfine Structure,
        S. Mrozowski,
        Acta Phys. Polon 2, 235-7

OH

(33.12) Isotope Effect,
H. L. Johnston and D. H. Dawson,
Naturwissenschaften 21, 495-6

(33.13) Excitation Conditions,
R. C. Johnson and E. G. Dunstan,
Philos. Mag. 16, 472-8

(33.14) New Lines,
K. Chamberlain and H. B. Cutter,
Phys. Rev. 43, 771-2

(33.15) New Bands,
K. Chamberlain and H. B. Cutter,
Phys. Rev. 44, 927-30

(33.16) Detailed Study,
D. H. Dawson and H. L. Johnston,
Phys. Rev. 43, 980-91

(33.17) (2,2), (1,1) and (0,0) Bands,
H. L. Johnston and D. H. Dawson,
Phys. Rev. 43, 580

(33.18) (1,2) Bands,
H. L. Johnston, D. H. Dawson, and M. K. Walker,
Phys. Rev. 43, 473-80

(33.19) Weak Bands,
T. Tanaka and Z. Koana,
Proc. Phys. Math. Soc. 15, 272-90

(33.20) Analysis,
T. Tanaka and M. Siraisi,
Proc. Phys. Math. Soc. 15, 195-209

(34.21) Lifetime,
O. Oldenberg,
J. Chem. Phys. 2, 713-4

(34.22) Electronic Isotope Effect,
H. L. Johnston,
Phys. Rev. 45, 79-81

(34.23) Irregular Rotation,
O. Oldenberg,
Phys. Rev. 46, 210-5

(34.24) Spectral Analysis,
T. Tanaka and Z. Koana,
Proc. Phys. Math. Soc. 16, 365-400

(34.25) W. D. Harkins,
Trans. Faraday Soc. 30, 221-7

(35.26) A. Naherniac,
C. R. Acad. Sci. 200, 1742-3

(35.27) Lifetime,
O. Oldenberg,
J. Chem. Phys. 3, 266-75

(35.28) Chemical Reaction,
A. A. Frost and O. Oldenberg,
Phys. Rev. 47, 788

(36.29) Kinetics,
A. A. Frost and O. Oldenberg,
J. Chem. Phys. 4, 642-8

(36.30) Kinetics,
A. A. Frost and O. Oldenberg,
J. Chem. Phys. 4, 781-4

(37.31) Discharge Flames,
L. Avramenko and V. Kondratjew,
Acta Physochim. U.R.S.S. 7, 567-80

(37.32) Flame Spectra,
V. Kondratjew and M. Ziskin,
Acta Physicochim. U.R.S.S. 6, 307-19

(37.33) Intensity Distribution,
V. Kondratjew and M. Ziskin,
Acta Physicochim. U.R.S.S. 7, 65-74

(37.34) Electronic Levels, Theory,
J. R. Stehn,
J. Chem. Phys. 5, 186-91

(37.35) Theory,
M. A. Kowner,
J. Exper. Theor. Phys. U.S.S.R. 7, 12-8

OH

(37.36) Theory,
O. Oldenberg,
J. Phys. Chem. 41, 293-7

(38.37) Lifetime Absorption,
V. Kondratjew,
Bull. Acad. Sci. U.R.S.S. Ser. Phys. 371-2

(38.38) Night Sky Spectrum,
G. Dejardin and R. Bernard,
C.R. Acad. Sci. 206, 1747-8

(38.39) Theory,
J. Rouvillois and H. Muraour,
C.R. Acad. Sci. 206, 1719-21

(38.40) Absorption,
O. Oldenberg and F. F. Rieke,
J. Chem. Phys. 6, 169

(38.41) Kinetics,
O. Oldenberg and F. F. Rieke,
J. Chem. Phys. 6, 439-47

(38.42) Kinetics,
O. Oldenberg and F. F. Rieke,
J. Chem. Phys. 6, 779-82

(38.43) Rotational Energy Distribution,
E. R. Lyman,
Phys. Rev. 53, 379-83

(38.44) Rotational Energy Distribution,
E. R. Lyman and F. A. Jenkins,
Phys. Rev. 53, 214

(38.45) Intensity,
O. Oldenberg and F. F. Rieke,
Phys. Rev. 53, 941

(38.46) Intensity Distribution,
H. Wakesima,
Proc. Phys. Math. Soc. 20, 374-7

(39.47) Stellar Spectra,
N. T. Bobravnioff,
Astrophys. J. 89, 301-10

(39.48) Flame Spectra. Intensities of Electronic Transitions,
R. S. Mulliken,
Astrophys. J. 89, 283-8

(39.49) Solar Spectrum,
F. E. Roach,
Astrophys. J. 89, 99-115

(39.50) Theory,
C. H. D. Clark and J. L. Stoves,
Philos. Mag. 27, 389-403

(39.51) Sensitized Fluorescence,
E. R. Lyman,
Phys. Rev. 55, 1126

(39.52) Sensitized Fluorescence,
E. R. Lyman,
Phys. Rev. 56, 466-70

(40.53) Flame Spectra,
V. Kondratjew,
Bull. Acad. Sci. U.R.S.S. Cl. Chim. Phys. 501-8

(40.54) Theory,
G. B. B. M. Sutherland,
J. Chem. Phys. 8, 161-4

(40.55) Discharge Across Flames,
F. L. Brown,
Phys. Rev. 57, 942

(40.56) Explosions,
O. Oldenberg, E. G. Schneider, and H. S. Sommers Jr.,
Phys. Rev. 58, 1121

(40.57) Theory,
C. H. D. Clark,
Trans. Faraday Soc. 36, 370-6

(41.58) Temperature Determinations,
G. D. Cristescu and R. Grigorovici,
Bull. Soc. Roumaine Phys. 42, 37-51

(41.59) Isotope Effect,
M. G. Sastry,
Current Sci. 10, 362-3

OH

(41.60) Theory,
H. M. Hulburt and J. O. Hirschfelder,
J. Chem. Phys. 9, 61-9

(41.61) Flame Spectra,
K. H. Geib and W. M. Vaidya,
Proc. Roy. Soc. A 178, 351-5

(41.62) Flame Spectra,
W. M. Vaidya,
Proc. Roy. Soc. A 178, 356-69

(41.63) Flame Spectra,
K. H. Geib,
Z. Elektrochem. 47, 275-6

(42.64) Discharge Across Flames,
L. I. Avramenko,
Acta Physiochim. U.R.S.S. 17, 197-210

(42.65) Astrophysics,
C. T. Elvey, P. Swings, and H. W. Babcock,
Astrophys. J. 95, 218-9

(42.66) Comet Spectra,
D. M. Popper and P. Swings,
Astrophys. J. 96, 156-7

(42.67) Raman and Infrared Spectra,
L. S. Druskina,
J. Exper. Theor. Phys. U.S.S.R. 12, 54-7

(42.68) Band Structure,
M. G. Sastry,
Indian J. Phys. 16, 27-34

(42.69) Photochemistry,
W. J. Blaedel, R. A. Ogg Jr., and P. A. Leighton,
J. Am. Chem. Soc. 64, 2499-500

(42.70) Force Constant; Internuclear Distance,
G. Glockler and G. E. Evans,
J. Chem. Phys. 10, 606

(42.71) Irregular Rotation,
H. Wakesima,
Proc. Phys. Math. Soc. 24, 367-74

(42.72) Flame Spectra,
A. G. Gaydon,
Proc. Roy. Soc. A 179, 439-50

(42.73) Comet Spectra,
P. Swings,
Publ. Astron. Soc. Pacific 54, 123-36

(42.74) Theory,
P. Swings,
Publ. Astron. Soc. Pacific 54, 232-6

(42.75) Comet Spectra,
P. Swings,
Rev. Mod. Phys. 14, 190-4

(43.76) Infrared Spectrum,
P. Barchewitz,
Ann. Physique 18, 167-89

(43.77) Infrared Spectrum,
P. Barchewitz,
Arch. Orig. Serv. Docum. 120, 77 pp

(44.78) Infrared Spectrum,
A. M. Vergnoux,
C. R. Acad. Sci. 219, 125-7

(45.79) Infrared Spectrum,
A. M. Vergnous,
Thesis, Paris

(47.80) Infrared Spectrum,
J. Louisfert,
J. Phys. Radium 8, 21-8

(48.81) Dissociation Energy,
G. Glockler,
J. Chem. Phys. 16, 602-4

(48.82) Continuous Spectra,
G. A. Hornbeck,
J. Chem. Phys. 16, 845-6

(48.83) Infrared Spectrum,
E. K. Plyler,
J. Res. Nat. Bur. Stand. 40, 113-20

OH

(48.84) Infrared Spectrum,
E. K. Plyler and C. J. Humphreys,
J. Res. Nat. Bur. Stand. 40, 449-56

(48.85) Continuous Spectra. Flame Absorption
G. Herzberg and L. Herzberg,
Nature 161, 283

(48.86) Flame Spectra,
A. G. Gaydon and H. G. Wolfhard,
Proc. Roy. Soc. A 194, 169-84

(50.87) Night Sky Emission,
A. B. Meinel,
Astrophys. J. 111, 207

(50.88) Night Sky Emission,
A. B. Meinel,
Astrophys. J. 111, 433-4

(50.89) Night Sky Emission,
A. B. Meinel,
Astrophys. J. 111, 555-64

(50.90) Analysis,
A. B. Meinel,
Astrophys. J. 112, 120-30

(50.91) G. Herzberg,
"Molecular Spectra and Molecular Structure I,"
2nd Ed. New York, D. Van Nostrand Co. (616 pgs)

(51.92) Vibration-Rotation, Infrared,
G. A. Hornbeck and R. C. Herman,
J. Chem. Phys. 19, 512

(51.93) Excitation in Fluorescence,
P. J. Dyne and D. W. G. Style,
Nature 167, 899

(51.94) Predissociation,
A. G. Gaydon and H. G. Wolfhard,
Proc. Roy. Soc. A 208, 63-75

(53.95) Infrared Band. Vibration-Rotation,
R. C. Herman and G. A. Hornbeck,
Astrophys. J. 118, 214-27

(53.96) Infrared Bands. Vibration-Rotation,
G. Dejardin, J. Janin and M. Peyron
Cah. Phys. 46, 3-16

(53.97) Infrared Spectrum,
W. S. Benedict, E. K. Plyler, and C. J. Humphreys,
J. Chem. Phys. 21, 398-402

(53.98) Rotational Intensities,
A. M. Bass and H. P. Broida,
Nat. Bur. Stand. Circ. 541, 22 pp

(53.99) Intensity, Rotational Structure,
H. P. Broida and W. R. Kane,
Phys. Pev. 89, 1053-9

(53.100) Infrared Bands, Vibration Rotation,
G. Dejardin, J. Janin, and M. Peyron,
Phys. Rev. 90, 359

(53.101) Microwave Spectra,
T. M. Sanders Jr., A. L. Schalow, G. C. Dousmanis, and C. H. Townes,
Phys. Rev. 89, 1158-9

(54.102) Visible Spectrum,
S. Benoist,
C. R. Acad. Sci. 238, 883-5

(54.103) Infrared Structure, Vibration Rotation,
G. Dejardin, J. Janin, and M. Peyron,
J. Phys. Radium 15, 222

(54.104) B-A Analysis,
H. Schüler, L. Reineback and A. Michel,
Z. Naturforsch A 9, 279-85

(55.105) B-A Analysis,
S. Benoist,
Ann. Physique 10, 363-76

(55.106) Astrophysics. Infrared Bands,
J. W. Chamberlain and F. L. Roesler,
Astrophys. J. 121, 541-7

OH

(55. 107) Pure Rotation,
R. P. Madden and W. S. Benedict,
J. Chem. Phys. 23, 408-9

(55. 108) Spectrum of B-A Emission,
S. Leach,
J. Chim. Phys. 52, 492-7

(55. 109) Hyperfine Structure,
G. D. Dousmainis, T. M. Sanders Jr., and C. H. Thownes,
Phys. Rev. 100, 1735-54

(56. 110) B-A Analysis,
R. F. Barrow,
Arkin Fyski 11, 281-90

(56. 111) Flames, Infrared,
T. M. Cawthon, Jr., and J. D. MacKinley,
J. Chem. Phys. 25, 585-6

(56. 112) B-A Constants,
R. F. Barrow and A. R. Downie,
Proc. Phys. Soc. A 69, 178-80

(56. 113) Franck-Condon Factors,
R. W. Nicholls,
Proc. Phys. Soc. A 69, 741-53

(56. 114) Franck-Condon Factors,
R. W. Nicholls and W. R. Jarmain,
Proc. Phys. Soc. A 69, 253-64

(56. 115) B-A Vibrational-Rotational Analysis,
H. Schüler and A. Michel,
Z. Naturforsch. A 77, 403-6

(57. 116) Solar Spectrum,
C. E. Moore and H. P. Broida,
Mem. Soc. Roy. Sci. 18, 252-63

(57. 117) Infrared Spectrum,
H. C. Allen Jr., L. R. Blaine, and E. K. Plyler,
Spectrochim. Acta 9, 126-32

(57. 118) B-A, C-A Vibrational-Rotational Analysis,
A. Michel,
Z. Naturforsch. A 12, 887-96

(58.119) LCAO-MO-SCF Calculations,
R. Gaspar and I. Tamassy-Lentei,
Acta Phys. Acad. Sci. 9, 105-13

(58.120) Valence-Band and LCAO-MO-SCF Calculations of $X^2\pi$ Using Hartree-Fock Atomic Orbitals,
A. J. Freeman,
J. Chem. Phys. 28, 230-43

(58.121) LCAO-MO-SCF Study of the Ground State,
M. Krauss,
J. Chem. Phys. 28, 1021-6

(58.122) Configuration Interaction Study Using Wave Function of (58.121),
M. Krauss and J. F. Wehner,
J. Chem. Phys. 29, 1287-97

(58.123) $^2\Sigma$ Rotation Population in High Frequency Discharge,
T. Carrington and H. P. Broida,
J. Mol. Spectrosc. 2, 273-86

(58.124) Flames,
M. Charton and A. G. Gaydon,
Proc. Roy. Soc. A 245, 84-92

(58.125) Study of Ground and Excited States by LCAO-MO-SCF-VB Calculations,
A. C. Hurley,
Proc. Roy. Soc. A 248, 118-35

(58.126) Franck-Condon Factors, A-X,
R. W. Nicholls, P. A. Fraser, and W. R. Jarmain,
"Transition Probabilities Parameters of Molecular Spectra Arising from Combustion Processes,"
Combustion and Flame 26, 13-38

(59.127) W. E. Kaskan,
"Abnormal Excitation of OH in $H_2/O_2/N_2$ Flames,"
J. Chem. Phys. 31, 944-56

(59.128) Electronic Quenching Cross-Sections, A State,
T. Carrington,
"Electronic Quenching of OH($^2\Sigma^+$) in Flames and its Significance in the Interaction of Rotational Relaxation,"
J. Chem. Phys. 30, 1087-95

OH

(59.129) $^2\Sigma^+$ Photodissociation,
T. Carrington,
J. Chem. Phys. 31, 1243-52

(59.130) Solar Spectrum,
C. E. Moore and H. P. Broida.
J. Res. Nat. Bur. Stand. A 63, 279-95

(59.131) Study of Ground and Excited States by LCAO-MO-SCF-VB Method,
A. C. Hurley.
Proc. Roy. Soc. A 249, 402-13

(60.132) Flash Photolysis,
W. D. MacGrath and R. G. W. Norrish,
Proc. Roy. Soc. A 254, 317-26

(60.133) Valence Band Configuration Interactions and LACO-MO-SCF-CI Calculations of $X^2\pi$,
A. J. Freeman,
Rev. Mod. Phys. 32, 273-4

(60.134) Vibration-Rotation: Night Sky,
G. Kifte,
Dept. of Physics and Meterology,
Agricultural College 11

(60.135) Oscillator Strength, A-X,
M. Lapp
"Shock-tube Measurements of the f-number for the (0.0) band of the OH $^2\Sigma \rightarrow ^2\pi$ transition,
Cal. Tech. TR #11

(61.136) Potential Curves . Calculations,
R. J. Fallon, I. Tobias and J. T. Vanderslice,
J. Chem. Phys. 34, 167-8

(61.137) Flames,
S. L. N. G. Krishnamachari and H. P. Broida,
J. Chem. Phys. 34, 1709-11

(61.138) Predissociation. Excitation by Electron Streams,
M. Horani and S. Leach,
J. Chim. Phys. 58, 825-9

(61.139) B-A, C-A, Rotational Vibrational Analysis,
L. Herman, P. Helenback and R. Herman,
J. Phys. Radium 22, 83-92

(61.140) Study of Ground and Excited States by LCAO—MO-SCF-VB Method,
A. C. Hurley,
Proc. Roy. Soc. A 261, 237-45

(61.141) Auroras, Night Sky,
J. W. Chamberlain,
Physics of the Aurora and Airglow,
New York, Acad. Press, 1961

(61.142) Meinel Bands,
J. P. Ostriker,
Yerkes Obs. Univ. Chicago,
Sci. Rept. 35

(62.143) Vibrational-Rotational Analysis. Infrared,
D. Garvin, H. P. Broida, and H. J. Kostkowski,
J. Chem. Phys. 37, 193

(62.144) Vibrational Rotational Band Analysis,
A. M. Bass and D. Garvin,
J. Mol. Spectrosc. 9, 114-23

(62.145) Complex A-X Analysis. Constants,
G. H. Dieke and H. M. Crosswhite,
J. Quant. Spectrosc. Radiative Trans. 2, 97-199

(62.146) Franck-Condon Factors,
R. W. Nicholls,
J. Quant. Spectrosc. Radiative Trans. 2, 433-9

(62.147) Intensities. Vibration-Rotation Interaction,
R. C. M. Learner,
Proc. Roy. Soc. A 269, 311-26

(62.148) G. Pannetier, L. Marsigny, and M. Ben-Caid,
"Spectroscopic Study of the Reaction of Atomic Oxygen or Active Nitrogen with Hydrocarbons of Chlorine or Bromine. Observation of the ($B^2\Sigma_u^+ - X^2\Sigma_g^+$) Transition of the N_2^+ Molecule in Certain Reactions,"
C. R. Acad. Sci. 12, 1270-71

(63.149) Vibrational-Rotational Analysis,
P. Felenbok,
Ann. Astrophys. 26, 393-428

OH

(63.150) Franck-Condon Factors,
P. Felenbok,
C.R. Acad. Sci. 256, 2334-7

(63.151) Oscillator Strength,
D. M. Golden, F. P. Delgreco, and F. Kaufman,
J. Chem. Phys. 39, 3034-41

(63.152) Hyperfine Structure Constants,
K. Kayama,
J. Chem. Phys. 39, 1507-13

(63.153) Shockwave Rotational Temperature,
H. Miyama and T. Takeyama,
J. Chem. Phys. 39, 351-2

(64.154) C-X Identification,
P. Felenbok and J. Czarny,
Ann. Astrophys. 27, 244-6

(64.155) Oscillator Strength of A-X,
R. G. Bennett and F. W. Dalby,
J. Chem. Phys. 40, 1414-6

(64.156) Valence Band Calculation of the Ground State,
D. M. Bishop and J. R. Hoyland,
Molecular Phys. 7, 161-4

(64.157) Infrared Absorption,
J. F. Ogilvie,
Nature 204, 572

(65.158) Valence-Band Calculations of the Ground State (STO),
K. Pecul,
Acta Phys. Polon. 27, 713-22

(65.159) A-X Rotational Analysis,
A. Stoebner, R. Delbourge, and P. Laffitte,
C.R. Acad. Sci. 261, 5044-6

(65.160) Stark Effect,
D. H. Phelps and F. W. Dalby,
Can. J. Phys. 43, 144-54

(65.161) Vibration-Rotation Interaction, Shock Wave Intensity,
R. Watson and W. R. Ferguson,
J. Quant. Spect. Radiative Transfer 5, 595-609

(65.162) C. C. McDonald and R. J. Goll,
"Studies of Gaseous Atom-Molecule Reactions by Electron Paramagnetic Resonance Spectroscopy,"
J. Phys. Chem. 69, 293-7

(66.163) P. J. T. Zeegers,
"Recombination of Radicals and Related Effects in Flames,"
Doctoral Thesis, Rotterdam 1966

(66.164) H. P. Hooymayers,
"Quenching of Excited Alkali Atoms and Hydrogen Radicals and Related Effects in Flames,"
Doctoral Thesis, Rotterdam 1966

(66.165) "A" Coefficient for Λ Doubling,
B. E. Turner,
Nature 212, 184-5

(66.166) Absorption,
G. Schulz and F. H. LeCerf,
Z. Physik 197, 228-45

(66.167) Absorption Without Flames,
R. Bleekrode,
Thesis, Amsterdam

(67.168) Astrophysics, Interstellar Matter,
B. J. Robinson and R. X. MacGee,
Annual Rev. Astron Astrophys. 5, 183-212

(67.169) Study of Band Characteristics Through Interpretation of Wave Function of (67.170),
R. F. W. Bader, I. Keaveny, and P. E. Cade,
J. Chem. Phys. 47, 3381-402

(67.170) LCAO-MO-SCF Calculation of Ground State. Hartree-Fock Limit Reached,
P. E. Cade and W. M. Hue,
J. Chem. Phys. 47, 614-48

(67.171) A-X Rotational Analysis,
A. Stoebner and R. Delbourgo,
J. Chim. Phys. 64, 1115-23

(67.172) Emission, Predissociation,
D. W. Naegeli and H. B. Palmer,
J. Mol. Spectrosc. 23, 44-52

OH

(67.173) MO-SCF Calculations,
R. N. Dixon,
Molecular Phys. 12, 83-90

(67.174) "A" Coefficient for Hyperfine Transition,
D. R. Lide Jr.,
Nature 213, 694-5

(67.175) "A" Coefficient for Hyperfine Transition,
B. E. Turner,
Nature 214, 379

(67.176) Transition Moment with Vibration Rotation Interaction,
J. Anketell and R. C. M. Learner,
Proc. Roy. Soc. A 301, 355-61

(67.177) Oscillator Strengths, Shockwaves,
J. Anketell and A. Pery-Thorne,
Proc. Roy. Soc. A 301, 343-53

(67.178) A-X System Through Recombination,
G. B. Spindler, S. Ticktin, and H. I. Schiff,
"The Chemiluminescent Reaction O + H $\rightarrow$ OH + hv,"
Nature 214, 1006-7

(67.179) Transition Probabilities, A-X,
J. Anketell and R. C. M. Learner,
"Vibrational Rotation Interaction in OH and He the Transition Moment,"
Proc. Roy. Soc. A 301, 355-61

(67.180) S. Ticktin, G. Spindler, and H. I. Schiff,
"Production of Excited OH ($A^2\Sigma^+$) Molecules by the Association of Ground State Oxygen and Hydrogen Atoms,"
Disc. Faraday Soc. 44, 218-55

(68.181) A-X System Through Recombination,
H. B. Palmer and D. W. Naegeli,
"Predissociation of Chemiluminescent OH and OD,"
J. Mol. Spectrosc. 28, 417-21

(68.182) B-A Rotational Analysis. Predissociation,
J. Czarny and P. Felenbok,
Ann. Astrophys. 31, 141-52

(68.183) R. M. Dagnal, D. H. Smith, K. C. Thompson, and T. S. West,
"Emission Spectra Obtained from the Combustion of Organic Compounds in Hydrogen Flames,"
Analyst 94, 871-8

(69.184) R. M. Dagnall, D. J. Smith, K. C. Thompson, and T. S. West,
"Emission Spectra Obtained from the Combination of Organic Compounds in Hydrogen Flames,"
Analyst 94, 871-878

(69.185) P. R. Schwartz and A. H. Barrett,
"Observations of the $^2\pi_{1/2}$, J = 5/2 Excited State of OH in W3,"
Astrophy. J. 157, L109-L110

(69.186) C, B → A Vibrational-Rotational Analysis,
C. Carbone and F. W. Dalby,
"Spectrum of the Hydroxyl Radical,"
Can. J. Phys. 47, 1945-57

(70.187) Radiative Lifetime, A-X,
W. H. Smith,
"Radiative and Predissociation Probabilities for the OH $A^2\Sigma^+$ State,"
J. Chem. Phys. 53, 792-93

(70.188) Vibration-Rotation Relaxation Time,
K. H. Welge, S. V. Filseth and J. Davenport,
"Rotation-Vibration Energy Transfer in Collisions Between $OH(A^2\Sigma^+)$ and Ar and N_2,"
J. Chem. Phys. 53, 502-7

(70.189) E. M. Bulewicz, P. J. Padley, and R. E. Smith,
"Spectroscopic Studies of C_2, CH, and OH Radicals in Low Pressure Acetylene Oxygen Flames,"
Proc. Roy. Soc. A 315, 129-148

(71.190) E. C. Shane and W. Brennen,
"Chemiluminescence of the Atomic Oxygen-Hydrazine Reaction,"
J. Chem. Phys. 55, 1479-80

(71.191) Vibrational Transition Probabilities,
R. E. Murphy,
"Infrared Emission of OH in the Fundamental and First Overtone Bands,"
J. Chem. Phys. 54, 4852-9

OH

(71.192) P. E. Charters, R. G. MacDonald and J. C. Polanyi,
"Formation of Vibrationally Excited OH by the Reaction $H + O_3$,"
Applied Optics 10, 1749-54

(71.193) E. A. Scarl and F. W. Dalby,
"High-Field Stark Effects on the Near Ultraviolet Spectrum of the Hydroxyl Radical,"
Can. J. Phys. 49, 2825-32

(71.194) Spectroscopic Analysis A-X,
E. A. Moore and W. G. Richards,
"A Reanalysis of the $A^2\Sigma^+ - X^2\pi$ System of OH,"
Physica Scripta 3, 223-30

(71.195) Predissociation of A State,
S. Durmez and J. N. Murrell,
"Predissociation by the Continuance of a Band State Without Curve Crossing,"
Trans. Faraday Soc. 67, 3395-8

(71.196) Radiative Lifetime, A-X,
R. L. De Zafra, A. Marshall, and H. Metcalf,
"Measurement of Lifetime and g Factors by Level Crossing and Optical Double Presonance in the OH and OD Free Radicals,"
Physical Review A 3, 1557-67

(71.197) Vibrational Laser Action,
A. B. Callear and H. E. van den Bergh,
"An Hydroxyl Radical Infrared Laser,"
Chem. Phys. Letters 8, 17-18

(71.198) Predissociation of A State,
P. S. Julienne and M. Krauss,
"Formation of OH Through Inverse Predissociation,"
Astrophys. J. 170, 65-70

(72.199) A-X System Through F + H, O,
G. Schatz and M. J. Kaufman,
"Chemiluminescent Reactions of Atomic Fluorine,"
Princeton University TR NR 092-531

(72.200) B. G. Elmergreen and W. H. Smith,
"Direct Measurements of the Lifetimes and Predissociation Probabilities for Rotational Levels of the OH and OD $A^2\Sigma^+$ States,
Astrophys. J. 178, 557-64

(72.201) Theoretical Study of Ground State Rotational Levels,
M. Mizushima,
"Molecular Parameters of OH Free Radicals,"
Physical Review A, 143-57

(73.202) R. A. Sutherland and R. A. Anderson,
"Radiative and Predissociative Lifetimes of the $A^2\Sigma^+$ State of OH,"
J. Chem. Phys. 58, 1226-34

(73.203) J. J. Ter Meulen and A. Dymanus,
"Beam-Maser Measurements of the Ground State Transition Frequencies of OH,"
Astrophys. J. 172, L21-L23

(73.204) Λ Doubling,
I. E. Valtz and V. A. Soglasnova,
"Λ-Doubling of the ^{17}OH Molecule at Microwave Frequencies,"
Astro. Physical Letters 13, 23-4

(71.L1) A. G. Gaydon and I. Kopp,
"Predissociation in the Spectrum of OH; a Reinterpretation,"
J. Phys. B (Atom. Mol. Phys.) 4, 752-8

(71.L2) K. P. Lee, W. G. Tam, R. Larouche, and G. A. Woonton,
"Electron Resonance of Vibrationally Excited OH Radicals,"
Can. J. Phys. 49, 2207-10

(71.L3) R. Nanes and D. W. Robinson,
"Magnetic Rotation Spectra of the $A^2\Sigma^+ - X^2\pi_i$ Transition of OH and OD,"
J. Chem. Phys. 55, 963-74

(71.L4) F. Penny,
"Time Resolved Spectroscopy of a Pulsed Discharge Through Water Vapor. Observations of Emissions from the $C^2\Sigma^+$ State of OH,"
Spectrosc. Letters 4, 319-27

(71.L5) F. F. Gardner and J. C. Ribes,
"Observations of the Excited Lines of OH Near 4700 MHz,"
Astrophys. Letters 9, 175-9

(72.L6) V. Bondybey, P. K. Pearson, and H. F. Schaefer III,
"Theoretical Potential Energy Curves for OH, HF^+, HF, HF^-, NeH^+ and NeH,"
J. Chem. Phys. 57, 1123-5

OH

(72. L7) R. K. Hinkley, J. A. Hall, T. E. H. Walker, and W. G. Richard
"Λ-Doubling in $^2\pi$ States of Diatomic Molecules,"
J. Phys. B (Atom. Mol. Phys.) 5, 204-12

(72. L8) B. E. Turner,
"OH and Formaldehyde Absorption in the Direction of Cygnus X-3,"
Nature 239, 132-3

(72. L9) B. Zuckerman, J. L. Yen, C. A. Gottlieb, and P. Palmer,
"Observations of the $^2\pi_{3/2}$, J = 5/2 State of Interstellar OH,"
Astrophys. J. 177, 59-78

(72. L10) R. Engleman, Jr.,
"Accurate wave Numbers of the $A^2\Sigma \rightarrow X^2\pi$ (0,0) and (1,0) Bands of OH and OD,"
J. Quant. Spectrosc. Radiative Transfer 12, 1347-50

(72. L11) N. Sokabe,
"Rotational Excitation of OH ($A^2\Sigma^+$, v' = 0) Resulting from Dissociative Collision of H_2O with Metastable Argon Atoms,"
J. Phys. Soc. 33, 473-82

(73. L12) R. K. Hinkley, T. E. H. Walker, and W. G. Richards,
"On the e.p.r. Spectrum of Vibrationally Excited Hydroxyl Radicals,"
Proc. Roy. Soc. A 331, 553-60

(73. L13) S. H. Knowles, K. J. Johnston, J. M. Moran, and J. A. Ball,
"Inteferometric Observations of the $^2\pi_{3/2}$, J = 5/2 State of Interstellar OH,"
Astrophys. J. 180, L117-L121

(73. L14) Preliminary Calculations on new $D^2\Sigma^-$ State,
I. Easson and M. H. L. Pryce,
"Calculated Potential Energy Curves of OH,"
Can. J. Phys. 51, 518-29

(73. L15) K. R. German, T. H. Bergman, E. M. Weinstock, and R. N. Zare,
"Zero-Field Level Crossing and Optical Radio-Frequency Double Resonance Studies of the $A^2\Sigma^+$ States of OH and OD,"
J. Chem. Phys. 58, 4304-18

PF

Methods of Production and Experimental Technique

Discharge in He + traces of PF_3.

Afterglow of microwave discharge into PF_5 + He (or Ar) (72.3).

BAND SYSTEMS

System	Transition	Sources	Wavelength Limits	Degrading	Band Head, $\nu_{0,0}$	Remarks	Bibliography
I	$B^3\Pi \to X^3\Sigma^-$	Discharge	4000-3000	R	29338.68 29481.80 29623.06		(62.1)
II	$b^1\Sigma^+ \to X^3\Sigma^-$	Discharge	7483-7444	V	13363.49(0,0)		(72.3)
III	$d^1\Pi \to a^1\Delta$	Discharge	4150-3320	R	28712.14		(62.1)
IV	$g^1\Pi \to a^1\Delta$	Discharge	2450-2200	V	44544.80		(62.1)
V	$d^1\Pi \to b^1\Sigma^+$	Discharge	4820-4218	R	22444.95		(62.1)
VI	$g^1\Pi \to b^1\Sigma^+$	Discharge	2800-2600	V	38277.77		(62.1)

PF

I. **B$^3\Pi \rightarrow$ X$^3\Sigma^-$ System**

Most intense bands in the region of 3300 Å. The structure of the bands is complex, caused by the transition: $^3\Pi$(case a) $\rightarrow$ $^3\Sigma^-$(case b).

II. **b$^1\Sigma^+ \rightarrow$ X$^3\Sigma^-$ System**

Band heads (72.3):

(v', v'')	(0, 0)	(1, 1)	(2, 2)	(3, 3)
λ	7483.08	7472.22	7461.77	7451.51

III. **d$^1\Pi \rightarrow$ a$^1\Delta$ System**

Band heads (62.1):

(v', v'')	(0, 1)	(0, 0)
λ	3587.64	3481.47

IV. **g$^1\Pi \rightarrow$ a$^1\Delta$ System**

Only a single progression is observed:

(v', v'')	(0, 1)	(0, 0)
λ	2288.21	2245.27

V. **d$^1\Pi \rightarrow$ b$^1\Sigma^+$ System**

Band heads:

(v', v'')	(0, 1)	(0, 0)	(1, 0)
λ	4630.26	4453.50	4373.04

VI. **g$^1\Pi \rightarrow$ b$^1\Sigma^+$ System**

Only a single progression observed:

(v', v'')	(0, 1)	(0, 0)
λ	2671.99	2612.00

SPECTROSCOPIC CONSTANTS

State	T_e	ω_e	$x_e\omega_e$	B_e	$\alpha_e \times 10^3$	$D_e \times 10^6$	r_e	Remarks	Bibliography
$g^1\Pi$	51644.95[a]	–	–	0.6186[c]	–		–		(62.1)
$d^1\Pi$	35812.29[a]	413[b]	–	0.4848	6.2		1.721		(62.1)
$B^3\Pi_2$	29623.06[a]	436[b]		0.4693	3.7		⎫		(64.2)
$B^3\Pi_1$	29481.80[a]	436[b]		0.4663	3.8		⎬ 1.752		(64.2)
$B^3\Pi_0$	29338.69[a]	436[b]		0.4632	4.0		⎭		(64.2)
$b^1\Sigma^+$	13353.91	866.14	4.51	0.5725	4.5		1.5812		(72.3, 62.1)
$a^1\Delta$	7090.41	858.79	4.438	0.5699	4.67		1.5849	$y_e\omega_e = 0.0147$ cm^{-1}	(62.1)
$X^3\Sigma^-$	0	846.75	4.489	0.5665	4.56		1.5896	$\lambda_o = 2.9623$ cm^{-1} $\gamma_o = 0.0018$ cm^{-1} $y_e\omega_e = 0.019$ cm^{-1}	(62.1)

[a] T_o, [b] $\Delta G_{1/2}$, [c] B_o.

Dissociation energy = 4.5 ± 1.0 eV, 104 kcal/mole, 36296 cm^{-1}.

PF

Perturbations and General Information

Vibrational perturbations of $d^1\Pi$ state (62.1).

Many perturbations $B^3\Pi_{2,1,0}$ states (64.2).

BIBLIOGRAPHY

(62.1) B → X, g → b, g → a, d → b, d → a Systems,
A. E. Douglas and M. Frackowiak,
Can. J. Phys. 40, 832-49

(64.2) Interpretation of B → X Multiplet Separations,
I. Kovacs,
Can. J. Phys. 42, 2180-4

(72.3) b → X System,
R. Colin, J. Devillers, and F. Prevot,
"The $b^1\Sigma^+$ - $X^3\Sigma^-$ Band Systems of PF",
J. Mol. Spectrosc. 44, 230-5

PH

Methods of Production and Experimental Technique

Absorption after flash photolysis of PH_3.

Emission from a discharge in compounds containing P and H, a flash photolysis and flames.

BAND SYSTEMS

System	Transition	Sources	Wavelength Limits	Degrading	Characteristic Bands, λ	Remarks	Bibliography
I	$A^3\Pi_i \rightleftarrows X^3\Sigma^-$	Absorption	3600-3180	R	3419.6(0,0)		(68.20, 64.14, 60.8)
II	$B^3\Pi \leftarrow X^3\Sigma^-$	Absorption	-	-	1435(0,0)		(68.20)
III	$^1\phi \leftarrow a^1\Delta$	Absorption	-	-	1625(0,0)		(68.20)
IV	$^1\Pi \leftarrow a^1\Delta$	Absorption	-	-	1595(0,0)		(68.20)

PH

I. $A^3\Pi_i \rightleftarrows X^3\Sigma^-$ System

Band heads:

(v', v'')	(0, 0)	(1, 0)
λ	3426.8(P_1)	3220(Q_1)
	3419.6(Q_1)	3195(R_3)

II. $B^3\Pi \leftarrow X^3\Sigma^-$ System

Weak band with complex structure.

III. $^1\phi \leftarrow a^1\Delta$ System

Band heads (68.20):

(v', v'')	(0, 0)	(1, 1)	(2, 2)
λ	1624.7	1624.6	1624.4

IV. $^1\Pi \leftarrow a^1\Delta$ System

Isolated band with weak intensity.

SPECTROSCOPIC CONSTANTS

State	T_o	ω_e	$x_e\omega_e$	B_o	$\alpha_e \times 10^3$	$D_o \times 10^4$	r_o	Remarks	Bibliography
$^1\Pi$	70387.7[a]			8.47		4.1	1.427		(68.20)
$B^3\Pi$	69587.8			7.3					(68.20)
$^1\phi$	69211.15[a]			8.60		5.4	1.416		(68.20)
$A^3\Pi$	29318.28			8.01		5.5	1.467		(68.20)
$a^1\Delta$	7662.4[a]			8.44		4.1	1.430		(68.20)
$X^3\Sigma^-$	0			8.41		4.3	1.432		(68.20)

[a] Estimated (68.21).

Dissociation energy = 3.5 ± 0.3 eV, 81 kcal/mole, 28230 cm^{-1}.

Perturbations and General Information

Λ doubling produces a perturbation of the $^3\Pi_o$ state (60.8).

Electron affinity = 0.93 eV (67.19).

Radiative lifetime:
$$A^3\Pi_i - X^2\Sigma^-$$

$\tau = 440 \pm 50$ μsec (64.15).

BIBLIOGRAPHY

(30.1) Observation,
R. W. B. Pearse,
Proc. Roy. Soc. A 129, 328-54

(36.2) Theory,
G. Gilbert,
Phys. Rev. 49, 619-24

(36.3) Intensities,
P. Nolan and F. A. Jenkins,
Phys. Rev. 50, 943-9

(36.4) Spincoupling,
M. Ishaque and R. W. B. Pearse,
Proc. Roy. Soc. A 156, 221-32

(36.5) Theory of the $A^3\Pi$ State,
A. Budo,
Z. Physik 98, 437-44

(37.6) Theory, Intensities,
A. Budo,
Z. Physik 105, 579-87

(39.7) Analysis,
M. Ishaque and R. W. B. Pearse,
Proc. Roy. Soc. A 173, 265-77

(60.8) Analytical Development,
F. Legay,
Can. J. Phys. 38, 797-805

(62.10) A-X System,
H. Guenebaut, B. Pascat,
"On the Behavior of PH_3, Where Diluted in Argon, in a Shock Tube,"
C. R. Acad. Sci. 255, 1741-3

(63.11) L. Thanh My and M. Peyron,
J. Chem. Phys. 41, 1442-9

(63.12) Production of $^3\Pi$ in Flames,
H. Guenebaut, B. Pascat,
"On the Reaction of Gaseous PH_3 with Atomic Nitrogen,"
C. R. Acad. Sci. 256, 2850-3

(64.13) Semi-Empirical Calculations,
P. C. Jordan,
J. Chem. Phys. 41, 1442-9

(64.14) Production by Flash Photolysis,
K. H. Becker, K. H. Welge,
"Fluorescent Investigation and Photochemical Primary Processes in the Vacuum U.V. of NH_3, N_2H_4, PH_3 and Reactions of the Electronically Excited Radicals $NH^*(^1\Pi)$, $NH^*(^3\Pi)$, and $PH^*(^3\Pi)$,"
Z. Naturforsch. A 19, 1006-15

(64.15) Radiative Lifetime, A-X,
E. Fink, K. H. Welge,
"Lifetime of the Electronic States $N_2(C^3\Pi_a)$, $N_2^+(B^2\Sigma_a^+)$, $NH(A^3\Pi)$, $NH(c^1\Pi)$, and $PH(^3\Pi)$,"
Z. Naturforsch. A 19, 1193-1201

(64.16) PH ($^3\Pi$) Produced in Flames,
H. Guenebaut, B. Pascat,
"On the Reaction of Gaseous PH_3 with Atomic Hydrogen: Observation in Emission of a New System of Bands,"
J. Chem. Phys. 61, 592-5

(65.17) PH ($^3\Pi$) Produced by Flash Photolysis. Deslandres Scheme,
D. Kley, K. H. Welge,
"Investigation of Photodissociation of PH_3 in the Quartz-U.V. through Absorption Spectroscopy Flash Photolysis,"
Z. Naturforsch. A 20, 124-131

(67.18) LCAO-MO-SCF Calculations of the Ground State,
P. E. Cade and W. M. Huo,
J. Chem. Phys. 47, 649-72

PH

(67.19) P. E. Cade,
"The Electron Affinities of the Diatomic Hydrides, CH, NH, SiH, and PH,"
Proc. Phys. Soc. 91, 842-54

(67.20) Observation of a Singlet System, Rotational Analysis,
W. J. Balfour, A. E. Douglas,
"The Absorption Spectrum of PH in the Vacuum Ultraviolet,"
Can. J. Phys. 46, 2277-80

(68.21) Estimation of a-X Separation,
P. E. Cade,
"Theoretical Predictions of the Singlet-Triplet Intercombination Separations for NH, OH^+, PH and SH^+,"
Can. J. Phys. 46, 1989-91

(69.22) Theory,
R. F. W. Bader, W. H. Henneker, and I. Keavens,
"Molecular Charge Distribution and Chemical Binding IV. The Second Row Diatomic Hydrides,"
J. Chem. Phys. 50, 5313-33

PN

Methods of Production and Experimental Technique

Absorption in phosphorus nitride at T > 450°C (39.4).

Emission from a discharge into (P + N_2) (34.3, 33.2).

Microwave discharge into PCl_3 + N_2 (73.12).

BAND SYSTEMS

System	Transition	Sources	Wavelength Limits	Degrading	Characteristic Bands, λ	Remarks	Bibliography
I	$A^1\Pi \rightleftarrows X^1\Sigma^+$	Discharge	3000-2375	R	2604.98(0,1) 2518.21(0,0)		(63.8, 63.9, 39.4)

PN

I. $A^1\Pi \rightleftarrows X^1\Sigma^+$

Band heads (Q), λ (intensity) (39.4, 33.2):

v', v''	0	1	2	3	4	5	6
0	2518.21 (20)	2604.98 (10)	2696.90 (3)				
1	2451.07 (8)	2532.99 (3)	2620.07 (8)	2712.07 (4)			
2	2388.24 (2)	2466.19 (7)		2635.15 (4)	2727.52 (4)	2825.05 (1)	
3		2403.26 (3)	2481.36 (3)		2650.85 (2)	2742.95 (3)	2840.56 (1)
4			2418.75 (3)		2579.45		2758.83 (2)
5				2434.36 (2)		2595.32 (2)	
6				2375.59 (1)			
7					2391.71 (1)		
8						2407.59 (1)	
9							2424.13 (1)

SPECTROSCOPIC CONSTANTS

State	T_e	ω_e	$x_e\omega_e$	B_e	$\alpha_e \times 10^3$	$D_e \times 10^6$	r_e	Remarks	Bibliography
$A\,^1\Pi$	39805.66	1103.09	7.222	0.73071	6.63	–	1.5424		(33.2)
$X\,^1\Sigma^+$	0	1337.24	6.983	0.78621	5.57	1.09	1.4869		(33.2)

Dissociation energy = 7.57 ± 0.03 eV, 174.6 kcal/mole, 61057 cm^{-1} (68.11).

PN

Purturbations and General Information

Relative transition probabilities $A^1\Pi - X^1\Sigma$ (73.12):

The 0,0 band probability is taken to be 0.58

v', v''	0	1	2	3	4
0	0.58	0.31	0.08	0.02	-
1	0.38	0.15	0.32	0.14	0.07

Franck-Condon factors — Morse potential (73.12):

v', v''	0	1	2	3	4
0	0.58	0.32	0.09	0.01	0.00
1	0.31	0.11	0.34	0.18	0.06
2	0.09	0.31	0.00	0.21	0.22

Potential energy curves — RKRV potential (66.10):

State	v	$U(cm^{-1})$	$R_{min}(\text{Å})$	$R_{max}(\text{Å})$
$X^1\Sigma^+$	0	663.4	1.442	1.545
	1	1986.7	1.410	1.588
	2	3296.0	1.389	1.620
	3	4591.3	1.372	1.647
	4	5872.4	1.359	1.672
	5	7139.8	1.347	1.695
	6	8394.8	1.336	1.716
	7	9633.3	1.327	1.737
	8	10859.2	1.318	1.757
$A^1\Pi$	0	546.1	1.494	1.606
	1	1634.6	1.458	1.655
	2	2708.8	1.436	1.691
	3	3770.6	1.419	1.723
	4	4814.1	1.404	1.751
	5	5844.5	1.392	1.778
	6	6861.1	1.381	1.803
	7	7858.7	1.372	1.827
	8	8849.7	1.363	1.850

BIBLIOGRAPHY

(33.1) J. Curry, L. Herzberg, and G. Herzberg,
J. Chem. Phys. 1, 749

(33.2) A-X System Analysis,
J. Curry, L. Herzberg, and G. Herzberg,
Z. Physik 86, 348-66

(34.3) P. N. Ghosh and A. C. Datta,
Z. Physik 87, 500-4

(39.4) H. Moureu, B. Rosen, and G. Wetroff,
C. R. Acad. Sci. 209, 207-9

(42.5) Dissociation Energy,
A. G. Gaydon,
Proc. Roy. Soc. A 179, 439-50

(44.6) Dissociation Energy,
R. F. Barrow,
Proc. Phys. Soc. 56, 211-2

(47.7) A. G. Gaydon,
"Dissociation Energies and Spectra of Diatomic Molecules,"
Chapman and Hall, London

(63.8) A-X Produced by Active Nitrogen,
H. Guenebaut, B. Pascat, C. Coulet, and L. Marsigny
"On the Emission of PN and PO in the Reations of PCl_3 and PCl_5 with Atomic Nitrogen,"
C. R. Acad. Sci. 257, 135-8

(63.9) A-X Produced by Active Nitrogen
H. Guenebaut and B. Pascat
"On the Reaction of Gaseous PH_3 with Atomic Nitrogen,"
C. R. Acad. Sci. 256, 2850-3

(66.10) Potential Energy Curves, A, X States,
R. B. Singh and D. K. Rai,
"Potential Curves for Some Diatomic Molecules: P_2, PN, SiN, NBr, BaO, BeF, SF, and SnF,"
Indian J. Pure. Appl. Phys. 4, 102-5

PN

(68.11) Dissociation Energy,
O. M. Uy, F. J. Kohl, and K. D. Carlson,
J. Phys. Chem. 72, 1611-6

(73.12) Relative Transition Probabilities, Franck-Condon Factors,
M. B. Moeller and S. J. Silvers,
"Fluorescence Spectra of PN and BF,"
Chem. Phys. Letters 19, 78-85

(71.L1) J. Raymonda and W. Klemperer,
"Molecular Beam Electric Resonance Spectrum of $^{31}P^{14}N$,"
J. Chem. Phys. 55, 232-3

(71.L2) F. C. Wyse, E. L. Manson, and W. Gordy,
"Millimeter Wave Rotational Spectrum and Molecular Constants of $^{31}P^{14}N$,"
J. Chem. Phys. 57, 1106-8

(72.L2) J. Hoeft, E. Tiemann, and T. Törring,
"Rotational Spectrum of PN,"
Z. Naturforsch. 27a, 703-4

PO

Methods of Production and Experimentation Technique

Emission from a high frequency discharge (68.26, 68.27, 68.28, 67.24, 67.23, 66.17, 66.18, 65.15, 62.8, 59.6, 58.4, 58.5).

High pressure discharge, Geissler tube; aluminum electrodes, He or Ar carrier gas (67.24, 66.17, 55.2, 55.3).

Luminescence from the reactions: [(P$_4$ vapor + He) + atomic oxygen (high frequency or high pressure discharge)], [(P + He) (high frequency discharge) + atomic oxygen (high frequency or high pressure discharge)] (67.24, 64.11, 64.12).

Hollow cathode (66.19, 66.20).

Discharge into POCl$_3$ + Argon (70.37).

BAND SYSTEMS

	System	Transition	Sources	Wavelength Limits	Degrading	Band Head, $v_{0,0}$	Remarks	Bibliography
β	I	$A^2\Sigma^+ \to X^2\Pi_r$	All sources	3750-3050	V and R	30695.24		(68.28, 68.30, 67.24, 66.19, 66.20, 65.16, 59.6)
	II	$A'^2\Pi_i \to X^2\Pi_r$	High frequency discharge	4800-3800	-	25760.1(0,6)		(70.37)
γ	III	$B^2\Sigma^+ \to X^2\Pi_r$	Fluorescense	2800-2250	V	40486.30		(67.23, 64.11, 64.12, 58.4)
	IV	$B'^2\Sigma^+ \to X^2\Pi_r$	Fluorescense	2635-2370	V	40406 40629		(67.23, 64.11, 64.12)
	V	$B''^2\Pi \to X^2\Pi_r$	Fluorescense	2600-2370	V	40427 40651		(67.23, 64.11, 64.12)
	VI	$C^2\Sigma^- \to X^2\Pi_r$	High frequency discharge	3010-2250	R	43435.20 43658.20		(71.40, 71.41, 66.17, 65.13, 65.15)
	VII	$C'^2\Delta \to X^2\Pi_r$	High frequency discharge	2195-1976	R	46779.8 47006.4		(71.40, 62.8, 55.2)
	VIII	$D^2\Pi_r \to X^2\Pi_r$	High frequency discharge	2170-2050	R	48615	$v_{0,0}$ from deperturbation	(68.26, 68.29, 68.31, 66.18, 58.5)

PO BAND SYSTEMS

System	Transition	Sources	Wavelength Limits	Degrading	Band Head, $\nu_{0,0}$	Remarks	Bibliography
IX	$D'^2\Pi_r \to X^2\Pi_r$	High frequency discharge	2110-2045	R	48835.4		(71.42, 68.26, 68.27, 68.31, 66.18)
X	$E^2\Delta \to X^2\Pi_r$	High voltage discharge	1930-1825	V	53100.7 53323.1		(70.46)
XI	$D^2\Pi_r \to A^2\Sigma^+$	High frequency discharge	6870-5583	R	17920	$\nu_{0,0}$ from deperturbation	(68.26, 68.27, 66.18, 58.5)
XII	$D'^2\Pi_r \to A^2\Sigma^+$	High frequency discharge	5925-5580	R	18106	$\nu_{0,0}$ from deperturbation	(68.26, 68.27, 66.18, 62.8, 58.5)
XIII	$F^2\Sigma^+ \to A^2\Sigma^+$	High frequency discharge	4445-4040	R	24744(head)		(66.18)
XIV	$G^2\Sigma^+ \to A^2\Sigma^+$	High frequency discharge	4520-3915	R	25513.76(head)		(69.33, 66.18)
XV	$H^2\Sigma^+ \to A^2\Sigma^+$	High frequency discharge	4150-3960	R	25219.33(head)		(70.34)
XVI	$I^2\Sigma^+ \to A^2\Sigma^+$	High frequency discharge	6642	-	15055.7		(72.43, 70.35)

System	Transition	Sources	Wavelength Limits	Degrading	Band Head, $\nu_{0,0}$	Remarks	Bibliography
XVII	$A^2\Sigma^+ \to B^2\Sigma^+$	High frequency discharge	6763.1, 7266.7	V	14786.3(7,4) 13761.4(7,5)		(72.43)
XVIII	$F^2\Sigma^+ \to B^2\Sigma^+$	High frequency discharge	6814.1, 7323.8	-	14675.4(0,4) 13654.1(0,5)		(72.43)
XIX	$N^2\Sigma^+ \to B^2\Sigma^+$	High frequency discharge	4770-4590	R	21397(v',0)		(70.36)
XX	$G^2\Sigma^+ \to B^2\Sigma^+$	High frequency discharge	4246-3877	-	25472(0,0)		(70.38, 70.39)
XXI	$I^2\Sigma^+ \to B^2\Sigma^+$	High frequency discharge	4768-4287	-	23275.4(v',0)		(70.38, 70.39)
XXII	$K^2\Sigma^+ \to B^2\Sigma^+$	High frequency discharge	4028-3988	V	-		(70.35)
XXIII	$L^2\Sigma^+ \to B^2\Sigma^+$	High frequency discharge	4077-3873	-	-	-	(70.38)
XXIV	$M^2\Sigma^+ \to B^2\Sigma^+$	High frequency discharge	4076-3728	-	28804.8(v',0)	-	(70.38)

I. $A^2\Sigma^+ \rightarrow X^2\Pi_r$ System (β)

Band heads, λ(intensity) (degrades) (68.29, 59.6):

v', v''	0	1	2	3	4	5	6
0	3270.47 (10)(V)	3405.69 (6)(V)	3551.39 (1)(V)				
	3246.21 (10)(V)	3386.37	3523.34 (1)(V)				
		3379.78 (6)(V)					
1	3280.47 (V)	3414.14 (5)(V)	3558.71 (1)(V)	3714.61			
	3255.28 (8)(V)	3387.92 (6)(V)	3530.46 (1)(V)	3683.93			
2	3043.58	3159.77	3290 (1)(V)	3424.64	3567.93	3722.68	
	-	3135.13	3266.64 (V)	3397.81 (4)(V)	3539.41 (1)(V)	3691.83	
3		3056.88	3173.43	3296.31 (1)(R)	3433	3579.31	3732.89
		3035.58	3149.71	3274	3409.80 (2)(V)	3550.35 (1)(V)	3701.80
4			3071.48	3188.07	3311.81 (5)(R)	3442.41	3590.5
			3050.24	3164.90	3300.8	3420	3563.47
			-	-	3285.79	-	-
5				3087.45	3204.19	3328.24 (7)(R)	3460.05 (1)(R)
				3066.31	3181.19	3319.7	3431.45
				-	-	3302.86 (5)(R)	-
6					3105.33	3222.07	3346.18 (7)(R)
					3083.13	3198.83 (2)(R)	3339.37
					-	-	3320.96 (7)(R)

PO

II. $A'^2\Pi_i \rightarrow X^2\Pi_r$ System

Band heads (P_2) (72.37):

(v', v'')	(0, 6)	(0, 8)	(0, 9)	(0, 10)	(1, 10)
λ	3882.0	4257.1	4469.3	4700.8	4540.5

III. $B^2\Sigma^+ \rightarrow X^2\Pi_r$ System (γ)

Band heads:

v', v''	λ	v', v''	λ	v', v''	λ
0, 2	2634.8	0, 0	2476.4	2, 0	2318.8
1, 3	2621.9	1, 1	2466.7	3, 1	2312.0
2, 4	2609.1	2, 2	2457.4	4, 2	2305.3
0, 1	2553.6	1, 0	2394.6		
1, 2	2542.4	2, 1	2386.4		
2, 3	2531.4				

IV. $B'^2\Sigma^+ \rightarrow X^2\Pi_r$ System

Band heads (P_2), λ (64.11, 64.12):

v', v''	λ	v', v''	λ
0, 2	2632.8	0, 0	2474.2
1, 3	–	1, 1	2465.0
0, 1	2551.0	2, 2	2455.4
		1, 0	2393.2
		2, 1	2384.6

V. $B''^2\Sigma^+ \rightarrow X^2\Pi_r$ System

Band heads (P_1), λ (64.11, 64.12):

v', v''	λ	v', v''	λ
1, 3	2601.0	0, 0	2459.2
0, 1	2535.6	1, 1	2448.8
1, 2	2523.8	1, 0	2378.2

VI. $C^2\Sigma^- \to X^2\Pi_r$ System

Band heads (Q_2), λ (66.17):

v', v''	λ	v', v''	λ	v', v''	λ
6, 13	3006.49	2, 8	2804.93	0, 4	2588.40
5, 12	2977.85	5, 10	2797.57	2, 5	2558.99
4, 11	2949.85	1, 7	2779.06	0, 3	2511.62
3, 10	2922.20	4, 9	2770.70	0, 2	2438.43
6, 12	2913.14	3, 8	2744.50	2, 3	2413.94
2, 9	2895.49	2, 7	2718.76	0, 1	2368.77
5, 11	2885.78	1, 6	2693.67	2, 2	2346.30
4, 10	2857.93	3, 7	2662.05	0, 0	2302.28
3, 9	2831.23	2, 6	2636.97	1, 0	2259.75

VII. $C'^2\Delta \to X^2\Pi_r$ System

Band heads (62.8, 55.2):

v', v''	0, 1	0, 0	1, 0	2, 0
λ	2194.27	2136.95	2103.80	2071.96
	2183.39	2126.67	2093.79	2062.22

VIII. $D^2\Pi_r \to X^2\Pi_r$ System

Band heads, λ (intensity) (55.2):

v', v''	0	1	2
λ	2062.0(10)	2115.2(8)	2170.5(6)
	2053.8(10)	2106.6(8)	2161.6(6)

X. $E^2\Delta \to X^2\Pi_r$ System

Band heads (P_2), λ (intensity) (55.2):

v', v''	λ	v', v''	λ
0, 1	1927.70(4)	0, 0	1883.42(10)
1, 2	1918.46(3)	1, 1	1875.05(5)
		1, 0	1833.20(2)

PO

XI. $D^2\Pi_r \rightarrow A^2\Sigma^+$ System

Band heads (68.27, 66.18):

v', v''	0, 0	0, 1	0, 2	0, 3
λ (Q_{11})	5592.99	5972.95	6396.98	6866.52
λ (R_{11})	5588.32	5967.14	6389.77	6857.38
λ (R_{22})	5584.45	5962.94	6385.15	-
λ (R_{21})	5578.04	5955.24	-	-

XIII. $F^2\Sigma^+ \rightarrow A^2\Sigma^+$ System

Band heads (66.18):

(v', v'')	(0, 2)	(0, 1)	(0, 0)
λ(intensity)	4443.31(3)	4236.20(4)	4040.20(3)

XIV. $G^2\Sigma^+ \rightarrow A^2\Sigma^+$ System

Band heads (69.33):

(v', v'')	(0, 1)	(0, 0)
λ(intensity)	4099.94(2)	3919.45(10)

XV. $H^2\Sigma^+ \rightarrow A^2\Sigma^+$ System

Band heads (70.34):

(v', v'')	(0, 1)	(0, 0)
λ(intensity)	4151.27(2)	3964.09(10)

XVI. $I^2\Sigma^+ \rightarrow A^2\Sigma^+$ System

Band heads, λ (72.43):

(v', v'')	(0, 0)	(1, 1)	(2, 2)
λ	6643.2	6642.7	6639.7

PO

XX. $I\,^2\Sigma^+ \to B\,^2\Sigma^+$ System

Band heads, λ (70.39):

(v', v'')	(v', 0)	(v', 1)	(v', 2)
λ	4296.5	4516.8	4754.9

XXI. $G\,^2\Sigma^+ \to B\,^2\Sigma^+$ System

Band heads, λ (70.38):

(v', v'')	(v', 0)	(v', 1)	(v' + 1, 1)
λ	3925.9	4109.3	3731.1

XXIII. $L\,^2\Sigma^+ \to B\,^2\Sigma^+$ System

Band heads, λ (70.38):

(v', v'')	(v', 0)	(v', 1)
λ	3878.5	4056.8

XXIV. $M\,^2\Sigma^+ \to B\,^2\Sigma^+$ System

Band heads, λ (70.38):

(v', v'')	(v', 0)	(v', 1)	(v', 2)
λ	3730.6	3896.1	4072.2

PO

SPECTROSCOPIC CONSTANTS

State	T_e	ω_e	$x_e\omega_e$	B_e	$\alpha_e \times 10^3$	$D_e \times 10^6$	r_e	Remarks	Bibliography
$H^2\Sigma^+$	56016	(1391.06)	(6.95)	(0.779)			(1.432)		(72.47) (71.L1)
$I^2\Sigma^+$	55458.90	1391.06	6.95	0.799	5.3		(1.431)		(72.47) (71.L1)
$E^2\Delta$	53092.4	1480.24	11.98	0.758	7.4	0.8	1.451		(72.46)
$G^2\Sigma^+$	52651	1260	5.40	0.770	12.7		(1.451)		(72.47) (71.L1)
$F^2\Sigma^+$	49899.65	835.84	5.77	0.616	5.1		(1.670)		(72.47) (71.L1)
$D'^2\Pi_r$	48801			0.437			1.912	$A = 32.3$ cm^{-1}	(68.26, 68.27, 66.18, 59.5)
$D^2\Pi_r$	48615	(1238)		0.749			1.460	$A = 25.1$ cm^{-1}	(71.L1, 68.26, 68.27, 66.18, 59.5)
$C'^2\Delta$	47250.6	745	3.5						(62.8, 55.2)
$C^2\Sigma^-$	45717.25	768.94	5.14	0.584	5.6		1.653	$\gamma = 0.0325$ cm^{-1}	(71.40, 71.41)
$B''^2\Pi$	40539[a]	1389[b]							(67.23, 64.11, 64.12)
$B'^2\Sigma$	40446	1375.2	2.8						(67.23, 64.11, 64.12)
$B^2\Sigma^+$	40407.6	1390.27	7.19	0.7801	5.4	1.0	1.431		(67.23, 64.11, 64.12)

SPECTROSCOPIC CONSTANTS

State	T_e	ω_e	$x_e\omega_e$	B_e	$\alpha_e \times 10^3$	$D_e \times 10^6$	r_e	Remarks	Bibliography
$A'^2\Pi_i$	32884.3			0.5396^c		1.0^d		$A_1 = -13.26$ cm^{-1}	(70.37)
$A^2\Sigma^+$	30731.1	1164.86	13.71	0.7473	8.8	1.4	1.466	$\gamma = -0.0076$ cm^{-1}	(68.28, 67.23, 66.19, 66.20)
$X^2\Pi_r$	0	1233.38	6.56	0.7337	5.5	1.3	1.473	$A = 223.92$ cm^{-1}	(72.43, 68.26, 68.27, 68.28, 67.23)

$^a T_o$, $^b \Delta G_{1/2}$, $^c B_o$, $^d D_o$.

Dissociation energy = 6.14 ± 0.08 eV, 141.7 kcal/mole, 49523 cm^{-1}.

PO

PO

Perturbations and General Information

Franck-Condon factors, r-centroids — Morse potential:

$A^2\Sigma^+ - X^2\Pi_r$ (66.20a)

v', v''	0	1	2	3	4	5
0	0.696 1.455 2477.9	0.244 1.400 2555.05	0.050 1.346 2636.3	0.007 1.290 2721.5	0.003	
1	0.258 1.518 2396.3	0.280 1.462 2468.3	0.318 1.407 2543.94	0.108 1.353 2623.42	1.297 2706.81	
2	0.041 1.584 2320.6	0.353 1.525 2387.94	0.185 1.469 2458.96	0.235 1.414 2533.00	0.027 1.360 2616.7	1.304 2692.4
3	0.004	0.094 1.592 2313.7	1.533 2379.9			
4	0.000		1.600 2306.9			
5				1.608 2300.4		

$C^2\Sigma^- - X^2\Pi_r$ (67.25)

v', v''	0	1	2	3	4
0	0.153 1.529 2290.66	0.286 1.560 2356.05	0.273 1.593 2425.04	0.172 1.626 2497.34	0.082 1.658 2573.11
1	0.267 1.503 2248.28	0.120 1.539 2311.74	0.000 - -	0.294 1.603 2447.66	- - -
2	0.098	0.001	0.089 1.550 2333.84	0.147 1.582 2400.83	
3	0.071	0.065	0.055		
4	0.054				

Top   = Frank-Condon factor
Middle = r-centroid
Bottom = Wavelength

Potential energy curves (66.17):

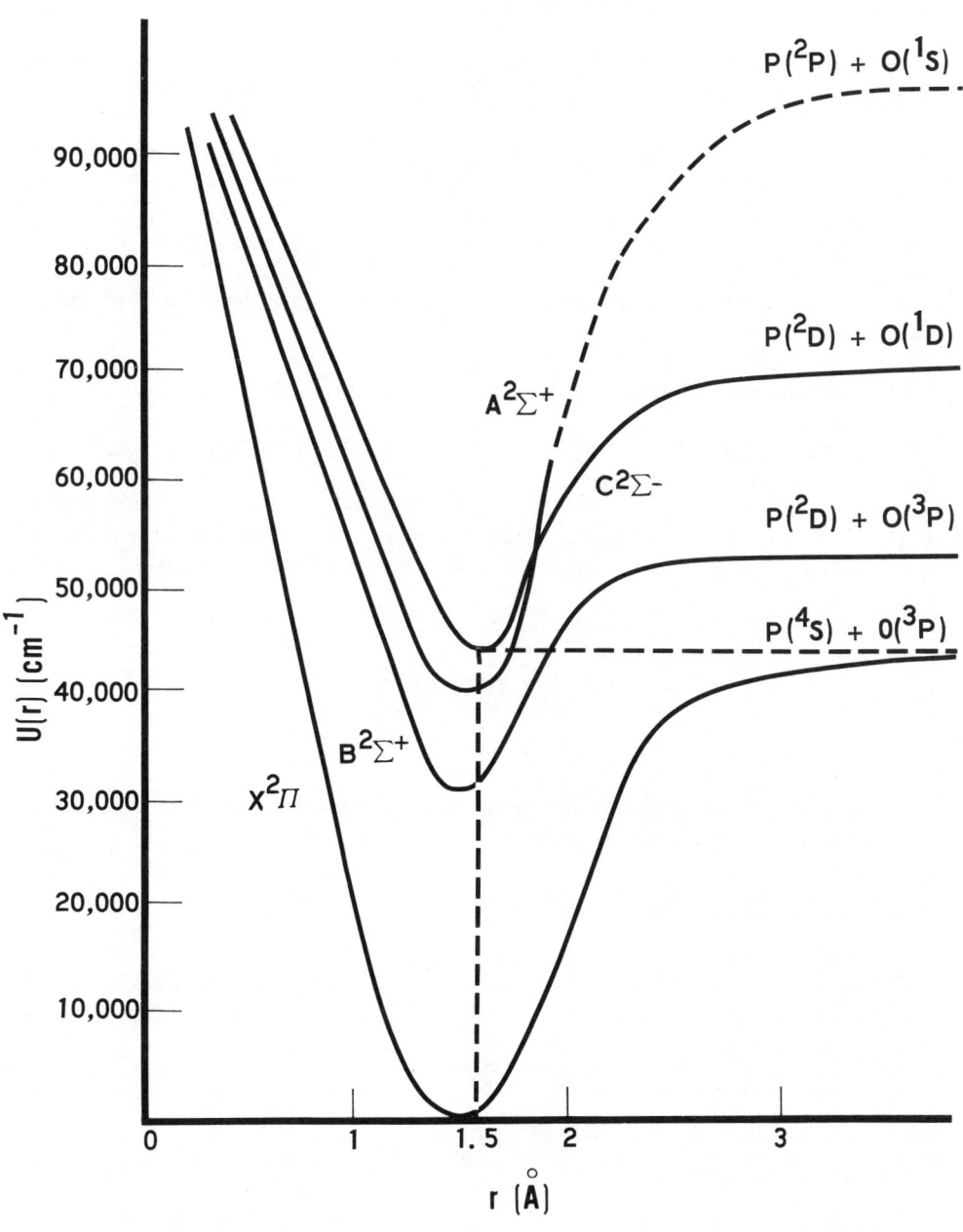

$A^2\Sigma^+$ state is perturbed in the v' = 6 level at J = 39.5 and 43.5 for the F_1 state and J = 38.5 and 43.5 for the F_2 states (67.24). It is also perturbed in the v' = 10 and 12 levels (68.27, 66.20).

$B^2\Sigma^+$ state is perturbed in the v' = 0 level between N = 10-17 and in the v' = 1 level between N = 30-35 (67.23, 58.4).

$D^2\Pi$ and $D'^2\Pi$ states homogeneously perturb each other in all levels of v = 0 with a maximum at J = 23.5. $D'^2\Pi$ also perturbed at J = 23.5-34.5 of the F_1 state and J = 31.5-33.5 of the F_2 state (71.42, 68.31). $D^2\Pi$ state predissociates for N > 33, yet N = 34 is observed (68.31).

$E^2\Delta$ state has local predissociations in its v' = 1 and 2 levels and is totally predissociated above v' = 2 (72.46).

$F^2\Sigma^+$ state (v = 0) homogeneously perturbs $A^2\Sigma^+$ (v = 7) (72.43).

$G^2\Sigma^+$ and $H^2\Sigma^+$ states display only one vibrational level each. All the rotational levels are perturbed.

$N^2\Sigma^+$ state is perturbed in all rotational levels. Predissociation occurs for N = 16-26 of the F_2 state and N = 16-30 of the F_1 state (70.36).

BIBLIOGRAPHY

(46.1) β System: Classification into the Distinct Transitions. Interpretation Abandoned,
W. Heilpern,
Helv. Phys. Acta 19, 245-65

(55.2) β System, Vibrational Analysis, Evaluation of Rotational Constants D-X, C'-X, E-X Systems, Vibrational Analysis, Evaluation of the Rotational Constants. Dissociation Energy,
K. Dressler,
Helv. Phys. Acta 28, 563-90

(55.3) K. Dressler and E. Miescher,
Proc. Phys. Soc. A 68, 542-4

(58.4) γ System, Vibrational and Rotational Analysis. Perturbations in $B^2\Sigma^+$,
K. S. Rao,
Can. J. Phys. 36, 1526-35

(58.5) D, D'-A, Vibrational Analysis,
K. K. Durga and P. T. Rao,
Indian J. Phys. 32, 223-9

(59.6) β System, Rotational Analysis of Several Degradations Toward the U.V. and I.R. Regions,
N. L. Singh,
Can. J. Phys. 37, 136-43

(60.7) Edlen Formula. Conversion Table,
C. D. Coleman, W. R. Bozman, and W. F. Meggers,
NBS Monograph 3, (1), 508 pp.

(62.8) C, C'-X Systems. Vibrational Analysis,
C. Santaram and P. T. Rao,
Z. Physik 168, 553-9

(63.9) Flame Studies in Active Nitrogen,
H. Guenebaut, B. Pascat, C. Couet, and L. Marsigny,
C. R. Acad. Sci. 257, 135-8

(63.10) C-X System, Evaluation of the Rotational Constants. Potential Energy Curves for $X^2\Pi$, $A^2\Sigma^+$, $C^2\Sigma^-$, Dissociation Energy,
C. Santharam and P. T. Rao,
Indian J. Phys. 37, 14-7

(64.11) γ Systems. Observation of the New B' and B"-X Transitions. Vibrational Analysis,
H. Guenebaut, C. Couet, and D. Houlon,
C. R. Acad. Sci. 258, 3457-60

(64.12) γ Systems. Observation of the New B' and B"-X Transitions. Vibrational Analysis,
H. Guenebaut, C. Couet, and D. Houlon,
C. R. Acad. Sci. 258, 6370-3

(65.13) C-X System. Vibrational and Rotational Analysis. H-A System. Band at 3964.11 Å,
R. K. Asundi,
J. Chem. Phys. 43, No. 10 (part 2) 24.S

(65.14) RKRV Potential Energy Curves, $X^2\Pi$, $A^2\Sigma^+$, $B^2\Sigma^+$. Dissociation Energy,
R. B. Singh and D. K. Rai,
J. Phys. Chem. 69, 3461-2

PO

(65.15) C-X System. Vibrational and Rotational Analysis,
N. A. Narasimham, M. N. Dixit, and V. Sethuraman,
Proc. Ind. Acad. Sci. A 62, 314-29

(65.16) β System: Classification into Two Distinct Transitions. Interpretation Abandoned,
G. Graff and G. Werth
Z. Physik 183, 223-33

(66.17) C-X System. Vibrational and Rotational Analysis,
H. Guenebaut, C. Couet, and B. Coquart,
J. Chim. Phys. 63, 969-82

(66.18) D, D'-A Systems. Found One Band of the D-A System which Has Not Fit this System Since. (68.26, 68.27). F and G-A Systems, Vibrational Analysis,
S. Mrozowski and C. Santaram,
J. Opt. Soc. Am. 56, 1174-8

(66.19) β System. New Bands. Vibration Classification. Perturbation at v = 0 of $A^2\Sigma^+$ State,
H. Meinel and L. Krauss,
Z. Naturforsch. A 21, 1520-2

(66.20) β System. New Bands. Vibration Classification. Perturbation at v = 0 of $A^2\Sigma^+$ State,
H. Meinel and L. Krauss,
Z. Naturforsch. A 21, 1878-83

(66.20a) S. Sankaranarayan,
Indian J. Phys. 40, 678-80

(67.21) Calculation of the Franck-Condon Factors for $B^2\Sigma - X^2\Pi$,
S. Sankaranarayan,
Indian J. Phys. 40, 678-80

(67.22) Calculation of a Molecular SCF, Wavefunction. Hamiltonian Matricies, Population,
D. B. Boyd and W. N. Lipscomb,
J. Chem. Phys. 46, 910-9

(67.23) γ System, Vibrational and Rotational Analysis. Perturbation in $B^2\Sigma^+$,
B. Coquart, C. Couet, N. T. Anh, and H. Guenebaut,
J. Chim. Phys. 64, 1197-208

(67.24) β System. Rotational Analysis of Several Bands Degrading Towards U.V. and I.R. Perturbations at v = 6 of $A^2\Sigma^+$,
B. S. Mohanty, K. N. Upadhya, R. B. Singh, and N. L. Singh,
J. Mol. Spectrosc. 24, 19-37

(67.25) D. S. Rai and B. S. Mohanty,
"Franck-Condon Factors and r-Centroids of the C-X System of the PO Molecule,"
Current Sci. 36, 231-2

(68.26) D-A System: The (1, 1) Band of (66.18) Does Not Fit in That System. Vibrational and Rotational Analysis of all the Systems. Homogeneous Pertubation Between D and D' (Maximum at J = 23.5). D Predissociates at v = 0 for N > 33. D' and E' of (62.8) are One and the Same State,
R. D. Verma and M. N. Dixit,
Can. J. Phys. 46, 2079-86

(68.27) D-A System: The (1, 1) Bands of (66.18) Does Not Appear in This System. Vibrational and Rotational Analysis of all Systems. Homogeneous Perturbation Between D and D' (Maximum at J = 23.5). D Predissociates at v = 0 for N > 33. D' and E' of (62.8) are One and the Same State,
C. Couet, B. Coquart, Ngo Tuan Anh, and H. Guenebaut,
J. Chim. Phys. 65, 1241-58

(68.28) β System. Rotational Analysis of Several Bands Degrading Towards the U.V. and I.R.,
C. Couet, Ngo Tuan Anh, B. Coquart, and H. Guenebaut,
J. Chim. Phys. 65, 217-26

(68.29) β, γ and C-X Systems. Isotope Displacement,
M. N. Dixit and N. A. Narasimham,
Proc. Indian Acad. Sci. A 68, 1-12

(68.30) A-X Rotational Analysis,
B. S. Mohanty, D. K. Rai, and K. N. Upadhya,
"Structure and Analysis of Some Bands of the β-System of the PO-Molecule,"
Proc. Indian Acad. Sci. A 68, 165-75

(68.31) C. Couet, B. Coquart, N. T. Ahn, and H. Guenebaut,
"Rotational Analysis of the Ultraviolet System $D^2\Pi - X^2\Pi$ and $D'^2\Pi - X^2\Pi$ of PO,"
C. R. Acad. Sci. B 266, 1219-21

(69.32) R. M. Dagnall, D. J. Smith, K. C. Thompson, and T. S. West,
"Emission Spectra Obtained from the Combustion of Organic Compounds in Hydrogen Flames,"
Analyst 94, 871-8

PO

(69.33) G-A System, Rotational Analysis of (v', 0) and v', 1). All the Levels are Perturbed,
B. Coquart, Ngo Tuan Anh, C. Couet, and H. Guenebaut,
C. R. Acad. Sci. B 269, 1242-4

(70.34) H-A System, Vibrational and Rotational Analysis of 2 Perturbed Bands,
B. Coquart, Ngo Tuan Anh, C. Couet, and H. Guenebaut,
C. R. Acad. Sci. C 270, 150-3

(70.35) K-A System,
B. Coquart, N. T. Ahn, C. Couet, and H. Guenebaut,
"Rotational Analysis of the New Visible Transition of the Rydberg-Non-Rydberg Type of the PO Radical,"
C. R. Acad. Sci. B 270, 1227-9

(70.36) N-B Rydberg Transition System,
B. Coquart, N. T. Ahn, and C. Couet,
"Evidence and Rotational Analysis of the Bands Pertaining to a New Visible Band System of the PO Radical,"
C. R. Acad. Sci. B 271, 708-11

(70.37) R. D. Verma,
"Emission Spectrum of the PO Molecule. I. A New $B'^2\Pi_i - X^2\Pi_r$ Transition,"
Can. J. Phys. 48, 2391-8

(70.38) M, G, L → β Systems,
B. Coquart, N. T. Ahn, C. Couet, and H. Guenebaut,
"Evidence and Rotational Analysis of the New Bands of the PO Radical in the Visible Region,"
C. R. Acad. Sci. C 270, 1702-5

(70.39) I → β System,
B. Coquart, N. T. Ahn, C. Couet, and H. Guenebaut,
"Rotational Analysis of a New Visible Electronic System of the PO Radical,"
C. R. Acad. Sci. C 270, 776-9

(71.40) B. Coquart, N. T. Ahn, M. Larzilliere, and C. Couet,
"New Spectroscopic Information on the $C^2\Sigma^-$ State of the PO Radical,
C. R. Acad. Sci. B 273, 384-7

(71.41) C. Couet, M. Larzilliere, and H. Guenebaut,
"Rotational Analysis of the Many Bands of the $C-X^2\Pi$ System of PO,
C. R. Acad. Sci. C 272, 425-7

(71.42) R. D. Verma,
"A Note on the D'$^2\Pi_r$ State of the PO Molecule,"
Can. J. Phys. 49, 279-80

(72.43) I-A, A, F-B Systems, Rotational Analysis,
S. Guha, S. S. Jois, and R. D. Verma,
"Emission Spectrum of the PO Molecule. Part III. The Spectrum of the Red Regions,"
Can. J. Phys. 50, 1579-86

(72.44) B. Coquart and J. C. Prudhomme,
"Reconsideration of the γ System of PO; Localization of the Rotational Perturbations in the First Vibrational Levels of the A$^2\Sigma^+$ State,
C. R. Acad. Sci. B 275, 383-5

(72.45) r-Centroids, A-X,
P. D. Singh and M. M. Shukla,
"Vibrational Transition Probabilities and r-Centroids for Some Diatomic Molecular Band Systems,"
J. Quant. Spectrosc. Radiat. Transfer 12, 1249-52

(72.46) B. Coquart, M. Larzilliere, and T. A. Ngo,
"Emission Spectrums of the PO Molecule: The Transition E$^2\Delta$ - X$^2\Pi_r$: Study of the Rydberg State E$^2\Delta$,
Can. J. Phys. 50, 2945-56

(72.47) Rotational Analysis F, G, H, and I States,
N. T. Ahn, M. Larzilliere, and J. C. Prudhomme,
"Rotational Analysis and Deperturbations of the Visible System of P^{16}O and P^{18}O,"
C. R. Acad. Sci. B 275, 423-6

(71.L1) A, F, G, H, I → X Systems,
R. D. Verma, M. N. Dixit, S. S. Jois, S. Nagaraj, and S. R. Singhal,
"Emission Spectrum of the PO Molecule Part II. $^2\Sigma$ - $^2\Sigma$ Transitions,"
Can. J. Phys. 49, 3180-3200

(72.L2) B. Coquart, C. Couet, H. Guenebaut, M. Larzilliere, and T. A. Ngo,
"Emission Spectra of the PO Molecule; C$^2\Sigma^-$ - X$^2\Pi_r$ Transition and the Nature of the C' State,"
Can. J. Phys. 50, 1014-22

PO

(72.L3) Theoretical,
F. Ackerman, H. Lefebvre-Brion, and A. L. Roche,
"Calculated Rydberg States of the PO Molecule,"
Can. J. Phys. 50, 692-9

(72.L4) B-X System, Rotational Analysis,
S. B. Rai, D. K. Rai, and K. N. Upadhya,
"Analysis of Some Bands of the β System of PO,"
J. Phys. B. (Atom. Mol. Phys.) 5, 1038-47

(73.L5) A-B System,
R. D. Verma, and S. S. Jois,
"Emission Spectrum of the PO Molecule. Part IV. Spectrum in the Region 7000-12000 Å,"
Can. J. Phys. 51, 322-33

PS

Methods of Production and Experimental Technique

Emission from a microwave discharge (69.3).

Discharge in a Geissler tube in P_4S_3 (55.1, 55.2).

Microwave discharge in neon containing trace amounts of compounds containing P and S (71.4).

BAND SYSTEMS

System	Transition	Sources	Wavelength Limits	Degrading	Band Head, $\nu_{0,0}$	Remarks	Bibliography
I	$B^2\Pi \rightarrow X^2\Pi$	Discharge	6000-4200	R			(69.3, 55.1, 55.2)
II	$C^2\Sigma \rightarrow X^2\Pi$	Discharge	3300-2700	R			(71.4, 69.3, 55.1, 55.2)

PS

I. $B^2\Pi \rightarrow X^2\Pi$ System

Band heads (R_1) of $P^{32}S$ (69.3), λ:

v', v''	0	1	2	3	4	5	6
0				4852.89	5027.75	5213.91	5412.55
1			4579.90	4736.43	4902.68	5079.86	
2			4476.70	4625.72			
3		4242.65	4378.81				

II. $C^2\Sigma \rightarrow X^2\Pi$ System

Band heads (R_1) of $P^{32}S$ (69.3), λ:

v', v''	0	1	2	3	4	5	6
0	2890.58	2953.16	3018.05	3085.27			
1	2847.06	2907.83		3035.79	3103.28	3173.20	
2	2805.37					3121.59	3191.65
3	2765.35						
4		2782.62					

SPECTROSCOPIC CONSTANTS

State	T_e	(c)/(a) ω_e	(c)/(a) $x_e\omega_e$	B_e	$\alpha_e \times 10^3$	$D_e \times 10^6$	r_e	Remarks	Bibliography
$C^2\Sigma$	34686.5 34365.7	527.2 534.8	3.2 3.31	0.2644	1.96		2.013	$\gamma_1 = 0.015$ cm^{-1}	(71.4, 69.3, 55.1)
$B^2\Pi$	22894.0 22666.9	505.1 512.2	2.1 2.15	—			1.900[b]		(69.3, 55.1)
$X^2\Pi$	321.9 0	728.5 739.1	2.9 2.96	0.29					(71.4, 69.3, 55.1)

[a] for P^{32}S, [b] r_o, [c] for P^{34}S.

Dissociation energy = 5.2 ± 0.1 eV, 120 kcal/mole, 41942 cm^{-1}.

BIBLIOGRAPHY

(55.1) C, B →X Systems, Vibrational Analysis,
K. Dressler,
Helv. Phys. Acta 28, 563-90

(55.2) C →X System, Vibrational Analysis,
K. Dressler and E. Miescher,
Proc. Phys. Soc. A 68, 542-4

(69.3) Isotope Effect,
N. A. Narasimham and T. K. B. Subramanian,
J. Mol. Spectrosc. 29, 294-304

(71.4) C →X System,
N. A. Narasimham and T. K. B. Subramanian,
"The Ultraviolet Bands of PS,"
J. Mol. Spectrosc. 37, 371-2

PbCl

Methods of Production and Experimental Technique

Absorption from a superheated, saturated vapor of $PbCl_2$ at $\geq 700°C$ (49.30, 36.57).

Emission from a high frequency discharge into $PbCl_2$ (36.3).

Fluorescence (irradiation of $PbCl_2$ by U.V. light) (32.1).

High frequency discharge into $PbCl_4$ (70.11).

Reaction of $Pb + Cl_2$ (n.p. 12).

BAND SYSTEMS

System	Transition	Sources	Wavelength Limits	Degrading	Band Head, $\nu_{0,0}$	Remarks	Bibliography
I	$A(^2\Sigma^+) \rightleftarrows X^2\Pi_{3/2}$	Discharge	8820-6260	R	13546.2		(70.11, 68.10, 67.9)
II	$A(^2\Sigma) \rightleftarrows X^2\Pi_{1/2}$	Discharge	5900-4100	R	21827.5		(70.11, 68.10, 67.9)
III	$B(^2\Sigma^+) \leftarrow X^2\Pi_{1/2}$	Discharge	2956-2591	V	34980.0		(66.7, 52.5, 49.4)

PbCl

I. $A(^2\Sigma^+) \rightleftarrows X^2\Pi_{3/2}$ System

Band heads of $Pb^{35}Cl$ (67.9):

v', v''	λ	Intensity	v', v''	λ	Intensity
0, 5	8360	8	3, 1	7191.3	10
0, 4	8146	7	4, 1	7079.0	8
1, 4	7998	6	3, 0	7029.6	9
1, 3	7800	6	5, 1	6970.6	7
1, 2	7611	8	4, 0	6921.9	8
2, 2	7483	6	5, 0	6818.7	7
3, 2	7359	7	6, 0	6718.8	7

II. $A(^2\Sigma) \rightleftarrows X^2\Pi_{1/2}$ System

Band system has single R heads.

Bands of greatest intensity in emission (int.$_e$) (36.3) or absorption (int.$_a$) (36.2):

v', v''	λ	Int.$_e$	Int.$_a$	v', v''	λ	Int.$_e$	Int.$_a$
6, 17	5471.9	8		0, 4	4846.0		7
5, 16	5455.7	8		0, 3	4777.5		8
4, 15	5439.2	8		1, 3	4725.9		7
5, 15	5374.5	8		0, 2	4710.2		4
4, 14	5358.4	8		1, 2	4660.3		10
3, 13	5342.1	8		2, 2	4611.6		4
3, 12	5263.1	8		1, 1	4596.0		8
2, 11	5246.9	8		2, 1	4548.9		10
2, 10	5169.5	10	0	3, 1	4503.0		7
1, 9	5133.4	10	0	2, 0	4487.2		6
1, 8	5078.6	9	2	3, 0	4442.3		5
0, 7	5062.0	10	1	4, 0	4399.0		6
0, 6	4988.2	10	2	5, 0	4356.4		7
0, 5	4916.4	10	4	6, 0	4315.3		4

III. $B(^2\Sigma^+) \leftarrow X^2\Pi_{1/2}$ System

Strong predissociation.

Bands of greatest intensity (66.7):

v', v''	λ (Int.)
0, 2	2907.64(4)
0, 1	2882.69(7)
0, 0	2857.94(10)
1, 0	2826.86(6)
2, 1	2820.28(4)
2, 0	2796.92(5)
3, 0	2767.22(4)
4, 0	2738.66(4)

PbCl

SPECTROSCOPIC CONSTANTS

State	T_e	ω_e	$x_e\omega_e$	B_e	$\alpha_e \times 10^3$	$D_e \times 10^6$	r_e	Remarks	Bibliography
$B(^2\Sigma^+)$	34937.5	386.3	1.36	-	-				(68.7)
$A^2\Sigma$	21863.1	228.8	0.795	0.10716	0.52		2.293	$y_e\omega_e = 0.006$ cm^{-1}	(70.11, 68.10, 67.9)
$X^2\Pi_{3/2}$	8274	321.6	0.30	-	-		-		(68.10, 67.9)
$X^2\Pi_{1/2}$	0	304.2	0.89	0.11862[a]	-		2.179[b]		(68.10, 67.9)

[a] B_o, [b] r_o.

Dissociation energy - 3.77 eV, 86.9 kcal/mole, 30408 cm^{-1} (70.11).

PbCl

Perturbations and General Information

Ionization potential = 7.5 ± 0.1 eV (67.8).

Electron affinity = 1.0 ± 0.2 eV (67.8).

BIBLIOGRAPHY

(32.1) A-X ($^2\Pi_{1/2}$) System in Fluorescence,
B. Popov and H. Neuimin,
Phys. Z. Soviet Union 2, 394-421

(36.2) A ← X ($^2\Pi_{1/2}$) System, Vibrational Analysis,
F. Morgan,
Phys. Rev. 49, 47-50

(36.3) A → X ($^2\Pi_{1/2}$) System, Vibrational Analysis,
G. D. Rochester,
Proc. Roy. Soc. A 153, 407-21

(49.4) B ← X System Vibrational Analysis,
K. Wieland and R. Newburgh,
Helv. Phys. Acta 22, 590-1

(52.5) B ← X ($^2\Pi_{1/2}$) System. Vibrational Analysis. Isotope Effect Pb$^{35/37}$Cl,
K. Wieland and R. Newburgh,
Helv. Phys. Acta 25, 87-106

(64.6) A → X ($^2\Pi_{1/2}$) System. Rotational Analysis Possibly Incorrect,
V. S. Rao and P. T. Rao,
Z. Physik 181, 58-66

(66.7) B ← X ($^2\Pi_{1/2}$) System. Vibrational Analysis. Isotope Effect Pb$^{35/37}$Cl,
H. Cordes and F. Gehrke,
Z. Phys. Chem. 51, 281-9

(67.8) Electronic Shock,
J. W. Hastie, H. Bloom, and J. D. Morrison,
J. Chem. Phys. 47, 1580-3

(67.9) Extension of the A → X ($^2\Pi_{1/2}$) and A → X ($^2\Pi_{3/2}$) Systems,
P. Deschamps,
Thesis, Paris

PbCl

(68.10) A-X, Rotational Analysis,
O. N. Singh and I. S. Singh,
"Rotational Analysis of the A-X Bands of the PbCl Molecule,"
Current. Sci. 37, 282-3

(70.11) Extension of A-X System,
S. P. Singh,
"A-X Band System of PbCl in the Visible Region,"
Indian J. Pure App. Phys. 8, 114

(n. p. 12) A-X Observation from Pb + Cl_2,
G. A. Capelle, R. S. Bradford, and H. P. Broida,
"Chemiluminescence and Laser Photoluminescence of Diatomic Metal Halides,"
(To Be Published)

PbF

Methods of Production and Experimental Technique

Absorption in PbF_2 at $T \sim 1500°C$.

Emission from a high frequency discharge or dc discharge in PbF_2.

BAND SYSTEMS

System	Transition	Sources	Wavelength Limits	Degrading	Band Head, $\nu_{0,0}$	Remarks	Bibliography
I	$A(\Omega=1/2) \to X_2\,^2\Pi_{3/2}$	Discharge	8300-7300	R	14198.3		(59.6)
II	$A(\Omega=1/2) \rightleftarrows X_1\,^2\Pi_{1/2}$	Discharge	5300-4100	R	22511.5		(36.1, 36.2)
III	$B\,^2\Sigma^+ \to X_2\,^2\Pi_{3/2}$	Discharge	3795-3565	V	27417		(72.13, 67.10, 38.3)
IV	$B\,^2\Sigma^+ \rightleftarrows X_1\,^2\Pi_{1/2}$	Discharge	2925-2665	V	35695.3		(67.10, 38.3)
V	$C \leftarrow X_1\,^2\Pi_{1/2}$	Absorption	2690-2570	V	38089	Uncertain classification (52.11)	(38.3)
VI	$D \leftarrow X_1\,^2\Pi_{1/2}$	Discharge	2305-2245	(V)	43863		(38.3)
VII	$E_2 \to X_2\,^2\Pi_{3/2}$	Discharge	2763-2684	-	37232.9		(67.10)
VIII	$E_1 \leftarrow X_1\,^2\Pi_{1/2}$	Absorption	2250-2120	V	45432		(38.3)
IX	$F \leftarrow X_1\,^2\Pi_{1/2}$	Absorption	2130-2060	(V)	47927		(38.3)
X	Continuum emission		3800-2900	Continuum	32800 (maximum)		(38.3)
XI	Continuum absorption		3200-2300	Continuum	41000 (maximum)		(38.3)

PbF

I. $\underline{A(\Omega = 1/2) \to X_2\ {}^2\Pi_{3/2}\ \text{System}}$

Band heads:

(v', v'')	(0, 4)	(0, 3)	(0, 2)	(0, 1)
λ	8237.6	7897.2	7588.6	7299.55

II. $\underline{A(\Omega = 1/2) \rightleftarrows X_1\ {}^2\Pi_{1/2}\ \text{System}}$

Intense P heads with Q heads ~ 7 cm^{-1} toward longer wavelength.

Band heads in emission (int.$_e$) and absorption (int.$_a$), λ (int.$_e$, int.$_a$)

v', v''	0	1	2	3	4	5	6
0	4441.2 (8, 10)	4542.5 (10, 10)	4647.7 (9, 8)	4756.7 (7, 5)			
1	4364.7 (10, 10)		4563.9 (7, 5)	4669.1 (9, 7)	4778.1 (7, 6)	4891.3 (6, 1)	
2	4291.5 (5, 8)		4484.0 (3, 5)		4690.6 (8, 4)	4799.6 (7, 5)	4912.7 (6, 2)
3		4312.7 (3, 6)		4505.4 (2, 5)			

III. $\underline{B\ {}^2\Sigma^+ \to X_2\ {}^2\Pi_{1/2}\ \text{System}}$

Observed P and Q heads:

v', v''	λ_P, λ_Q	Intensity
0, 2	3792.8, 3792.0	2
1, 3	3780.6, 3779.6	2
0, 1	3718.9, 3717.8	5
1, 2	3707.7, 3706.6	3
0, 0	3647.3, 3646.2	10
1, 0	3568.8, 3567.6	3

IV. $B\,^2\Sigma^+ \rightleftarrows X_1\,^2\Pi_{1/2}$ System

P, Q head separation ~ 7 cm^{-1}.

Band heads (P):

v', v''	λ(Intensity)
0, 2	2882.0(3)
1, 3	2872.6(3)
0, 1	2841.4(8)
0, 0	2801.2(9)
1, 0	2754.5(10)
2, 1	2747.2(5)
2, 0	2709.8(7)

V. $C \leftarrow X_1\,^2\Pi_{1/2}$ System

Weak system with a superimposed continuum. $v_Q - v_P \sim 10$ cm^{-1}.

Possibly this system can be represented by the B-X$_1$ system with strong perturbations of v' > 4 (52.5). Both interpretations are given:

v', v'' B ← X$_1$	v', v'' C ← X$_1$	λ	Intensity
5, 3	1, 3	2688.6	0
5, 2	1, 2	2653.4	1
6, 3	2, 3	2646.9	2
7, 4	3, 4	2640.6	1
8, 5	4, 5	2634.4	0
5, 1	1, 1	2618.7	0
6, 2	2, 2	2612.8	2
7, 3	3, 3	2606.9	2
8, 4	4, 4	2601.2	1
9, 5	5, 5	2595.3	0
6, 1	2, 1	2579.2	0
7, 2	3, 2	2574.2	0

VI. $D \leftarrow X_1\,^2\Pi_{1/2}$ System

Bands with simple heads:

(v', v'')	(0, 1)	(0, 0)	(1, 0)
λ(Intensity)	2305.6(1)	2279.1(2)	2248.5(2)

PbF

VII. $E_2 \to X_2\ ^2\Pi_{3/2}$ System

Three double-headed bands. Band heads (Q):

(v', v'')	(0, 0)	(0, 1)	(0, 2)
λ	2684.5(5)	2723.2(3)	2762.9(2)

VIII. $E_1 \leftarrow X_1\ ^2\Pi_{1/2}$ System

System appears to predissociate at v' = 0 (38.3):

v', v''	λ_Q(Int.), λ_P
0, 2	2249.7(0), 2250.8
0, 1	2224.8(3), 2225.8
1, 2	2221.6(1)
2, 3	2218.8(0)
0, 0	2200.4(8), 2201.2
1, 0	2173.6(10)
2, 0	2147.4(5)
3, 0	2121.5(1)

IX. $F \leftarrow X_1\ ^2\Pi_{1/2}$ System

Line-like bands:

(v', v'')	(0, 2)	(0, 1)	(0, 0)	(1, 0)
λ(Intensity)	2130.5(0)	2108.1(2)	2085.8(2)	2058.9(1)

SPECTROSCOPIC CONSTANTS

State	T_e	ω_e	$x_e\omega_e$	B_e	$\alpha_e \times 10^3$	$D_e \times 10^6$	r_e	Remarks	Bibliography
F	~47866	628[b]	-	-	-				(38.3)
E	~45404	562[b]	-	-	-				(67.10, 38.3)
D	~43818	597[b]	-	-	-				(38.3)
C	38046	594.0	2.50	-	-				(52.5, 38.3)
$B\ ^2\Sigma^+$	35643	612.8	3.42	0.2510	1.5	0.279[d]	1.964	γ is small	(72.13, 67.10, 38.3)
$A(\Omega=1/2)$	22566.6	379.8	1.77	0.2070[c]	-		2.163[e]	$P_o \sim 0.619$ cm^{-1}	(36.1, 36.2)
$X_2\ ^2\Pi_{3/2}$	8264.7[a]	531.1	1.50	0.2338	2.0	0.132[d]	2.035		(72.13, 59.6, 38.3)
$X_1\ ^2\Pi_{1/2}$	0	507.2	2.30	0.2281[c]	-	-	2.061[e]	$P_o \sim -0.138$ cm^{-1}	(36.1, 36.2)

[a] $A_v(X^2\Pi) = 8264.7 + 26.0\ (v+1/2)$, [b] ω_o, [c] B_o, [d] D_o, [e] r_o.

Dissociation energy = 3.0 ± 0.5 eV, 69 kcal/mole, 24197 cm^{-1}.

PbF

Perturbations and General Information

Absence of emission bands of the $B \rightarrow X_1, X_2$ System for $v' > 1$ in addition to the diffuseness of the absorption for $v' > 1$ indicates predissociation in this region (49.4).

$C \leftarrow X$ system can also be interpreted as the $B \leftarrow X$ system with strong perturbations for $v' > 4$ (52.5).

Predissociation of the B ($v' > 1$) and E ($v' > 0$) states is caused by the crossing of the A state (67.10).

BIBLIOGRAPHY

(36.1) $A \leftarrow X$, System,
F. Morgan,
Phys. Rev. 49, 47-50

(36.2) $A \rightarrow X$, System,
E. C. W. Smith,
Proc. Roy. Soc. A 155, 173-82

(38.3)   All Systems,
G. D. Rochester,
Proc. Roy. Soc. A 167, 567-80

(49.4)   Predissociation. Dissociation Energy,
K. Wieland and R. Newburgh,
Helv. Phys. Acta 22, 590-1

(52.5)   Detailed $C \leftarrow X$ System. Predissociation. Dissociation Energy,
K. Wieland and R. Newburgh,
Helv. Phys. Acta 25 87-106

(59.6) $A \rightarrow X_2$ System,
R. F. Barrow, D. Butler, J. W. C. Johns, and J. L. Powell,
Proc. Phys. Soc. 73, 317-20

(64.7)   A-X System Rotational Analysis Possibly Incorrect,
K. M. Rao and P. T. Rao,
Can. J. Phys. 42, 690-5

(67.8) B-X_1, B-X_2 Systems, Rotational Analysis Possibly Incorrect,
Y. P. Reddy and P. T. Rao,
Proc. Int. Conf. Spectr., Bombay 1, 129-32

(67.9) A-X System, Rotational Analysis Possibly Incorrect,
O. N. Singh and I. S. Singh,
Proc. Int. Conf. Spectr., Bombay 1, 133-7

(67.10) B → X_1, X_2 Systems, Vibrational Analysis,
S. P. Singh,
"Emission Spectrum of PbF Molecule,"
Ind. J. Pure Appl. Phys. 5, 292-4

(68.11) Dissociation Energy, Ionization Potential,
K. Zmbov, J. W. Hastie, and J. L. Margrave,
Trans. Faraday Soc. 64, 861-7

(69.12) A-X System Rotational Analysis Possibly Incorrect,
O. N. Singh, M. P. Srivastava, and I. S. Singh,
Can. J. Phys. 47, 1639-41

(72.13) B → X_2 System, Rotational Analysis,
O. N. Singh, I. S. Singh, and O. N. Singh,
"Rotational Analysis of B-X_2 System of ^{208}PbF,"
Can. J. Phys. 50, 2206-10

PbO

Methods of Production and Experimental Technique

Absorption.

Emission from an arc (C and Ca) + (Pb, plus a salt of Pb); discharge (particularly microwaves).

BAND SYSTEMS

System	Transition	Sources	Wavelength Limits	Degrading	Band Head, $v_{0,0}$	Remarks	Bibliography
I	$AO^+ \rightleftarrows X^1\Sigma^+$	Absorption, discharge	6750-4750	R	19724.5		(63.12, 61.10, 36.5, 30.1, 30.2)
II	$B1 \rightleftarrows X^1\Sigma^+$	Absorption, discharge	Emission - 5770-4140 Absorption - 5170-3880	R	22174.2		(62.11, 61.10, 36.5)
III	$CO^+ \rightleftarrows X^1\Sigma^+$	Absorption, discharge	4200-3600	R	~23725		(63.12, 61.10, 36.5)
IV	$C'1 \rightleftarrows X^1\Sigma^+$	Absorption, discharge	3650-3400	R	~24835		(63.12, 61.10, 36.5)
V	$D1 \rightleftarrows X^1\Sigma^+$	Absorption, discharge	Emission - 3700-3200 Absorption - 3670-2950	R	30098		(63.12, 61.10, 36.5)
VI	$EO^+ \rightleftarrows X^1\Sigma^+$	Absorption, discharge	Emission - 3500-3130 Absorption - 3060-2780	R	34320.5		(62.11, 61.10, 47.7)
VII	$q \leftarrow X^1\Sigma^+$	Absorption	2045-1860	R	51072		(63.2)
VIII	$r \leftarrow X^1\Sigma^+$	Absorption	2055-1865	R	51570		(63.2)
IX	$p \leftarrow X^1\Sigma^+$	Absorption	1825-1775	R	56323		(63.2)
X	$F \leftarrow X^1\Sigma^+$	Absorption	1810-1738	R	56513		(63.2)
XI	$O \leftarrow X^1\Sigma^+$	Absorption	~1750	-	-	Continuum	(63.12)

I. $AO^+ \rightleftarrows X^1\Sigma^+$ System

Band heads in emission (30.1):

v', v''	λ	Intensity	v', v''	λ	Intensity
3, 8	6433.63	3	0, 4	5910.74	6
0, 6	6427.73	3	0, 3	5677.78	6
2, 7	6342.01	3	2, 4	5617.65	3
1, 6	6250.75	5	0, 2	5459.38	6 d.
0, 5	6160.52	4	1, 2	5331.11	3
			1, 1	5138.18	3

II. $B1 \rightleftarrows X^1\Sigma^+$ System

Band heads in emission (30.37) (30.1):

v', v''	λ	Intensity	v', v''	λ	Intensity
0, 5	5353.82	3	0, 1	4657.98	5
0, 4	5162.31	6	1, 1	4553.71	6
0, 3	4983.79	6	1, 0	4410.38	5
0, 2	4816.90	6	2, 0	4317.06	4
			3, 0	4229.01	4

III. $CO^+ \rightleftarrows X^1\Sigma^+$ System

Band heads in absorption (36.5):

v', v''	λ	Intensity	v', v''	λ	Intensity
2, 1	4156.20	3	4, 0	3877.85	8
2, 0	4037.63	3	6, 1	3838.24	2
4, 1	3987.70	4	5, 0	3804.94	6
3, 0	3955.04	7	6, 0	3735.94	4
5, 1	3910.30	3	7, 0	3669.63	3
			8, 0	3607.35	1

IV. $C'1 \rightleftarrows X^1\Sigma^+$ System

In absorption, the band heads form a single intense progression with v'' = 0 (30.37). v' numbering is deduced from isotopic studies (63.12):

(v', v'')	(6, 0)	(7, 0)
λ	3612.8	3554.8

PbO

V. $\quad$ D1 $\rightleftarrows$ X$^1\Sigma^+$ System

Band heads in emission (30.1):

(v', v'')	(0, 2)	(0, 1)	(1, 1)	(1, 0)	(2, 0)
λ(Intensity)	3485.68(6)	3401.92(5)	3341.83(2)	3264.36(2)	3209.22(2)

VI. $\quad$ EO$^+$ $\rightleftarrows$ X$^1\Sigma^+$ System

In emission, bands with v' > 1 are not observed.
Band heads in absorption:

v', v''	λ	Intensity	v', v''	λ	Intensity
1, 3	3062.67	4	3, 2	2925.64	3
2, 3	3023.38	2	2, 1	2900.21	4
1, 2	2998.52	4	4, 2	2894.21	1
2, 2	2960.73	3	3, 1	2866.17	5
1, 1	2936.19	2	4, 1	2836.57	1
			3, 0	2808.5	1

VIII. $\quad$ q, r ← X$^1\Sigma^+$ Systems

Weak, superimposed systems observed only in absorption (63.12).

IX. $\quad$ p ← X$^1\Sigma^+$ System

Only an intense v' = 0 progression (63.12):

(v', v'')	(0, 2)	(0, 1)	(0, 0)
λ	1820.7	1797.6	1775.5

X. $\quad$ F ← X$^1\Sigma^+$ System

Intense system with diffuse bands implying predissociation.
Band maxima:

(v', v'')	(0, 2)	(0, 1)	(0, 0)	(1, 0)	(2, 0)
λ	1809.7	1787.5	1769.5	1753.8	1738.2

PbO

SPECTROSCOPIC CONSTANTS

State	T_e	ω_e	$x_e\omega_e$	B_e	$\alpha_e \times 10^3$	$D_e \times 10^6$	r_e	Remarks	Bibliography
O	56500							Repulsive	(63.12)
F	(56513)	500							(63.12)
p	(56323)								(63.12)
r	51661	540.5	6						(63.12)
q	51153	558.5	3						(63.12)
EO^+	34455	454.1	6.95	0.2421	2.6		2.165		(62.11, 61.10)
D1	30194	530.4	2.9	0.2710	2.8	0.283	2.047		(62.11, 61.10)
C'1	24947	494	3.0	0.2491	1.8	0.25	2.135		(62.11, 61.10)
CO^+	23820	532	3.9	0.2545	2.1	0.25	2.112		(62.11, 61.10)
B1	22289	(489)	-	0.2648	2.6	0.30	2.071		(62.11, 61.10)
AO^+	19862.3	444.2	0.46	0.2588	1.4	0.33	2.095		(62.11, 61.10)
$X^1\Sigma^+$	0	721.46	3.53	0.307519	1.9167	0.22	1.92181		(64.13, 63.12, 62.11, 61.10)

Dissociation energy - 3.87 ± 0.05 eV, 89.3 kcal/mole, 31214 cm^{-1}.

PbO

Perturbations and General Information

B1 state, v' = 1-4 levels are all perturbed; probably by states with $\Omega = 1$ or 2.

D1 state is perturbed by a state with $\Omega = 2$.

EO^+ state perturbed in v = 1-4.

Potential energy curves - RKRV potential (65.15), (71.L1):

State	v	$U(cm^{-1})$	$r_{min}(\text{Å})$	$r_{max}(\text{Å})$	$T_e + U(cm^{-1})$
XO^+	0	360.0	1.8683	1.9811	360.0
	1	1070.6	1.8329	2.0288	1070.6
	2	1771.0	1.8102	2.0637	1771.0
	3	2475.4	1.7925	2.0938	2475.4
	4	3169.3	1.7781	2.1208	3169.3
	5	3855.1	1.7654	2.1462	3855.1
	6	4534.6	1.7543	2.1700	4534.6
	7	5206.8	1.7443	2.1929	5206.8
	8	5870.9	1.7352	2.2150	5870.9
AO^+	0	225.0	2.027	2.169	20088.3
	1	670.0	1.980	2.228	20533.3
	2	1114.0	1.950	2.271	20977.3
B1	0	248.5	2.008	2.143	22533.4
	1	743.2	1.967	2.202	23028.1
	2	1232.6	1.942	2.246	23517.5
	3	1719.1	1.923	2.283	24004.0
	4	2212.1	1.907	2.318	24497.0
	5	2684.6	1.893	2.351	24969.5
	6	3155.3	1.881	2.382	25440.2
D1	0	264.5	1.986	2.117	30463.2
	1	788.5	1.947	2.175	30987.2
	2	1308.5	1.923	2.220	31507.2
	3	1814.9	1.905	2.258	32013.6
	4	2321.5	1.891	2.293	32520.2
	5	2822.0	1.880	2.325	33020.7
	6	3321.5	1.870	2.356	33520.2
	7	3808.6	1.862	2.385	34007.3
	8	4314.7	1.855	2.414	34513.4

PbO

Franck-Condon factors, r-centroids — Morse potential:

$AO^+ - X^1\Sigma^+$ (58.9)

v', v''	0	1	2	3	4	5	6
0	5068.8		5459.4	5677.8	5910.7	6160.5	6427.7
	0.02	0.09	0.17	0.21	0.14	0.08	0.04
1		5138.2	5311.1				6550.7
	0.07	0.16	0.13	0.02	0.00	0.07	0.14
2					5617.6		
	0.14	0.13	0.01	0.03	0.10	0.06	0.00
3	0.17	0.04	0.02	0.09	0.02	0.15	0.08
4	0.17	0.00	0.08	0.03	0.01	0.07	0.02
5	0.14	0.02	0.07	0.00	0.07	0.01	0.02

Top = Wavelength
Bottom = Franck-Condon factors

$B1 - X^1\Sigma^+$ (70.17)

v', v''	0	1	2	3	4	5
0	0.05154	0.15781	0.23807	0.23597	0.17418	0.10266
	1.988	2.02	2.052	2.092	2.116	2.148
	4510	4660	4819	4987	5164	5355
1		0.14216	0.19151			
		1.972	2.002			
		4412	4555			
2		0.20794				
		1.964				
		4319				
3		0.00001				
		1.948				
		4230				
4		0.07133				
		1.924				
		4144				

Top = Franck-Condon factor
Middle = r-centroid
Bottom = Wavelength

PbO

$D1 - X^1\Sigma^+$ (58.9)

v', v''	0	1	2	3	4	5	6
0	3320.7	3401.9	3485.7				
	0.13	0.27	0.28	0.18	0.08	0.00	0.00
1	3264.4	3341.8			3594.2		
	0.24	0.14	0	0.10	0.20	0.17	0.08
2	3209.2			3442.8			
	0.25	0.00	0.12	0.09	0	0.09	0.18
3	0.18	0.05	0.10	0.00	0.12	0.04	0.01
4	0.10	0.14	0.00	0.10	0.02	0.04	0.10

Top   = Wavelength
Bottom = Franck-Condon factors

Chemiluminescence Studies (71.19)

$Pb + O \rightarrow PbO$ ($X^1\Sigma^+$, $A0^+$, $B1$ and $C0^+$).

Relative emission intensities: $B > A > C$.

Radiative lifetime:
  $B1 \rightarrow X^1\Sigma^+$

  $\tau = 1500 \pm 300$ nsec (71.19).

BIBLIOGRAPHY

(30.1)   A, B, D → X Systems, Vibrational Analysis,
         S. Bloomenthal,
         Phys. Rev. 35, 34-45

(30.2)   A, D → X Systems, Rotational Analysis,
         A. Christy and S. Bloomenthal,
         Phys. Rev. 35, 46-50

(35.3)   Preliminary Note to (35.4),
         E. N. Shawhan and F. Morgan,
         Phys. Rev. 47, 199

(35. 4) D, C ← X Systems,
E. N. Shawhan and F. Morgan,
Phys. Rev. 47, 377-8

(36. 5) B, C, D, E ← X Systems,
H. G. Howell,
Proc. Roy. Soc. A 153, 683-98

(40. 6) Dissociation Energy,
R. F. Barrow,
Trans. Faraday Soc. 36, 1053-5

(47. 7) E ← X System,
E. E. Vago and R. F. Barrow,
Proc. Phys. Soc. 59, 449-57

(49. 8) Thermochmistry,
L. Brewer and D. F. Mastick,
UCRL No. 571

(58. 9) Franck-Condon Factors, A, D-X,
R. W. Nicholls, P. A. Fraser, and W. R. Jarmain,
"Transition Probability Parameters of Molecular Spectra
Arising from Combustion Processes,"
Combustion and Flame 26, 13-38

(61.10) Rotational Analysis,
B. F. Barrow, J. L. Deutsch, and D. N. Travis,
Nature 191, 374-5

(62. 11) B, E-X Systems, Rotational Analysis,
J. L. Deutsch,
Theises, Oxford

(63. 12) A, C, C', D-X Systems, Rotational Analysis. Absorption Bands
in the Vacuum Ultraviolet,
D. N. Travis,
Thesis, Oxford

(64. 13) Microwave Spectra,
T. Törring,
E. Naturforsch 19, 1426-8

(65. 14) Dissociation Energy,
J. Drowart, R. Colin, and G. Exsteen,
Trans. Faraday Soc. 61, 1376-83

PbO

(65.15)   K. P. R. Nair, R. B. Singh, and D. K. Rai,
          "Potential Energy Curves and Dissociation Energies of Oxides
          and Sulfides of Group IVA Elements,"
          J. Chem. Phys. 43, 3570-4

(69.16)   Stark Effect,
          J. Hoeit, F. J. Lovas, E. Tiemann, R. Tischer, and T. Törring,
          Z. Naturforsch. A 24, 1222-6

(70.17)   Franck-Condon Factors, r-Centroids, B-X System,
          P. S. Dube, K. N. Upadhya, and D. K. Rai,
          "Electronic Transition Moment Variation in the $B1 - X^1\Sigma^+$
          System of PbO and Determination of the Effective Vibrational
          Temperature,"
          J. Quant. Spectrosc. Radiative Transfer 10, 1191-4

(71.18)   P. S. Dube,
          "Einstein Coefficient Oscillator Strength and the Lifetime Measurement in the $B1 - X^1\Sigma$ System of PbO,"
          Current Sci. 40, 32

(71.19)   Radiative Lifetime, Chemiluminescence,
          S. E. Johnson,
          Thesis, Univ. of Calif.: Santa Barbara

(71.L1)   J. Singh and K. P. R. Nair,
          "Potential Energy Curves and Dissociation Energies of Silicon,
          Germanium and Lead Monoxides,"
          Indian J. Pure Appl. Phys. 9, 130-1

PbS

Methods of Production and Experimental Technique

Observation of all the systems in absorption.

In emission appears as the bands of the D-X system and some others in the positive column of a discharge in a high density stream.

BAND SYSTEMS

System	Transition	Sources	Wavelength Limits	Degrading	Characteristic Bands, λ	Remarks	Bibliography
I	$a1 \leftarrow X^1\Sigma^+$	Absorption	7700-6600	R	7614.2, 7081.8, 6943.8, 6876.4, 6745.4		(63.5, 35.1)
II	$AO^+ \leftarrow X^1\Sigma^+$	Absorption	7670-4545	R	5499.81, 5422.49, 5228.51, 5158.95, 5047.54, 4982.69		(63.5, 38.2, 35.1)
III	$B1 \leftarrow X^1\Sigma^+$	Absorption	5080-3950	R	4622.11, 4563.50, 4506.39, 4421.29, 4368.11, 4316.50		(63.5, 35.1)
IV	$C?O^+ \leftarrow X^1\Sigma^+$	Absorption	4400-3675	R	3923.51, 3881.45, 3840.40		(35.1)
V	$C'?1 \leftarrow X^1\Sigma^+$	Absorption	4080-3500	R	3796.0, 3757.4, 3720.0		(35.1)
VI	$D1 \leftrightarrows X^1\Sigma^+$	Absorption	3660-3130	R	3530.6, 3478.7, 3443.4, 3428.3, 3393.9, 3313.2		(63.5, 35.1)
VII	$E?O^+ \leftarrow X^1\Sigma^+$	Absorption	3200-2750	R			(47.3)

PbS

BAND SYSTEMS

System	Transition	Sources	Wavelength Limits	Degrading	Band Head, $\nu_{0,0}$	Remarks	Bibliography
VIII	$F \leftarrow X\,^1\Sigma^+$	Absorption	2175-2060	R	2151.89, 2132.51, 2115.95, 2097.24, 2078.7		(47.3)

PbS

I. <u>$a1 \leftarrow X^1\Sigma^+$ System</u>

Weak, superimposed on the A-X system (63.5, 35.1).

II. <u>$AO^+ \leftarrow X^1\Sigma^+$ System</u>

Band heads, λ (intensity):

v', v''	0	1	2	3	4	5	6
0			5579.59 (5)	5714.79 (8)	5852.74 (8)	5999.09 (10)	6152.11 (9)
1		5373.76 (5)	5499.81 (10)	5630.24 (10)	5765.15 (8)	5906.72 (1)	6055.37 (1)
2	5182.59 (3)	5299.91 (8)	5422.49 (9)	5549.15 (10)			5961.38 (3)
3	5113.98 (8)	5228.51 (10)	5347.51 (7)			5731.11 (5)	5871.09 (2)
4	5047.54 (10)	5158.95 (10)	5274.19 (0)		5519.21 (8)		
5	4982.69 (10)	5091.35 (8)		5320.59 (4)	5441.70 (2)		5698.31 (0)
6	4919.89 (10)	5025.24 (1)	5134.81 (4)	5248.60 (1)		5490.0 (1)	
7	4858.49 (10)		5068.20 (2)				
8	4798.90 (10)						
9	4741.09 (8)						
10	4684.99 (7)	4780.61 (6)					

III. <u>$B1 \leftarrow X^1\Sigma^+$ System</u>

Band heads, λ (intensity):

v', v''	0	1	2	3
0			4777.85(1)	4876.45(2)
1	4532.50(3)	4622.11(8)	4714.51(3)	4810.40(1)
2	4474.30(8)	4563.50(10)	4653.61(1)	
3	4421.29(10)	4506.39(8)		
4	4368.11(10)	4450.90(1)	4536.35(0)	4625.34(0)
5	4316.50(10)		4480.84(0)	

PbS

IV. $C?0^+ \leftarrow X^1\Sigma^+$ System

Band heads. v' numbering is uncertain:

v', v''	λ	Intensity	v', v''	λ	Intensity
(3), 0	4157.53	4	(8), 0	3923.51	8
(4), 0	4107.39	8	(9), 0	3881.45	9
(5), 0	4059.51	8	(10), 0	3840.40	8
(6), 0	4012.90	8	(11), 0	3800.60	8
(7), 0	3967.59	8	(12), 0	3761.90	8
			(13), 0	3724.50	6

V. $C'?1 \leftarrow X^1\Sigma^+$ System

Band heads. v' numbering is uncertain:

v', v''	λ	Intensity	v', v''	λ	Intensity
(2), 0	3918.46	6	(8), 0	3683.47	7
(3), 0	3876.30	6	(9), 0	3648.20	4
(4), 0	3835.49	6	(10), 0	3614.00	5
(5), 0	3795.99	8	(11), 0	3580.77	3
(6), 0	3757.40	8	(12), 0	3548.20	5
(7), 0	3720.00	8	(13), 0	3517.10	5

VI. $D1 \rightleftarrows X^1\Sigma^+$ System

Band heads, λ (intensity):

v', v''	0	1	2	3	4	5
0	3378.90(3)	3428.31(10)	3478.70(10)	3530.60(10)	3583.65(8)	3638.10(–
1	3345.54(6)	3393.90(10)	3443.40(10)	3494.30(7)	3546.10(5)	3599.34(?
2	3313.20(9)	3360.49(6)	3408.97(0)	3458.90(3)	3509.71(5)	3561.89(!
3	3281.50(8)	3328.00(5)		3424.39(6)		
4	3250.80(8)		3343.37(3)	3391.20(0)		
5	3221.10(8)		3311.80(4)			
6	3192.10(7)	3236.30(4)				
7	3164.40(6)	3207.55(4)				
8	3137.00(5)	3179.60(5)				

VII. $E?0^+ \leftarrow X^1\Sigma^+$ System

System not analyzed.

VIII. $F \leftarrow X^1\Sigma^+$ System

Band heads, λ:

v', v''	0	1	2	3	4	5	6
0	2094.47	2113.31	2132.51	2151.89	2171.4		
1	2078.7	2097.24	2115.95	2134.83	2154.2		
2	2064.2	2082.5		2119.82	2138.53	2157.8	
3		2068.6	2086.9	2105.4			2162.5
4			2073.4	2091.70			

PbS

SPECTROSCOPIC CONSTANTS

State	T_o	ω_e	$x_e\omega_e$	B_e	$\alpha_e \times 10^4$	$D_e \times 10^6$	r_e	Remarks	Bibliography
F	47730	370	(7.8)	–	–		–		(63.5)
E?O$^+$	34000	–	–	–	–		–		(63.5)
D1	29587.4	297.8	1.36	0.1016	6.4		2.477		(63.5)
C'?1	24952.3	283.9	1.171	–	–		–		(63.5)
C?O$^+$	23150.7	303.9	1.43	–	–		–		(63.5)
B1	21774.5	282.1	0.85	0.0999	6.0		2.467		(63.5)
AO$^+$	18768.9	260.8	0.36	0.0963	2.6		2.513		(63.5)
a1	14821.9	285.9	(0.88)	0.0926	3.7		2.562		(63.5)
X$^1\Sigma^+$	0	429.40	1.30	0.1163	4.2		2.287		(63.5)

Dissociation energy = 3.50 ± 0.1 eV, 80.8 kcal/mole, 28230 cm^{-1}.

Perturbations and General Information

Dipole moment = 4.02 ± 0.07 D (69.9).

Potential energy curved — Morse potential (65.7):

State	v	U(cm^{-1})	r_{min}(Å)	r_{max}(Å)	Te+U(cm^{-1})
XO$^+$	0	214.4	2.235	2.342	214.4
	1	641.3	2.200	2.385	641.3
	2	1066.1	2.177	2.416	1066.1
	3	1487.3	2.158	2.443	1487.3
	4	1903.9	2.143	2.466	1903.9
	5	2319.7	2.130	2.488	2319.7
	6	2735.3	2.118	2.508	2735.3
	7	3147.1	2.107	2.527	3147.1
	8	3557.3	2.097	2.546	3557.3
	9	3963.6	2.088	2.563	3963.6
	10	4368.9	2.079	2.580	4368.9
	11	4770.2	2.071	2.597	4770.2
	12	5168.2	2.064	2.613	5168.2
	13	5565.7	2.056	2.629	5565.7
	14	5959.2	2.050	2.645	5959.2
	15	6348.7	2.043	2.661	6348.7
	16	6733.7	2.036	2.676	6733.7
AO$^+$	0	130.0	2.446	2.583	18981.6
	1	390.1	2.400	2.637	19241.4
	2	649.7	2.370	2.676	19501.0
	3	907.5	2.346	2.708	19758.8
	4	1166.0	2.325	2.737	20017.3
	5	1423.6	2.308	2.763	20274.9
	6	1689.0	2.292	2.787	20532.3
	7	1937.3	2.277	2.811	20788.6
	8	2193.0	2.264	2.832	21044.3
	9	2446.7	2.251	2.853	21298.0
	10	2699.3	2.239	2.874	21550.6
	11	2951.3	2.228	2.893	21802.6
	12	3203.3	2.218	2.912	22054.6
	13	3453.3	2.208	2.930	22304.6
	14	3702.7	2.199	2.948	22554.0
	15	3953.7	2.191	2.965	22805.0
	16	4205.9	2.182	2.982	23057.2

PbS

State	v	U(cm^{-1})	r$_{min}$(Å)	r$_{max}$(Å)	Te+U(cm^{-1})
B1	0	141.0	2.405	2.536	21988.7
	1	422.2	2.363	2.591	22270.0
	2	700.6	2.336	2.632	22548.3
	3	977.2	2.316	2.666	22824.9
	4	1253.1	2.298	2.697	23100.8
	5	1526.4	2.284	2.725	23374.1
	6	1798.0	2.271	2.752	23645.7
	7	2068.0	2.259	2.778	23915.7
	8	2336.6	2.249	2.802	24184.3
	9	1603.5	2.239	2.826	24451.2
	10	2867.8	2.231	2.849	24715.5
	11	3131.0	2.223	2.872	24978.7
	12	3393.0	2.215	2.894	25241.3
	13	3652.4	2.208	2.915	25500.1
	14	3911.3	2.201	2.937	25759.0
	15	4166.2	2.195	2.957	26013.9
	16	4420.6	2.189	2.978	26268.3
	17	4674.6	2.184	2.998	26522.3
	18	4925.6	2.179	3.018	26773.3
E1	0	148.6	2.386	2.514	29799.1
	1	443.7	2.346	2.568	30094.2
	2	736.2	2.320	2.608	30386.7
	3	1020.3	2.299	2.642	30677.8
	4	1314.1	2.283	2.673	30964.6
	5	1598.5	2.267	2.703	31249.0
	6	1870.4	2.254	2.730	31520.9
	7	2145.9	2.244	2.755	31796.4
	8	2420.8	2.233	2.780	32071.3

BIBLIOGRAPHY

(35. 1)   All Systems,
G. D. Rochester and H. G. Howell,
Proc. Roy. Soc. A 148, 157-70

(38. 2)   A ← X System,
H. Bell and A. Harvey,
Proc. Phys. Soc. 50, 427-35

(47. 3)   E, F ← X Systems,
E. E. Vago and R. F. Barrow,
Proc. Phys. Soc. 59, 449-57

(62. 4)   Dissociation Energy,
R. Colin and J. Drowart,
J. Chem. Phys. 37, 1120-5

(63. 5)   Rotational Analysis, Resumé and Discussion,
R. F. Barrow, P. W. Fry and R. C. LeBargy,
Proc. Phys. Soc. 81, 697-704

(64. 6)   Microwave Spectra,
J. Hoeft,
Z. Naturforsch. A 19, 1134-6

(65. 7)   K. P. R. Nair, R. B. Singh, and D. K. Rai,
"Potential Energy Curves and Dissociation Energies of Oxides and Sulfides of Group IVA Elements,"
J. Chem. Phys. 43, 3570-4

(66. 8)   Dissociation Energy,
J. Drowart and P. Goldfinger,
Quarterly Rev. 20, 545-57

(69. 9)   Stark Effect,
A. N. Murty and R. F. Curl Jr.,
J. Mol. Spectrosc. 30, 102-10

(69. 10)   Stark Effect,
J. Hoelt, F. J. Lovas, E. Tiemann, R. Tischer, and T. Törring,
Z. Naturforsch A 24, 1222-6

PbSe

Methods of Production and Experimental Technique

Absorption.

Emission of the D → X system in a positive column discharge with high current density.

Ground state studied by microwave techniques.

BAND SYSTEMS

System	Transition	Sources	Wavelength Limits	Degrading	Characteristic Bands, λ	Remarks	Bibliography
I	$A \leftarrow X^1\Sigma^+$	Absorption	6100-4800	R	5503.11, 5372.60, 5325.12, 5278.73, 5202.55, 5158.61		(38.1)
II	$B \leftarrow X^1\Sigma^+$	Absorption	5410-4160	R	4921.22, 4855.89, 4813.13, 4791.33, 4749.00, 4708.38		(38.1)
III	$C \leftarrow X^1\Sigma^+$	Absorption	4500-4050	R	4197.75, 4165.93, 4134.87, 4104.90		(38.1)
IV	$D \rightleftarrows X^1\Sigma^+$	Absorption	3850-3250	R	3593.16, 3557.94, 3534.16, 3500.00, 3477.05, 3454.78		(47.4, 44.2)
V	$F \leftarrow X^1\Sigma^+$	Absorption	2300-2150	R	2253.13, 2239.33, 2228.11, 2214.55, 2203.56, 2190.05		(47.4, 44.2)

PbSe

I. $A \leftarrow X^1\Sigma^+$ System

Band heads, λ (intensity):

v', v''	0	1	2	3	4	5
0					5694.16 (3)	5784.30 (5)
1		5388.37 (5)	5470.20 (2)	5554.29 (2)	5640.28 (8)	5728.88 (7)
2	5263.36 (4)	5340.57 (3)	5421.46 (5)	5503.11 (8)	5587.60 (8)	5674.31 (9)
3	5217.10 (6)	5293.80 (7)	5372.60 (10)	5453.40 (7)	5535.96 (5)	5621.70 (3)
4	5172.51 (4)	5248.16 (8)	5325.12 (10)	5403.88 (7)	5485.36 (1)	
5	5128.65 (5)	5202.55 (10)	5278.73 (10)	5356.51 (5)		5518.15 (3)
6	5085.94 (7)	5158.61 (10)	5233.27 (4)		5387.57 (1)	
7	5043.93 (6)	5114.90 (8)				

II. $B \leftarrow X^1\Sigma^+$ System

Band heads, λ (intensity):

v', v''	0	1	2	3
0		4834.71(2)	4899.11(7)	4966.23(4)
1		4791.33(10)	4855.89(10)	4921.22(10)
2	4687.45(6)	4749.00(10)	4813.13(10)	4877.76(4)
3	4647.81(8)	4708.38(10)	4771.14(5)	4834.73(2)
4	4609.03(7)	4668.87(6)		
5	4570.88(9)	4629.62(1)		
6	4533.71(8)		4650.69(1)	
7	4497.21(8)		4611.66(1)	

PbSe

III. $C \leftarrow X^1\Sigma^+$ System

Band heads, λ (intensity), v' numbering is uncertain:

v', v''	0	1	2	0
0	4296.06(1)			3523.37
1	4263.09(2)	4314.64(2)	4365.15(2)	3500.00
2	4230.06(3)	4280.67(3)	4332.10(3)	3477.05
3	4197.75(4)	4247.29(2)	4297.93(2)	3454.78
4	4165.93(6)		4264.40(1)	3432.90
5	4134.87(7)	4182.52(2)		3411.12
6	4104.90(6)	4151.52(2)		
7	4074.73(2)	4120.82(3)		3368.62

IV. $D \rightleftarrows X^1\Sigma^+$ System

Band heads, λ :

v', v''	1	2
0	3557.94	3593.16
1	3534.16	3568.94
2	3510.80	3544.78
3	3488.34	3521.99
4	3466.01	3499.06
5	3443.86	
6	3421.99	
7	3400.49	

V. $F \leftarrow X^1\Sigma^+$ System

Band heads, λ :

v', v''	0	1	2
0	2211.94	2225.58	2239.33
1	2201.02	2214.55	2228.11
2	2190.0	2203.5	2217.15
3	2179.4	2192.7	
4	2169.0	2182.1	
5	2158.7		
6			
7			

SPECTROSCOPIC CONSTANTS

State	T_o	ω_e	$x_e\omega_e$	B_e	$\alpha_e \times 10^3$	$D_e \times 10^6$	r_e	Remarks	Bibliography
F	45194.5	224.8	0.50					$y_e\omega_e = 0.004$ cm^{-1}	(47.4, 46.3)
D	28374.4	190.4	0.53						(47.4, 44.2)
C	23268.5	183.0	0.25						(38.1)
B	20959.4	184.8	0.43						(38.1)
A	18661.5	166.9	0.14						(38.1)
X$^1\Sigma^+$	0	277.6	0.51	0.05059983	1.29		2.4022		(66.6, 38.1)

Dissociation energy = 2.5 ± 0.1 eV, 61.5 kcal/mole, 20164 cm^{-1}.

PbSe

Perturbations and General Information

Dipole moment $^{208}Pb^{80}Se$ = 3.28 ± 0.10 D (70.7).

BIBLIOGRAPHY

(38.1) A, B, C ← X Systems,
J. W. Walker, J. W. Straley, and A. W. Smith,
Phys. Rev. 53, 140-5

(44.2) D → X System,
D. Sharma,
Proc. Nat. Acad. Sci. A 14, 133-41

(46.3) F ← X System,
D. Sharma,
Nature 157, 663

(47.4) D-X, F ← X Systems,
E. E. Vago and R. F. Barrow,
Proc. Phys. Soc. 59, 449-57

(61.5) Dissociation Energy,
R. F. Porter,
J. Chem. Phys. 34, 583-7

(66.6) Microwave Spectra,
J. Hoeft and K. Manns,
Z. Naturforsch. A 21, 1884-9

(70.7) Stark Effect,
J. Hoeft, F. J. Lovas, E. Tiemann, and T. Törring,
"Electrical Dipole Moments of FeSe, GeSe, PbSe, and PbTe,"
Z. Naturforsch. 25, 539-41

PdB

Dissociation energy = 3.37 ± 0.22 eV, 77.7 kcal/mole, 27181 cm^{-1} (70.1).

BIBLIOGRAPHY

(70.1) A. V. Auwera-Mahieu, R. Peters, N. S. McIntyre, and J. Drowart, "Mass Spectrometric Determination of Dissociation Energies of the Borides and Silicides of some Transition Metals," Trans. Faraday 66, 809-16

PdH

Methods of Production and Experimental Technique

Absorption in a King furnace, $T > 3000^{\circ}C$.

Band Systems

Absorption is observed $6400 > \lambda > 3100 \text{Å}$, degrading R. Characteristic band at 4145. Bands are strongly perturbed.

BIBLIOGRAPHY

(64.1) A. Lagerqvist, H. Neuhaus, and R. Scullman, Proc. Phys. Soc. 83, 498-9

PdO

Dissociation energy = 2.91 ± 0.3 eV, 67 kcal/mole, 23471 cm^{-1} (68.3, 65.2, 64.1).

BIBLIOGRAPHY

(64.1) J. H. Norman, H. G. Staley, and W. E. Bell, J. Phys. Chem. 68, 662-3

(65.2) J. H. Norman, H. G. Staley, and W. E. Bell, J. Phys. Chem. 69, 1373-6

(68.3) J. H. Norman, H. G. Staley, and W. E. Bell, Advances Chem. Ser., U.S.A. 72, 101-14

PdSi

PdSi

Dissociation energy = 3.21 ± 0.14 eV, 74.0 kcal/mole, 25891 cm^{-1} (70.1).

BIBLIOGRAPHY

(70.1) A. V. Auwera-Mahieu, R. Peters, N. S. McIntyre, and J. Drowart, "Mass Spectrometric Determination of Dissociation Energies of the Borides and Silicides of some Transition Metals," Trans. Faraday Soc. 66, 809-16

PmS

Dissociation energy (estimated) = 4.38 ± 0.22 eV, 101 kcal/mole, 35328 cm^{-1} (69.1).

BIBLIOGRAPHY

(69.1) S. Smoes, P. Coppens, C. Bergman, and J. Drowart,
Trans. Faraday Soc. 65, 682-7

PrO

PrO

Dissociation energy = 7.73 ± 0.15 eV, 62348 cm^{-1}.

BIBLIOGRAPHY

(72.1) M. Shafi, C. L. Beckel, and R. Ergelke,
"Diatomic Molecule Ground State Dissociation Energies,"
J. Mol. Spectrosc. 42, 578-581

Dissociation energy = 5.21 ± 0.15 eV, 41222 cm^{-1}.

BIBLIOGRAPHY

(67.1) E. D. Cater, B. A. Holler, and J. A. Fries,
T. R. No. COO-1182-15

(72.2) M. Shafi, C. L. Beckel, and R. Ergelke,
"Diatomic Molecule Ground State Dissociation Energies,"
J. Mol. Spectrosc. 42, 578-581

PtB

Dissociation energy = 4.91 ± 0.18 eV, 113.3 kcal/mole, 39603 cm^{-1} (68.1).

BIBLIOGRAPHY

(68.1) N. S. McIntyre, A. V. Auwera-Mahieu, and J. Drowart, Trans. Faraday Soc. 64, 3006-10

PtC

Methods of Production and Experimental Technique

Absorption in a King furnace.

BAND SYSTEMS

System	Transition	Sources	Wavelength Limits	Degrading	Characteristic Bands, λ	Remarks	Bibliography
I	$A^1\Pi \leftarrow X^1\Sigma$	King furnace	6400-3100	R	5399.5 (0,0)		(66.2, 65.1)
II	$B^1\Sigma \leftarrow X^1\Sigma$	King furnace	6400-3100	R	3160.1 (0,1)	Perturbed	(66.2, 65.1)
III	$A'^1\Pi \leftarrow X^1\Sigma$	King furnace	8500-6000	R	7570 (0,0)		(67.4)
IV	$A''^1\Sigma \leftarrow X^1\Sigma$	King furnace	8500-6000	R	7899 (0,0)		(n.p. 5)

PtC

I. $A\,^1\Pi \leftarrow X\,^1\Sigma$ System

Band heads, λ:

v', v''	0	1	2
0	5399.5	5721.1	6079.8
1	5173.8		5795.5
2	4968.9	5240.1	
3			
4			

II. $B\,^1\Sigma \leftarrow X\,^1\Sigma$ System

Band heads, λ:

v', v''	0	1	2	3	4
0	3059.4	3160.1	3266.6	3379.5	
1	2984	3078		3285.7	3398.8
2	2912				

III. $A'\,^1\Pi \leftarrow X\,^1\Sigma$ System

Band heads, λ:

v', v''	0	1
0	7570	8217
1	7084	

IV. $A''\,^1\Sigma \leftarrow X\,^1\Sigma$ System

Band heads, λ:

v', v''	0	1	2	3
0	7899	8606		
1	7357	7968	8680	
2		7423	8038	8835
3			7490	
4				7559

SPECTROSCOPIC CONSTANTS

State	T_o	ω_e	$x_e\omega_e$	B_e	$\alpha_e \times 10^3$	$D_e \times 10^6$	r_e	Remarks	Bibliography
$A''^1\Sigma$	-	932[a]	-	0.5078[b]	-	0.62	-		(65.1)
$A'^1\Pi$	-	905[a]	-	0.5038[b]	-	0.55[c]	1.171[d]		(67.4)
$B^1\Sigma$	-	844[a]	-	0.4680	-	0.7	1.785	Perturbations	(66.2)
$A^1\Pi$	18510	818.8	5.5	0.4802	4.0	0.7	1.762		(66.2)
$X^1\Sigma$	0	1051.18	4.87	0.5303	3.2	0.55	1.677		(66.2)

[a] $\Delta G_{1/2}$, [b] B_o, [c] D_o, [d] r_o.

Dissociation energy = 6.30 ± 0.06 eV, 145.3 kcal/mole, 50814 cm^{-1} (67.3).

Perturbations and General Information

Perturbations of the $B^1\Sigma$ state:

        strongest $-$ $v' = 0$, $J' < 45$ and $J' > 85$

                               $v' = 1$, $J' > 55$

        weak $-$ $v' = 0$, $J' = 27$

                               $v' = 1$, $J' = 37$

        extremely weak $-$ $v' = 0$, $J' = 47, 55, 68, 71,$ and 93.

BIBLIOGRAPHY

(65. 1)   H. Neuhaus, R. Scullman, and B. Yttermo,
               Z. Naturforsch. A 20, 162

(66. 2)   R. Schullman and B. Yttermo,
               Arkiv. Fysik. 33, 231-54

(67. 3)   Dissociation Energy by Mass Spectra,
               A. Van der Auwera-Mahieu and J. Drowart,
               Chem. Phys. Letters 1, 311-3

(67. 4)   O. Appelblad, R. F. Barrow, and R. Schullman,
               Proc. Phys. Soc. 91, 260-1

(n. p. 5)   O. Appelblad and R. Schullman,
               (unpublished)

PtF

Methods of Production and Experimental Technique

Exploding a Pt wire in F_2 (73.1).

Lasing action has been observed through the above method between $11.1 \leq \lambda \leq 24$ μm (73.1).

BIBLIOGRAPHY

(73.1) W. W. Rice and W. H. Beattie,
"Metal Atom Oxidation Lasers,"
Chem. Phys. Letters 19, 82-5

PtH

Methods of Production and Experimental Technique

Absorption in a King furnace (T > 3000°C).

Exploding wire (64.1).

Discharge into a hollow cathode of platinum foil in hydrogen and argon (71.5).

BAND SYSTEMS

System	Transition	Sources	Wavelength Limits	Degrading	Characteristic Bands, λ	Remarks	Bibliography
I	$A^2\Delta_{5/2} - X^2\Delta_{5/2}$	King furnace	6400-3100	R	4547.16(0,0)		(65.3, 64.2)
	$A^2\Delta_{3/2} - X^2\Delta_{3/2}$	Hollow cathode	6400-3100	R	5088.58(0,0)		(71.5)
II	$C^2\phi_{7/2} - X^2\Delta_{5/2}$	Hollow cathode	4640-3958	R	4194.66(0,0)		(71.5)
III	$B^2\Delta_{5/2} - X^2\Delta_{5/2}$	King furnace	6400-3100	R	3751.67(0,0)		(65.3, 64.2)

I. $A^2\Delta - X^2\Delta$ System

$^2\Delta_{5/2} - {}^2\Delta_{5/2}$

Band heads (64.2):

(v', v'')	(0, 1)	(0, 0)	(1, 0)	(2, 0)
λ	5076.3	4547.16	4245.0	4004.45

$^2\Delta_{3/2} - {}^2\Delta_{3/2}$

Band heads (71.5):

(v', v'')	(0, 1)	(0, 0)	(1, 0)
λ	5720.93	5088.58	4729.45

II. $C^2\phi_{7/2} - X^2\Delta_{5/2}$ System

Band heads (71.5):

(v', v'')	(0, 1)	(0, 0)	(1, 0)
λ	4640.88	4194.66	3958

III. $B^2\Delta_{5/2} - X^2\Delta_{5/2}$ System

Band heads (64.2):

(v', v'')	(0, 1)	(0, 0)	(1, 0)
λ	4104.38	3751.67	3546.63

PtH

SPECTROSCOPIC CONSTANTS

State	T_o	ω_e (a)	$x_e\omega_e$	B_e	α_e	$D_e \times 10^4$	r_e	Remarks	Bibliography
$B^2\Delta_{5/2}$	26606.76	1546.97		6.004	0.301	3.15	1.673		(65.3, 64.2)
$A^2\Delta_{5/2}$	23856.85	1425.9		5.595[b]		3.6[c]			(71.5)
$C^2\phi_{7/2}$	21951.60	1566.21	72.85	5.670	0.262	3.55	1.722		(65.3, 64.2)
$A^2\Delta_{3/2}$	19364.69	1498.93		5.97[b]		4.8[c]			(71.5, 65.3, 64.2)
$X^2\Delta_{3/2}$	615	2176.50		7.18[b]		3.1[c]			(65.3, 64.2)
$X^2\Delta_{5/2}$	0	2293.93		7.198	0.198	2.63	1.528	$A = -615$ cm^{-1}	(65.3, 64.2)

[a]$\Delta G_{1/2}$, [b]B_o, [c]D_o.

Dissociation energy = 3.6 ± 0.4 eV, 83 kcal/mole, 29037 cm^{-1}.

Perturbations and General Information

$B^2\Delta_{5/2}$ state in the $v' = 0$ level, $J' = 13.5, 15.5, 16.5$ levels are divided into two components.

$B^2\Delta_{5/2}$ state there are predissociations for $v' = 0$, $J' = 17.5, 18.5, > 21.5$; $v' = 1$, $J' \geq 7.5$.

BIBLIOGRAPHY

(64.1)   V. A. Loginov,
Optics Spectr. 16, 220-3

(64.2)   A, B-X Systems,
H. Neuhaus and R. Scullman,
Z. Naturforsch. A 19, 659-60

(65.3)   A, B-X Systems,
R. Scullman,
Arkiv. Fysik. 28, 255-65

(65.4)   V. A. Loginov,
"Absorption Spectra of PtH and PtD,"
Optics Spectr. 18, 88

(71.5)   B. Kaving and R. Scullman,
"Two New Subsystems of PtH in the Region 3500-5800 Å,"
Can. J. Phys. 49, 2264-75

PtO

Methods of Production and Experimental Technique

Emission from a hollow cathode discharge, high voltage arc.

Band Systems

$A^1\Sigma - X^1\Sigma$ system observed $7000 > \lambda > 4000$ Å, degrading R.

Band heads, λ:

v', v''	0	1	2	3
0	5902.3	6210.6	6548.5	
1	5663.0	-	6255	6594
2	5445.6	5707		
3	5248	5489		

SPECTROSCOPIC CONSTANTS

State	T_e	ω_e	$x_e\omega_e$	B_e	$\alpha_e \times 10^3$	$D_e \times 10^6$	r_e	Remarks	Bibliography
$A^1\Sigma$	16932.99	727.07	5.42	0.35385	2.91	0.327			(n. p. 5)
$X^1\Sigma$	0	851.09	4.97	0.38223	2.83	0.296	1.727		(n. p. 5)

Dissociation energy = 3.84 ± 0.17 eV, 88 kcal/mole, 30972 cm^{-1} (68.4, 67.3).

BIBLIOGRAPHY

(50. 1)   Arc Spectra,
M. W. Feast,
Proc. Phys. Soc. A 63, 549-56

(65. 2)   Arc Spectra,
V. Raziunas, G. J. Macur, and S. Katz,
J. Chem. Phys. 43, 1010-5

(67. 3)   Dissociation Energy,
J. H. Norman, H. G. Staley, and W. E. Bell,
J. Phys. Chem. 71, 3686-9

(68. 4)   Dissociation Energy,
J. H. Norman, H. G. Staley, and W. E. Bell,
Advances Chem. Ser. 72, 101-14

(n.p. 5)   Emission Spectra,
C. Nilsson, R. Scullman, and N. Mehendale,
(Unpublished)

PtSi

Dissociation energy = 5.15 ± 0.19 eV, 118.8 kcal/mole, 41538 cm^{-1} (70.1).

BIBLIOGRAPHY

(70.1) A. V. Auwera-Mahieu, R. Peters, N. S. McIntyre, and J. Drowart, "Mass Spectrometric Determination of Dissociation Energies of the Borides and Silicides of some Transition Metals," Trans. Faraday Soc. 66, 809-16

PuF

PuF

Dissociation energy = 5.54 ± 0.30 eV, 127.7 kcal/mole, 44684 cm^{-1} (68.1).

BIBLIOGRAPHY

(68.1) R. A. Kent,
J. Am. Op. Soc. 90, 5657-9

ReO

Band Systems

Bands have been observed in both a.c. arcs (57.1) and d.c. arcs (65.2). Most bands are degraded R and are in the region $8800 > \lambda > 5750$Å. Analysis uncertain.

Band heads (65.2, 57.1).

$$\lambda \mid 7119.1 \mid \sim 7110 \mid 6443.8 \mid 6092.6 \mid 6077.6 \mid 6060.9$$

BIBLIOGRAPHY

(57.1)   λ Listing,
A. Gatterer, J. Junks, E. W. Salpeter, and B. Rosen,
"Molecular Spectra of Metallic Oxides,"
Ed. Specola Vaticana (1957)

(65.2)   Wavelength Ordering,
V. Raziunas, G. J. Macur, and S. Katz,
J. Chem. Phys. 43, 1010-5

RhB

Dissociation energy = 4.89 ± 0.22 eV, 112.8 kcal/mole, 39441 cm^{-1} (70.1).

BIBLIOGRAPHY

(70.1) A. V. Auwera-Mahieu, R. Peters, N. S. McIntyre, and J. Drowart, "Mass Spectrometric Determination of Dissociation Energies of the Borides and Silicides of some Transition Metals," Trans. Faraday Soc. 66, 809-16

RhC

Methods of Production and Experimental Technique

Absorption in a King furnace (T> 3000°C).

BAND SYSTEMS

System	Transition	Sources	Wavelength Limits	Degrading	Characteristic Bands, λ	Remarks	Bibliography
I	$A^2\Pi_r \leftarrow X^2\Sigma^+$	Ne matrix, King furnace	12250-6600	R	$\Pi_{1/2}$: 10488(0,0) $\Pi_{3/2}$: 9698(0,0)		(71.L1) (69.5)
II	$B^2\Sigma^+ \leftarrow X^2\Sigma^+$	King furnace	6600-4000	R	4695.5((0),0)	B and C states perturb each other strongly	(66.2, 65.1)
III	$C^2\Sigma^+ \leftarrow X^2\Sigma^+$	King furnace	6600-4000	R	4658.7(0,0)		(66.2, 65.1)
IV	$?^2\Sigma \leftarrow X^2\Sigma$	Ne matrix			4440(0,0)		(71.L1)
V	$?^2\Sigma \leftarrow X^2\Sigma$	Ne matrix			4487(n,0)	Could be possible d-X transition	(71.L1)

I. $A^2\Pi_r \leftarrow X^2\Sigma^+$ System

$^2\Pi_{1/2} - {}^2\Sigma$

Band heads, λ:

v', v''		0	1	2	3	4
0	Q	10617.4	11936			
	R	10591.1	11899			
1	Q	9655.3	10733.1	12068		
	R	9636.5	10706.8	12030		
2	Q	8861	9761.0	10852		
	R	-	9742.2	-		
3	Q		8961	9870		
	R		8947	9850		
4	Q		8287	9058		
	R		-	9047		
5	Q			8379		
	R			8369		
6	Q				8477	
7	Q					8573

$^2\Pi_{3/2} - {}^2\Sigma$

Band heads, (Q) λ:

v', v''	0	1	2	3	4
0	9810.7	10925.2			
1	8991.9	9920	11046		
2	8306.7	9093		11173	
3		8400	9194		11307
4			8493		
5				8592	

II. $B^2\Sigma^+ \leftarrow X^2\Sigma^+$ System

Band heads (R):

(v', v'')	(0, 0)	(0, 1)
λ	4695.4	4936.4

III. $C^2\Sigma^+ \leftarrow X^2\Sigma^+$ System

v', v''	0	1	2	3
0	4658.7	4895.8		
1	4484.9		4943.7	
2	4317.7	4520.8		
3	4165.7			4781.1

IV. $?^2\Sigma \leftarrow X^2\Sigma$ System

v', v''	0
0	4440
1	4315
2	4173

V. $?^2\Sigma \leftarrow X^2\Sigma$ System

(n, 0)	4487

SPECTROSCOPIC CONSTANTS

State	T_e	ω_e	$x_e\omega_e$	B_e	$\alpha_e \times 10^3$	$D_e \times 10^6$	r_e	Remarks	Bibliography
$D^2\Sigma^-$	≤21623	782[a]		0.482[b]	-	-	1.804[b]	$\gamma \sim -1.6$ cm^{-1}	(66.2)
$C^2\Sigma^+$	21452.0	927.8	13.7	0.5510	6.0	~1.0	1.692	$\gamma = -0.03$ cm^{-1}	(66.2)
$B^2\Sigma^+$	21285.0	-		0.5067[c]	-	~0.8	1.760[d]	$\gamma = 1.03$ cm^{-1}	(66.2, 65.1)
$A^2\Pi_{3/2}$	10187.24	932.12	5.48	0.57149	4.28	0.832	-	$\beta_e = 0.32$ cm^{-1} $y_e\omega_e = 0.021$ cm^{-1}	(69.5, 65.1)
$A^2\Pi_{1/2}$	9412.60	949.41	5.357	0.57329	4.26	0.826	-	$\beta_e = -0.19$ cm^{-1}	(69.5, 65.1)
$X^2\Sigma^+$	0	1049.87	4.94	0.6027	3.96	0.78	1.614	$\gamma = -0.065$ cm^{-1}	(69.5, 65.1)

[a] $\Delta G_{1/2}$, [b] B_e and r_e are for the lowest observed vibrational level, [c] B_0, [d] r_0, [e] state is observed through perturbations of the C state.

Dissociation energy = 6.01 ± 0.06 eV, 138.5 kcal/mole, 48475 cm^{-1} (67.3).

Perturbations and General Information

Lower vibrational levels of the B and C states perturb each other strongly (66.2).

Spectroscopic constants of the D state are derived from the perturbation of the vibrational levels of the C state assuming a $^2\Sigma^+ - {}^2\Sigma^-$ perturbation (66.2).

Franck-Condon factors, r-centroids — Morse potential (67.4).

$C^2\Sigma^+ - X^2\Sigma^+$:

v', v''	0	1	2	3
0	0.353 1.658	0.402 1.699		
1	0.349 1.624		0.281 1.708	
2	0.176 1.593	0.070 1.622		
3	0.086 1.677			0.160 1.677

Top = Franck-Condon factors
Bottom = r-centroids

BIBLIOGRAPHY

(65.1) A, B-X Systems,
A. Lagerqvist, H. Neuhaus, and R. Scullman,
Z. Naturforsch. A 20, 751-2

(66.2) C, D-X Systems,
A. Lagerqvist and R. Scullman,
Arkiv. Fysik. 32, 479-508

(67.3) Dissociation Energy,
A. V. Auwera-Mahieu and J. Drowart,
Chem. Phys. Letters 1, 311-3

RhC

(67.4) Franck-Condon Factors, r-Centroids, C-X,
V. M. Korwar,
Proc. Phys. Soc. 92, 523

(69.5) A-X System,
B. Kaving and R. Scullman,
J. Mol. Spectrosc. 32, 475-500

(71.L1) J. M. Brom, Jr., W. R. M. Graham, and W. Weltner, Jr.,
"ESR and Optical Spectroscopy of the RhC Molecule at 4°K,"
J. Chem. Phys. 57, 4116-24

RhO

Band Systems

Bands degrading R are observed in a dc arc (65.2):

λ | 6592.0 | 6563.5 | 6373.3 | 6268.1 | 6228.9 | 6173.3 | 6165.6 | 5878.5

Dissociation energy = 4.20 ± 0.3 eV, 97 kcal/mole, 33876 cm^{-1} (68.3, 64.1).

BIBLIOGRAPHY

(64.1) Thermochemistry,
J. H. Norman, H. G. Staley, and W. E. Bell,
J. Phys. Chem. 68, 662-3

(65.2) Arc Spectra,
V. Raziunas, G. J. Macur, and S. Katz,
J. Chem. Phys. 43, 1010-5

(68.3) Dissociation Energy,
J. H. Norman, H. G. Staley, and W. E. Bell,
Advances Chem. Ser. 72, 101-14

RhSi

Dissociation energy = 4.05 ± 0.19 eV, 93.4 kcal/mole, 32666 cm^{-1} (70.1).

BIBLIOGRAPHY

(70.1) A. V. Auwera-Mahieu, R. Peters, N. S. McIntyre, and J. Drowart, "Mass Spectrometric Determination of Dissociation Energies of the Borides and Silicides of some Transition Metals," Trans. Faraday Soc. <u>66</u>, 809-16

RuB

Dissociation energy = 4.59 ± 0.22 eV, 105.9 kcal/mole, 37022 cm^{-1} (70.1).

BIBLIOGRAPHY

(70.1) A. V. Auwera-Mahieu, R. Peters, N. S. McIntyre, and J. Drowart, "Mass Spectrometric Determination of Dissociation Energies of the Borides and Silicides of some Transition Metals," Trans. Faraday Soc. **66**, 809-16

RuC

Methods of Production and Experimental Technique

Absorption and emission in a King furnace (T ~ 3000°C).

Band Systems

Emission has been observed in two regions: $8700 > \lambda > 6000$ Å and $4800 > \lambda > 4100$ Å.

The red system is composed of eight subsystems, while the violet is composed of three. Both systems have a common lower level and the violet system is probably a triplet-triplet transition.

Band heads — red system:

v', v''	0	1	2
0	7224		
1	6764	7269	
2		6809	7316

Band heads — violet system:

(v', v'')	(1,0)	(0,0)	(0,1)	(0,2)	
λ	4182	4317	4518	4736	4317 Å system
	4239	4383	4588		4383 Å system
	4201				4337 Å system

SPECTROSCOPIC CONSTANTS

State	T_e	ω_e	$x_e\omega_e$	B_o	$\alpha_e \times 10^3$	$D_e \times 10^6$	r_e	Remarks	Bibliography
4317Å				0.5200		0.95			(72.3)
4337Å		775.25[a]		0.5133		0.97			(72.3)
4383Å				0.5273		1.0			(72.3)
X		1038.77	4.64	0.5860		0.75			(72.3)

[a] $\Delta G_{1/2}$.

Dissociation energy = 6.54 ± 0.14 eV, 151 kcal/mole, 52750 cm^{-1}.

BIBLIOGRAPHY

(68. 1)   N. S. MacIntyre, A. V. Auwera-Mahieu, and J. Drowart,
          Trans. Faraday Soc. 64, 3006-10

(71. 2)   R. Scullman and B. Thelin,
          Physica Scripta 3, 19

(72. 3)   R. Scullman and B. Thelin,
          "The Spectra of RuC in the Region 4100-4800 Å,"
          Physica Scripta 5, 201-8

RuO

Bands observed 6710 > λ > 5300Å, degrading R (65.1).

Most intense band (0, 0) 5525.85; however, incomplete analysis.

Possibly $^3\Sigma$ - $^3\Sigma$ transition.

RuO

SPECTROSCOPIC CONSTANTS

State	T_e	ω_e	$x_e\omega_e$	B_o	$\alpha_e \times 10^3$	$D_e \times 10^6$	r_o	Remarks	Bibliography
?	18094 + a	878.1	22.4	0.39			1.76		(65.1)
?	0	880.8	13.1	0.42			1.70		(65.1)

Dissociation energy = 5.33 ± 0.30 eV, 123 kcal/mole, 42990 cm^{-1} (68.2).

BIBLIOGRAPHY

(65.1) Spectral Reproduction, Incomplete Analysis,
V. Raziunas, G. J. Macur, and S. Katz,
J. Chem. Phys. 43, 1010-5

(68.2) Dissociation Energy,
J. H. Norman, H. G. Staley, and W. E. Bell,
Advances Chem. Ser. 72, 101-14

RuSi

RuSi

Dissociation energy = 4.07 ± 0.22 eV, 93.9 kcal/mole, 32827 cm^{-1} (70.1).

BIBLIOGRAPHY

(70.1)   A. V. Auwera-Mahieu, R. Peters, N. S. McIntyre, and J. Drowart, "Mass Spectrometric Determination of Dissociation Energies of the Borides and Silicides of some Transition Metals," Trans. Faraday Soc. 66, 809-16

SF

Methods of Production and Experimental Technique

Reaction of COS with F atoms produced by microwave discharge in CF_4 (70.3).

BAND SYSTEMS

	System	Transition	Sources	Wavelength Limits	Degrading	Characteristic Bands, λ	Remarks	Bibliography
	I	$A^2\Pi - X^2\Pi$	Absorption	4000-3300	-	$^2\Pi_{3/2}$: 24991 (0, 0)		(70.3)

I. $A^2\Pi \leftarrow X^2\Pi$ System

$^2\Pi_{3/2} - {}^2\Pi_{3/2}$

Band heads:

(v', v'')	(0, 0)	(1, 0)	(2, 0)	(3, 0)
λ	24991	25477.0	25951.5	26420.7

$^2\Pi_{1/2} - {}^2\Pi_{1/2}$

Band heads:

(v', v'')	(1, 0)	(2, 0)	(3, 0)	(4, 0)
λ	25676	26141	26622.4	27085.9

SPECTROSCOPIC CONSTANTS

State	T_o	ω_e	$x_e\omega_e$	B_o	$\alpha_e \times 10^3$	$D_e \times 10^6$	r_o	Remarks	Bibliography
$A^2\Pi_{1/2}$	25351	483	2.6					$A = -599$ cm^{-1}	(70.3)
$A^2\Pi_{3/2}$	24752.0	488	3.1						(70.3)
$X^2\Pi$	0			0.5527			1.599	$A = -398$ cm^{-1}	(70.3, 69.2)

Dissociation energy = 3.51 ± 0.09 eV, 81 kcal/mole, 28310 cm^{-1} (73.5)

BIBLIOGRAPHY

(67.1) ESR Spectra,
A. Carrington, G. N. Currie, P. N. Dyer, D. H. Levy, and T. A. Miller,
Chemical Comm. 641-2

(69.2) ESR Spectra,
A. Carrington, G. N. Currie, T. A. Miller, and D. H. Levy,
"Gas-Phase Electron Resonance Spectra of SF and SeF,"
J. Chem. Phys. 50, 2726-32

(70.3) A-X Systems,
G. Dilonarde and A. Trombetti,
"Spectrum of SF,"
Trans. Faraday Soc. 66, 2694-8

(70.4) SCF Calculations of Physical Observiables,
P. A. G. O'Hare and A. C. Wahl,
"Hartree-Fock Wavefunctions and Computed Properties for the Ground ($^2\Pi$) States of SF and SeF and their Positive and Negative Ions. A Comparison of the Theoretical Results with Experimental,"
J. Chem. Phys. 53, 2834-46

(73.5) Dissociation Energy,
D. L. Hildenbrand,
J. Phys. Chem. 77, 897

SO

Methods of Production and Experimental Technique

Microwave discharge into O_2 and the resulting species reacted with OCS (71.36).

Absorption in flash-photolysis of SO_2, SO_3 ($H_2S + O_2$), ($CS_2 + O_2$) (69.33, 64.20, 62.17, 58.15, 57.13), ($COS + O_2$) (69.33).

Emission from a noncondensed discharge (69.33, 34.8) with microwave discharge into a stream of SO_2 flames (47.10) chemiluminescence (57.14).

Afterglow from a microwave discharge into ($COS + O_2$) (68.32).

Microwave discharge through oxygen and sulfur vapor (68.31).

Flash-photolysis of SO_2 in He (71.39).

BAND SYSTEMS

System	Transition	Sources	Wavelength Limits	Degrading	Characteristic Bands, λ	Remarks	Bibliography
I	$b^1\Sigma^+ \rightarrow X^3\Sigma^-$	Microwave afterglow	12400-9300	R	10469.34(0,0) 11523.51(1,0)		(72.L2, 68.32)
II	$B^3\Sigma^- \rightleftarrows X^3\Sigma^-$	Discharge	4570-1900	R	3271.0(0,10) 3164.7(0,9) 3064.1(0,8)		(69.33, 64.20, 63.19, 62.17, 58.15, 57.13, 57.14, 47.10)
III	$A^3\Pi \rightleftarrows X^3\Sigma^-$	Discharge	2630-2400	R	2578.2 2568.0 (2,0) 2558.5		(69.33, 34.8)
IV	$D \leftarrow X^3\Sigma^-$	Absorption				Rydberg-state [a]	(69.35)
V	$E \leftarrow X^3\Sigma^-$	Absorption	~1470			Rydberg-state [a]	
VI	$F \leftarrow X^3\Sigma^-$	Absorption	1420.4-1388.6			Rydberg-state [a]	(71.39)

SO

I. $b^1\Sigma^+ \to X^3\Sigma^-$ System

Band heads (68.32):

(v', v'')	(0, 1)	(0, 0)	(1, 1)	(2, 2)
λ	10708	9549.08	9626.13	9707.4

II. $B^3\Sigma^- \rightleftarrows X^3\Sigma^-$ System

Band heads, λ (69.33):

v', v''	0	1	2	3	4	5	6
0			2555.5	2630.1	2708.6	2791.33	2877.66
1			2516.4	2589.0	2664.8	2744.02	2827.41
2			2477.7	2548.6	2622.2	2699.10	2779.79
3	2313.64	2376.34	2441.55	2509.94	2581.2	2655.64	
4	2282.52	2343.40	2406.95	2473.27			
5	2252.31	2311.55	2373.28				
6	2223.51	2281.42	2341.49	2404.13			
7	2195.8						
8	2169.62						
9	2144.04						
10	2119.68						

III. $A^3\Pi \rightleftarrows X^3\Sigma^-$ System

Band heads λ (69.33):

	$^3\Pi_0 - X^3\Sigma^-$		$^3\Pi_1 - X^3\Sigma^-$		$^3\Pi_2 - X^3\Sigma^-$	
v', v''	0	1	0	1	0	1
0	2634.70		2623.33		2613.37	
1	2605.50		2595.09		2585.40	
2	2578.07		2568.02		2558.54	
3	2551.42	2627.73	2541.71	2617.44	2532.53	
4	2525.51	2600.21	2516.11	2590.31	2507.21	
5		2573.49	2491.25	2564.06	2482.59	
6		2547.98	2467.29	2538.35	2458.73	2529.25
7		2523.81		2513.83		2505.66
8				2492.77		2485.34

SPECTROSCOPIC CONSTANTS

State	T_e	ω_e	$x_e\omega_e$	B_e	$\alpha_e \times 10^3$	$D_e \times 10^6$	r_e	Remarks	Bibliography
$C^3\Pi_i$	~42200	~170		0.501	-	24.2	~2.2		(70.37, 69.33, 68.31, 32.6)
$B^3\Sigma^-$	41628.7	630.4[b]	4.8[b]	0.502	6.2	1.28	1.775	$\lambda_0 = 3.5$ cm^{-1} $\gamma_0 = 0.010$ cm^{-1} [a]	(68.31, 58.15, 32.6)
$A^3\Pi_2$	38616.5	412.7	1.7	0.6164	20.4	4.8[c]	-	$\gamma_e = 0.0009$ cm^{-1}	(69.33)
$A^3\Pi_1$	38455.2	413.3	1.6	0.6107	19.4	4.0[c]	1.6094	$\gamma_e = 0.0009$ cm^{-1}	(69.33)
$A^3\Pi_0$	38292.5	415.2	1.6	0.6067	19.4	3.7[c]	-	$\gamma_e = 0.0009$ cm^{-1}	(69.33)
$b^1\Sigma^+$	10509.97	1068.66	7.25	0.70262	6.35	1.2	1.5004		(72.L2, 71.41, 68.32)
$a^1\Delta$	~6350			0.7098	-	-	1.494		(71.40, 71.44, 70.38, 68.32, 66.25)
$X^3\Sigma^-$	0	1149.22	5.63	0.7208	5.74	1.07	1.49198		(71.41, 67.28, 67.29, 58.15)

[a] derived from perturbations in $B^3\Sigma^-$ state, [b] approximation, [c] for v = 2.

Dissociation energy = 3.357 ± 0.003 eV, 123.57 kcal/mole, 27077 cm^{-1}.

SO

SO

Perturbations and General Information

$B^3\Sigma^-$ state strongly perturbed by the $C^3\Pi$ state (70.37, 69.33, 64.20).

$A^3\Pi$ state perturbed at v' = 0 and v' > 4 (69.33).

Predissociation of v = 0-3 levels of B state caused by C state (70.37, 69.33).

Ionization potential = 10.34 ± 0.02 eV (71.42).

Rate of production of SO ($a^1\Delta$) (72.45):

$$O_2(a^1\Delta) + SO(X^3\Sigma^-) \rightarrow SO(a^1\Delta) + O_2(X^3\Sigma^-)$$

$$k = (2.12 \pm 0.22) \times 10^8 \text{ liter/mole-sec.}$$

Franck-Condon factors, r-centroids — Morse potential:

$b^1\Sigma^+ - X^3\Sigma^-$ (70.36)

v', v''	0	1	2	3
0	0.937 1.497 9549.08	0.062 1.645 10708.00	0.002 1.810	0.000 1.975
1	0.062 1.508	0.815 1.508 9625.13	1.654	1.805
2	0.003 1.320	0.111 1.320	0.700 1.519 9707.40	1.659
3	0.000		1.358	0.287 1.529

Top = Franck-Condon factor
Middle = r-centroid
Bottom = Wavelength

$B^3\Sigma^- - X^3\Sigma^-$ - RKR potential (71.43)

v'	0	1	2	3
v'' = 0	0.337-4 1.620	0.251-3 1.612	0.976-3 1.605	0.264-2 1.598
1	0.120-3 1.609	0.807-3 1.620	0.287-2 1.612	0.713-2 1.603
2	0.869-3 1.645	0.487-2 1.636	0.142-1 1.627	0.287-1 1.618

v'	4	5	6	7
v'' = 0	0.556-2 1.591	0.978-2 1.584 0.728	0.149-1 1.577 1.160	0.203-1 1.569 1.790
1	0.140-1 1.595	0.229-1 1.588	0.327-1 1.580	0.419-1 1.573
2	0.450-1 1.610	0.581-1 1.602	0.638-1 1.594	0.608-1 1.586

v'	8	9	10
v'' = 0	0.253-1 1.562 2.13	0.293-1 1.554 2.57	0.322-1 1.547 2.92
1	0.488-1 1.566	0.527-1 1.560	0.531-1 1.553
2	0.505-1 1.578	0.364-1 1.571	0.222-1 1.563

Top = Franck-Condon factor
Bottom = r-centroid

SO

Radiative lifetimes — $A^3\Pi - X^3\Sigma^-$ — $\tau = 12.4 \pm 2.5$ nsec (72. L1),

$B^3\Sigma^- - X^3\Sigma^-$ — $\tau = 16.2$ nsec for $v' = 2$ (69. 34):

Franck-Condon Factors and f-Values for Absorption from the $v'' = 0$ Level of $X^3\Sigma^-$ to the $A^3\Pi$ State (72. L1).

$v'' = 0$

v'	$^3\Pi_0$	$^3\Pi_1$	$^3\Pi_2$
0	0.107	0.127	0.154
1	0.177	0.190	0.206
2	0.195	0.200	0.201
3	0.173	0.170	0.162
4	0.133	0.127	0.115
5	0.091	0.081	0.074
6	0.057	0.050	0.043
7	0.033	0.029	0.024
8	0.018	0.015	0.012
9	0.009	0.007	0.006
f_{00}	0.007	0.009	0.010
f_{10}	0.016	0.017	0.018
f_{20}	0.017	0.018	0.018
f_{30}	0.014	0.014	0.013
f_{40}	0.009	0.009	0.008
f_{50}	0.005	0.004	0.004
f_{60}	0.002	0.002	0.002

BIBLIOGRAPHY

(06. 1) Emission,
F. Lowater,
Astrophys. J. 23, 324-42

(20. 2) Emission,
W. H. Bair
Astrophys. J. 52, 301-16

(29. 3) Emission,
W. Kessel,
C. R. Soc. Polan. Phys. 4, 175-82

(29. 4) Emission,
V. Henri and F. Wolf,
J. Phys. Radium 10, 81-106

(31. 5) Preliminary Note to (32.6),
E. V. Martin and F. A. Jenkins,
Phys. Rev. 37, 226

(32. 6) Emission,
E. V. Martin,
Phys. Rev. 41, 167-93

(32. 7) Preliminary Note to (32.6),
E. V. Martin and F. A. Jenkins,
Phys. Rev. 39, 549

(34. 8) Emission,
B. N. Bhaduri,
Thesis, Londres

(36. 9) Thermochemistry,
P. Golfinger, W. Jeunehomme, and B. Rosen,
Nature 138, 205-6

(47. 10) Flames,
A. G. Gaydon and G. Whittingham,
Proc. Roy. Soc. A 189, 313-25

(54. 11) Thermochemistry,
G. Saint-Pierre and J. Chipman,
J. Am. Chem. Soc. 76, 4787-91

SO

(56.12) Thermochemistry,
G. Saint-Pierre and J. Chipman,
J. Metals 8, 1474-83

(57.13) Absorption,
A. L. Myerson, F. R. Taylor, and P. L. Hanst,
J. Chem. Phys. 26, 1309-20

(57.14) Chemiluminescence,
A. D. Walsh,
"The Threshold of Space,"
Publ. Symposium Publications Division, Pergamon Press, 165-7

(58.15) Absorption,
R. G. W. Norrish and G. A. Oldershaw,
Proc. Roy. Soc. A 249, 498-512

(58.16) Thermochemistry,
E. W. Dewing and F. D. Richardson,
Trans. Faraday Soc. 54, 679-84

(62.17) Preliminary Note to (64.20),
W. D. MacGrath and J. J. MacGarvey,
J. Chem. Phys. 37, 1574-5

(63.18) Comment on (64.20),
D. Abadie and R. Herman,
C. R. Acad. Sci. 257, 2820-2

(63.19) Absorption,
J. Akriche,
J. Chim. Phys. 60, 732-6

(64.20) Absorption,
J. J. MacGarvey and W. D. MacGrath,
Proc. Roy. Soc. A 278, 490-504

(64.21) Thermochemistry,
R. Colin, P. Goldfinger, and M. Jeunehomme,
Trans. Faraday Soc. 60, 306-16

(65.22) Potential Energy Curves,
A. N. Singh and D. K. Rai,
J. Chem. Phys. 43, 2151-2

(65.23) C. C. McDonald and R. G. Goll,
"Studies of Gaseous Atom-Molecule Reactions by Electron Paramagnetic Resonance Spectroscopy,"
J. Phys. Chem. 69, 293-7

(66.24) ESR Spectra,
J. M. Daniels and P. B. Dorain,
J. Chem. Phys. 45, 26-34

(66.25) ESR Spectrum,
A. Carrington, D. H. Levy, and T. A. Miller,
Proc. Roy. Soc. A 293, 108-16

(66.26) Review of Dissociation Energy,
J. Drowart and P. Goldfinger,
Quarterly Rev. 20, 545-57

(67.27) Theory,
K. Kayama and J. C. Baird,
J. Chem. Phys. 46, 2604-18

(67.28) Microwave Spectra,
I. Amano, E. Hirota, and Y. Morino,
J. Phys. Soc. 22, 399-412

(67.29) ESR Spectra,
A. Carrington, D. H. Levy, and T. A. Miller,
Proc. Roy. Soc. A 298, 340-58

(67.30) F. Jenc,
"Ground State Reduced Potential Curves of Diatomics Molecules,"
J. Mol. Spectrosc. 24, 284-9

(68.31) K. V. S. R. Apparao and N. A. Narasimham,
"Isotope Shift Studies of the $B^3\Sigma$-$X^3\Sigma^-$ Bands of SO,"
Proc. Ind. Acad. Sci. 68, 173-7

(68.32) Emission,
R. Colin,
Can. J. Phys. 46, 1539-46

(69.33) Absorption,
K. Dressler,
Can. J. Phys. 47, 547-61

SO

(69.34) Radiative Lifetime, B-X,
W. H. Smith,
"Absolute Transition Probabilities for Some Electronic States of CS, SO and S_2,"
J. Quant. Spectrosc. Radiative Transfer 9, 1191-9

(69.35) Rydberg States,
R. J. Donovan, D. Husain, and P. T. Jackson,
Trans. Faraday Soc. 65, 2930

(70.36) M. M. Shukla and I. D. Singh,
"Franck-Condon Factors and r-Centroids for the b-X System of SO Molecules,"
Current Science 39, 345-6

(70.37) D. Abadie,
"Perturbation Analysis of the Rotational Structure of the v=1 Level of the $B^3\Sigma^-$ State of the SO Molecule. Identification and Constants of the State Perturbed by $C^3\Pi_i$,"
Ann. Physique 5, 227-38

(70.38) ESR Spectrum of $a^1\Delta$ State,
S. Saito,
"Microwave Spectrum of the SO Radical in the First Electronically Excited State, $^1\Delta$,"
J. Chem. Phys. 53, 2544-5

(71.39) Rydberg States,
R. S. Donovan and D. J. Little,
"Vacuum Ultraviolet Spectrum of the SO Radical,"
Spectrosc. Letters 4, 213-215

(71.40) ESR Spectrum of $a^1\Delta$ State,
T. A. Miller,
"Sulfur-33 Hyperfine Interactions in the Gas-Phase Electron Resonance Spectra of ($^2\Pi$) SH and ($^1\Delta$) SO,"
J. Chem. Phys. 54, 1658-64

(71.41) b-X Spectrum in Flames,
A. M. Bouchoux, J. Marchand, and J. Janin,
"Contribution to the Study of the $b^1\Sigma^+ - X^3\Sigma^-$ System of the SO Molecule,"
Spectro. Chim. Acta A 27, 1909-15

(71.42) Ionization Potential,
N. Jonathan, D. J. Smith, and K. J. Ross,
"The High Resolution Photoelectronic Spectra of Transient Species: Sulfur Monoxide,"
Chem. Phys. Letters 9, 217-8

(71.43) W. H. Smith, H. S. Liszt,
"Franck-Condon Factors and Absolute Oscillator Strengths for NH, SiH, S_2 and SO,"
J. Quant. Spectrosc. Radiative Transfer 11, 45-54

(71.44) ESR Spectrum of $a^1\Delta$ State,
H. Uehara,
"Gas Phase E.P.R. Spectra of SO in the $^1\Delta$ State,"
Molecular Phys. 21, 407-15

(72.45) $a^1\Delta$ Production,
W. H. Breckenridge and T. A. Miller,
"Kinetic Study by EPR of the Production and Decay of SO ($^1\Delta$) in the Reaction of $O_2(^1\Delta_g)$ with SO($^3\Sigma^-$),"
J. Chem. Phys. 56, 465-474

(72.46) A. M. Bouchoux, J. Marchand, and J. Janin,
"Vibrational Temperature of the SO Radical in Hydrogen Sulfide-Oxygen Flames,"
C.R. Acad. Sci. B 274-256-8

(72.L1) A-X Franck-Condon Factors, Oscillator Strengths, Radiative Lifetimes,
W. H. Smith,
"The Oscillator Strengths of the SO $A^3\Pi - X^3\Sigma^-$ Band Systems,"
Astrophys. J. 176, 265-6

(72.L2) A. M. Bouchoux and M. J. Marchand,
"Rotational Structure of the (2,0), (3,1) and (4,2) Bands of the $b^1\Sigma^+ - X^3\Sigma^-$ System of SO,"
Spectro. Chim. Acta 28A, 1771-3

SbBr

SbBr

Methods of Production and Experimental Technique

Emission from a microwave discharge in SbBr$_3$ vapor.

BAND SYSTEMS

System	Transition	Sources	Wavelength Limits	Degrading	Band Head, $v_{0,0}$	Remarks	Bibliography
I	A($^3\Sigma^-$) → X($^3\Sigma^-$)	Microwave discharge	5340-4905	R			(63.1)
II	B$_1$($^3\Pi$) → X($^3\Sigma^-$)	Microwave discharge	3340-3050	R	32354(0,0)		(71.2, 63.1)
	B$_2$($^3\Pi$) → X($^3\Sigma^-$)	Microwave discharge	3020-2950	R	33719(0,0)		

II. $\underline{B(^3\Pi) \rightarrow X(^3\Sigma^-)}$ System

$\underline{B_1 \rightarrow X}$

Band heads (71.2):

(v', v'')	(1, 0)	(0, 0)	(0, 1)	(1, 1)
λ	3070.4	3089.9	3113.1	3093.3

$\underline{B_2 \rightarrow X}$

Band heads (71.2):

(v', v'')	(1, 0)	(0, 0)	(0, 1)	(1, 1)
λ	2947.2	2964.8	2986.1	2968.3

SbBr

SPECTROSCOPIC CONSTANTS

State	T_o	ω_e	$x_e\omega_e$	B_e	$\alpha_e \times 10^3$	$D_e \times 10^6$	r_e	Remarks	Bibliography
$B_2(^3\Pi)$	33719	201	0.3						(71.2, 63.1)
$B_1(^3\Pi)$	32354	207.55	0.85						(71.2, 63.1)
$A(^3\Sigma)$	19736	215.8	0.35						(63.1)
$B(^3\Sigma)$	0	242.9	0.53						(71.2, 63.1)

Dissociation energy = 3.2 ± 0.6 eV, 74 kcal/mole, 24197 cm^{-1}.

BIBLIOGRAPHY

(63.1) $A_1, B_1, B_2 \rightarrow X$ Systems, Vibrational Analysis,
N. L. Singh and M. N. Avasthi,
Ind. J. Pure. Appl. Phys. 1, 197-8

(71.2) $B_1, B_2 \rightarrow X$ System,
M. N. Avasthi,
"Electronic Spectra of Antimony Monobromide,"
Z. Naturforsch. A 26, 250-4

SbCl

Methods of Production and Experimental Technique

Absorption in flash photolysis (68.2).

Emission (active nitrogen + SbCl$_3$) (40.1).

BAND SYSTEMS

System	Transition	Sources	Wavelength Limits	Degrading	Band Head, $\nu_{0,0}$	Remarks	Bibliography
I	$A(^3\Sigma^-) \rightarrow X_1(^3\Sigma^-)$	Active N$_2$ + SbCl$_3$	5335-4480	R	22332.5		(70.3, 40.1)
II	$b(^1\Sigma^+) \rightarrow X_2(^1\Sigma^+)$	Active N$_2$ + SbCl$_3$	4825-4100	R	25790		(70.3, 40.1)
III	$B \leftarrow X_1(^3\Sigma^-)$	Absorption	2421-2350	V	41652(0,0)		(72.5, 71.4)
IV	$C \leftarrow X_1(^3\Sigma^-)$	Absorption	2380-2274	V	43098(0,0)		(72.5, 71.4, 68.2)
V	$D \leftarrow X_1(^3\Sigma^-)$	Absorption	2228-2168	V	45245(0,0)		(72.5, 71.4)

I. $A(^3\Sigma^-) \to X_1(^3\Sigma^-)$ System

Band heads (R) — vibrational analysis is uncertain:

v', v''	0, 10	0, 9	0, 8	0, 7	0, 6
λ	5338.5	5236.1	5141.0	5048.6	4959.7
Intensity	5	4	4	4	4

v', v''	1, 6	0, 5	1, 5	0, 2
λ	4902.0	4872.9	4816.1	4680.8
Intensity	4	3	3	3

II. $b(^1\Sigma^+) \to X_2(^1\Sigma^+)$ System

Band heads (R) — vibrational analysis is uncertain:

v', v''	0, 11	0, 10	0, 9	0, 8	0, 7	0, 6
λ	4575.0	4503.6	4433.9	4365.4	4298.4	4233.0
Intensity	4	4	4	4	4	3

III. $B \leftarrow X_1(^3\Sigma^-)$ System

Band heads (72.5):

(v', v'')	(1, 0)	(0, 0)	(0, 1)
λ(Intensity)	2375(1)	2400.1(0)	2421.3(0)

IV. $C \leftarrow X_1(^3\Sigma^-)$ System

Band heads (72.5):

(v', v'')	(1, 0)	(0, 0)	(0, 1)
λ(Intensity)	2296.5(8)	2319.6(10)	2339.5(1)

V. $D \leftarrow X_1(^3\Sigma^-)$ System

Band heads (72.5):

(v', v'')	(1, 0)	(0, 0)	(0, 1)
λ(Intensity)	2188.5(4)	2209.5(8)	2227.8(2)

SbCl

SPECTROSCOPIC CONSTANTS

State	T_e	ω_e	$x_e\omega_e$	B_e	$\alpha_e \times 10^3$	$D_e \times 10^6$	r_e	Remarks	Bibliography
D	45211	440	-						(72.5)
C	43063	440	3.3						(72.5, 68.2)
B	41612	448	4						(72.5)
b($^1\Sigma^+$)	25855+a	240.2	2.2						(72.5, 40.1)
A($^3\Sigma^-$)	22395	244.4	2.3						(72.5, 40.1)
$X_2(^1\Sigma^+)$	a	370	1.00						(72.5, 40.1)
$X_1(^3\Sigma^-)$	0	368.0	0.8						(72.5, 40.1)

Dissociation energy = 3.7 ± 0.5 eV, kcal/mole, 29840 cm^{-1}.

BIBLIOGRAPHY

(40. 1) A_1, $A_2 \rightarrow X$ Systems,
W. F. C. Ferguson and I. Hudes,
Phys. Rev. 57, 705-7

(68. 2) $C \leftarrow X$ System,
N. Basco and K. K. Yee,
Spectroscopy Letters 1, 19-22

(70. 3) A_1, $A_2 \rightarrow X$ Systems,
M. N. Avasthi,
"Electronic Spectrum of SbCl,"
Spectroscopy Letters 3, 291-5

(71. 4)   B, C, D $\leftarrow$ X Systems,
N. Danon, A. Chatalic, and B. Pannetier,
"Absorption Spectra of Antimony Monochloride,"
C. R. Acad. Sci. C 272, 1411-1414

(72. 5)   B, C, D $\leftarrow$ X Systems,
A. G. Briggs and R. J. Kemp,
"New Electronic Transitions of Antimony Monochloride,"
J. Chem. Soc. Faraday Trans. II 68, 1083-7

SbF

Methods of Production and Experimental Technique

Emission from a high frequency discharge into SbF_3. Active nitrogen + SbF_3.

BAND SYSTEMS

System	Transition	Sources	Wavelength Limits	Degrading	Band Head, ν_e	Remarks	Bibliography
I	$A'\,^3\Sigma \to X_3\,^3\Sigma^-$	h.f. discharge	5450-4050		22588.6		(73.L3, 71.10, 37.1)
II	$A''\,^3\Sigma \to X_3\,^3\Sigma^-$	h.f. discharge	5770-4730	R	20004.7		(73.L3, 71.10, 37.1)
III	$A_1\,^1\Delta \to X_2\,^1\Delta$	h.f. discharge	5000-3200		21885.2		(73.L3, 71.10, 37.1)
IV	$A_2\,^1\Sigma \to X_1\,^1\Sigma^+$	h.f. discharge	5000-3200	R	23993.5		(73.L3, 71.10, 37.1)
V	$A_3\,^1\Sigma \to X_1\,^1\Sigma^+$	h.f. discharge	5000-3200		27910.4		(73.L3, 71.10, 37.1)
VI	$B\,^1\Pi \to X_1\,^1\Sigma^+$	h.f. discharge	3340-3140	V	31099.3		(73.L3, 62.3)
VII	$C_1\,^1\Pi \to X_2\,^1\Delta$	Discharge	2550-2750	V	37937.6		(73.L3, 71.8, 70.6, 69.4)
VIII		Discharge			43517.7		(73.L3)
IX	$C_2\,^3\Pi_0 \to X_1\,^1\Sigma^+$	Discharge	2250-2450	V	43514.1		(73.L3, 71.10)
X	$C_3(^3\Pi) \to X_3\,^3\Sigma^-$	Discharge	2050-2250	V	44747.8		(73.L3, 71.10)

II. $\underline{A''\,^3\Sigma \to X_3\,^3\Sigma^-\text{ System}}$

Band heads, λ (73. L3):

(v', v'')	(0, 2)	(0, 1)	(0, 0)	(1, 0)	(2, 0)
λ	5343.94	5178.34	5021.38	4919.29	4822.46

SbF

SPECTROSCOPIC CONSTANTS

State	T_e	ω_e	$x_e\omega_e$	B_e	$\alpha_e \times 10^3$	$D_o \times 10^6$	r_e	Remarks	Bibliography
$C_3(^3\ )$	50334.1	700.9	2.80	0.3265^a		0.42	1.77^b	$A = -2.5$ cm^{-1}	(73.L3, 72.L2, 71.8, 39.2, 37.1)
$C_2\ ^3\Sigma^-_0$	44747.8	701.8	3.00	0.3285^a		0.5			(73.L3, 72.L2, 70.6, 69.5, 39.2, 37.1)
$^3\Pi_1$		701.4	2.93						(73.L3)
$C_1\ ^1\Pi$	39181.6	697.6	1.37	0.3225^a		0.25	1.78^b		(73.L3, 72.L2, 71.10, 69.4, 69.5, 39.2, 37.1)
$B\ ^1\Pi$	37919.3	701.0	2.89						(73.L3, 62.3)
$A_3\ ^1\Sigma$	34730.4	412.15	1.88						(73.L3)
$A_2\ ^1\Sigma$	30813.5	420.37	1.76						(73.L3, 70.7, 39.2, 37.1)
$A_1\ ^1\Delta$	23129.2	415.36	2.44						(73.L3, 69.5, 39.2, 37.1)
$A'\ ^3\Sigma$	22588.6	417.79	2.42						(73.L3)
$A''\ ^3\Sigma$	20004.7	418.9	2.64						(73.L3)
$X_1\ ^1\Sigma^+$	6820	612.55	2.58	0.3047^a	2	0.25	1.83		(73.L3, 72.L2, 69.5, 39.2, 37.1)

SbF

SPECTROSCOPIC CONSTANTS

State	T_e	ω_e	$x_e\omega_e$	B_e	$\alpha_e \times 10^3$	$D_o \times 10^6$	r_e	Remarks	Bibliography
$X_2\ ^1\Delta$	1244	615.62	2.57	0.2746		0.30			(73.L3, 72.L2, 71.8, 69.5, 39.2, 37.1)
$X_3\ ^3\Sigma^-$	0	608.83	2.59	0.30825	1.5		1.8		(73.L3, 72.L2, 70.7, 69.5, 39.2, 37.1)

$^a B_o$, $^b r_o$.

Dissociation energy = 4.5 ± 1 eV, 104 kcal/mole, 36300 cm^{-1}.

BIBLIOGRAPHY

(37.1) $A_{1,2} \to X$ System, Vibrational Analysis, $(C \to X)$ System,
G. D. Rochester,
Phys. Rev. 51, 486-90

(39.2) $A_3 \to X$ and $C_{1,2,3} \to X$ Systems, Vibrational Analysis,
H. G. Howell and G. D. Rochester,
Proc. Phys. Soc. 51, 329-34

(62.3) $B \to X$ System, Vibrational Analysis,
T. A. P. Rao and P. T. Rao,
Indian J. Phys. 36, 85-92

(69.4)   C. Sivaji and P. T. Rao,
"Rotational Analysis of the C_1 System of SbF,"
Current Sci. 38, 432-3

(69.5) $C \to X$ System,
M. M. Patel and K. C. Abraham,
"Vibrational Analysis of Ultraviolet Bands of SbF,"
Indian J. Pure. Appl. Phys. 7, 641-3

(70.6)   K. C. Abraham and M. M. Patel,
"Rotational Analysis of the C_3 System of the SbF Molecule,"
J. Phys. B 3, 1183-7

(70.7) $A_2 \to X_3$ System,
K. C. Abraham and M. M. Patel,
A New Band System of the SbF Molecule in the λ 4050-5450 λ Å Region,"
J. Phys. B 3, 881-2

(71.8)   R. Shanker and I. S. Singh,
"Rotational Analysis of the C_2 System of SbF Molecule,"
Current Sci. 40, 341-2

(71.9)   P. S. Dube, D. K. Rai, and N. L. Singh,
"Determination of the Effective Vibrational Temperature in the A_2 System of the SbF Molecule,"
Indian J. Pure Appl. Phys. 9, 484-5

(71.10)   K. C. Abraham and M. M. Patel,
"Fine Structure Analysis of the C_1 System ($\lambda 2600 - \lambda 2700$ Å) of the SbF Molecule,"
J. Phys. B 4, 1398-1406

(72. L1)   R. Shanker and I. S. Singh,
"Rotational Structure of the C_3 System of SbF Molecule,"
Indian J. Pure Appl. Phys. 10, 395-7

(72. L2)   R. Shanker, S. C. Srivastava, and I. S. Singh,
"Rotational Analysis of the 2550-2750 Å Band System of SbF Molecule,"
Indian J. Pure Appl. Phys. 10, 541-4

(73. L3)   M. Chakravorty, K. C. Abraham, and M. M. Patel,
"On the Emission Spectrum of Molecular SbF,"
J. Phys. B. (Atom. Mol. Phys.) 6, 757-60

SbO

Methods of Production and Experimental Technique

Emission from a carbon arc + Sb, Cu + Sb, Cu + Cu/Sb.

Discharge in a hollow cathode + Sb_2O_3; microwave discharge + SbI_3 (only for systems B → X, C → X, D → X).

Absorption by an arc flame.

BAND SYSTEMS

System	Transition	Sources	Wavelength Limits	Degrading	Band Head, $\nu_{0,0}$	Remarks	Bibliography
I	$A^2\Pi_r \rightleftarrows X^2\Pi_r$	Arc	6800-4500	R			(39.2, 31.1)
II	$B^2\Sigma \rightleftarrows X^2\Pi_r$	Discharge	4500-3400	R	24209.98(0,0) 26474.07(0,0)		(60.5, 60.6, 39.2, 31.1)
III	$C^2\Delta_r \rightleftarrows X^2\Pi_r$	Discharge	3600-2800	R	29622.9(0,0) 29567.3(0,0)		(60.4, 60.5)
IV	$D^2\Pi \rightarrow X^2\Pi_r$	Discharge	3260-2780	R	34081.8(0,0)		(70.8, 60.4, 60.5)
V	$E^2\Sigma \rightleftarrows X^2\Pi_r$	Arc, Absorption	2910-2450	V	39802(0,0) 37531(0,0)		(60.4, 43.3)
VI	$F(^2\Delta) \rightleftarrows X^2\Pi_r$	Arc, Absorption	2900-2570	R	38868.1(0,0) 38856.1(0,0)		(60.4)
VII	$G? \rightleftarrows X^2\Pi_r$	Arc, Absorption	2800-2560	V	38960.5(v,1) 38939.3(v,1)		(60.4)

SbO

I. $A^2\Pi_r \rightleftarrows X^2\Pi_r$ System

Band heads, λ (39.2):

Subsystem 1

v', v''	0	1	2	3	4	5
0		5065.05	5277.70	5505.57	5750.70	6015.20
1	4736.89	4926.18	5126.30	5341.25	5572.80	
2	4617.25	4795.80	4985.92	5189.90	5406.37	5638.80
3	4504.70	4675.03			5252.15	
4		4562.04				5318.80
5						
6						5042.36

Subsystem 2

v', v''	0	1	2	3	4	5
0		5679.46	5949.45	6240.19	6559.13	
1			5757.93	6029.99	6325.58	6647.03
2		5344.68		5837.69	6113.15	6412.03

II. $B^2\Sigma \rightleftarrows X^2\Pi_r$ System

Band heads, λ (39.2):

Subsystem 1

v', v''	0	1	2	3	4	5	6
0	3776.00	3894.84					
1	3696.45	3810.62	3939.64				
2	3621.71	3730.80		3965.28	4091.00		
3	3552.00	3656.80	3766.32				4259.36
4		3587.57					
5		3521.58	3623.15				
6		3460.60					

SbO

Subsystem 2

v', v''	0	1	2	3	4	5	6
0	4130.22	4272.74					
1	4035.00	4171.44	4314.81				
2	3946.71	4076.72		4358.09		4669.21	4361.10
3	3864.02	3987.90	4119.05				
4		3905.88					
5		3829.03	3949.14				
6		3755.95	3871.81				

III. $C^2\Delta_r \rightleftarrows X^2\Pi_r$ System

Band heads, λ(intensity) (60.4, 60.5):

$$^2\Delta_{3/2} \rightarrow {}^2\Pi_{1/2}$$

v', v''	0	1	2	3	4
0	3374.8(6)	3469.1(5)	3568.3(5)	3672.3(6)	
1	3311.4(4)	3402.4(4)	3497.9(4)		
2	3251.1(4)		3430.8(3)	3527.3(4)	3627.1(3)
3	3194.1(4)				
4	3140.0(4)				
5	3087.9(4)	3167.0(3)			

$$^2\Delta_{5/2} \rightarrow {}^2\Pi_{3/2}$$

v', v''	0	1
0	3580.7(5)	3687.3(5)
1	3510.6(4)	
2	3443.7(3)	
3	3380.1(3)	
4	3318.0(3)	
5	3260.4(4)	
6	3205.1(3)	

IV. $D^2\Pi \rightarrow X^2\Pi_{1/2}$ System

Band heads, λ(intensity) (60.4):

v', v''	0	1	2	3	4	5	6
0	2865.3(3)	2933.2(4)	3003.6(5)	3077.0(5)	3153.0(4)	3230.8(2)	
1	2825.2(2)	2891.4(4)	2959.8(4)	3030.9(3)			3259.5(3)
2	2786.8(2)						

V. $E^2\Sigma \rightleftarrows X^2\Pi_r$ System

Band heads (Q), λ (intensity in emission, absorption) (60.4):

$^2\Sigma \rightleftarrows {}^2\Pi_{1/2}$

v', v''	0	1	2	3	4
0	2511.8(3, 10)	2563.2(3, 8)	2617.5(5, 6)	2672.9(3, 3)	
1		2509.4(3, 5)	2560.9(2, 4)	2614.1(2, 2)	
2			2506.8(2, 2)	2557.8(1, 2)	2610.2(2, 2)
3				2504.7(2, 3)	
4					2501.8(1, 2)

$^2\Sigma \rightleftarrows {}^2\Pi_{3/2}$

v', v''	0	1	2	3	4
0	2663.7(5, 5)	2722.2(7, 4)	2782.5(7, 3)	2844.7(3, 0)	
1	2605.2(2, 2)	2661.1(3, 2)		2778.8(3, 1)	
2			2658.0(2, 1)	2715.2(2, 1)	2774.1(2, 0)
3				2655.5(3, 3)	

VI. $F(^2\Delta) \rightleftarrows X^2\Pi_r$ System

Band heads (Q), λ (intensity in emission, absorption) (60.4):

Subsystem 1

v', v''	0	1	2
0	2573.6(3, 10)	2628.3(4, 5)	2684.8(3, 2)
1		2588.4(2, 1)	2643.2(1, 1)

SbO

                      Subsystem 2

v', v''	0	1	2
0	2733.4(5, 3)	2794.9(5, 2)	2858.7(4, 1)
1	2689.9(1, 1)	2749.6(2, 0)	2811.3(2, 0)

VII. $\underline{G? \rightleftarrows X^2\Pi_r \text{ System}}$

Diffuse bands.

Band heads, λ (intensity in emission, absorption) (60.4):

v', v''	Head	Subsystem 1 λ	Subsystem 2 λ
v', 3	P	2680.7(2, 1)	
	Q	2677.5(3, 2)	
v', 2	P	2625.6(2, 1)	2791.3(2, 0)
	Q	2621.8(4, 3)	2787.8(4, 1)
v', 1	P	2571.4(2, 2)	2731.0(3, 1)
	Q	2568.1(3, 7)	2727.2(5, 2)

SbO

SPECTROSCOPIC CONSTANTS

State	T_e	ω_e	$x_e\omega_e$	B_e	$\alpha_e \times 10^3$	$D_e \times 10^6$	r_e	Remarks	Bibliography
$E^2\Sigma$	39785.6	849.4	2.9	–	–	–	–		(60.4)
$F(^2\Delta)$	38960	595	–	–	–	–	–		(60.4)
$D^2\Pi$	35049.8	499.9	3.0	0.268	–	0.4[b]	2.06	$A \approx 2273.4$ cm^{-1}	(70.8, 68.7, 60.4)
$C^2\Delta_r$	30316.8 29754.6	570.6	3.4	0.292	0.00	0.33	2.018		(68.7, 60.5)
$B^2\Sigma$	26594.0	582	6.5	0.321[a]	–	0.57[b]	1.927[c]		(68.7, 60.6)
$A^2\Pi_r$	20800.5 20667.5	569.0	5.0	–	–	–	–		(39.2)
$X^2\Pi_r$	2274.3 0	817.4	4.9	0.351[a]	1.00	–	1.841		(68.7, 60.4, 60.6)

[a]B_0, [b]D_1, [c]r_0.

Dissociation energy = 4 ± 1 eV, 92 kcal/mole 32263 cm^{-1}.

BIBLIOGRAPHY

(31.1) A, B → X Systems, Spectral Reproduction,
B. C. Mukherji,
Z. Physik 70, 552-8

(39.2) A, B → X Systems, Vibrational Analysis,
A. K. SenGupta,
Indian J. Phys. 13, 145-7

(43.3) E → X System, Vibrational Analysis,
A. K. SenGupta,
Indian J. Phys. 17, 216-22

(60.4) C, D → X Systems. E, F, G ⇌ X Systems, Vibrational Analysis,
M. Shimauchi,
Science Light 9, 109-33

(60.5) C, D → X Systems, Rotational Analysis,
S. V. J. Lakshman,
Z. Physik 158, 367-73

(60.6) B − X System, Rotational Analysis,
S. V. J. Lakshman,
Z. Physik 158, 386-91

(68.7) B, C, D → X Systems, Rotational Analysis,
D. V. K. Rao and P. T. Rao,
"Fine Structure Analysis and Isotope Effect Studies in B-X, C-X and D-X Systems of the SbO Molecule,"
Current Sci. 37, 310

(70.8) B. D. Rai, K. N. Upadhya, and D. K. Rai,
"Rotational Analysis of the D-X Band System of the SbO Molecule,"
J. Phys. B 3, 1374-9

ScCl

Methods of Production and Experimental Technique

Emission from a discharge.

Band Systems

Band heads (0, 0) (69.2):

System	Head	Deg.	λ	Remarks
$A^1\Sigma^+ \to X^1\Sigma^+$	R	R	8066.59	
$b^3\phi \to a^3\Delta$	R_3	R	7973.82	$Sc^{37}Cl$ head
	R_2	R	7954.78	
	R_1	R	7936.72	$Sc^{37}Cl$ head
$C^3\Delta \to a^3\Delta$	R_3	R	7636.91	
	R_2	R	7635.92	
	R_1	R	7633.76	
$B^1\Pi \to X^1\Sigma^+$	R	R	5687.81	
$D^1\Pi \to X^1\Sigma^+$	R	R	4653.37	
$e^3\Pi \to d^3\Sigma$	N_1	V	4495.04	$Sc^{37}Cl$ head
	P_1	V	4491.83	$Sc^{37}Cl$ head
	Q_1	V	4490.48	$Sc^{37}Cl$ head
	Q_1	V	4489.55	
	P_2	V	4470.71	$Sc^{37}Cl$ head
	P_3	V	4450.91	$Sc^{37}Cl$ head
$E^1\Sigma^+ \to X^1\Sigma^+$	R	R	3696.24	
?	P	V	3675.49	$Sc^{37}Cl$ head
	Q	V	3675.22	
$F^1\Pi \to X^1\Sigma^+$	R	R	3203.33	

ScCl

SPECTROSCOPIC CONSTANTS

State	T_e	ω_e	$x_e\omega_e$	B_o	$\alpha_o \times 10^3$	$D_e \times 10^6$	r_o	Remarks	Bibliography
$e^3\Pi$	y+22260.0	312.5	0.55	-	-	-	-	Constants of $Sc^{35}Cl$	(67.1)
$d^3\Sigma$	y	297.3	0.61	-	-	-	-		(67.1)
$c^3\Delta$	x+13113.8	355.9	2.18	-	-	-	-		(67.1)
$b^3\phi$	x+12567.6 (a)	-	-	-	-	-	-		(67.1)
$a^3\Delta$	x	398.3	1.36	-	-	-	-		(67.1)
$F^1\Pi$	31249.9	364.7	1.0	-	-	-	-		(67.1)
$E^1\Sigma^+$	27033.3	472.1	1.32	-	-	-	-		(67.1)
$D^1\Pi$	21521.1	373.1	1.6	0.1569	-	-	2.337		(67.1)
$B^1\Pi$	17576.6[a]	374.3	2.3	0.1551	-	-	2.351		(67.1)
$A^1\Sigma^+$	12431.2	373.9	0.9	0.1574	-	-	2.333		(67.1)
$X^1\Sigma^+$	0	447.4	1.8	0.1720	1.0	-	2.232		(67.1)

[a] T_o.

Dissociation energy = 3.4 eV (estimated), 78.4 kcal/mole, 27423 cm^{-1} (69.2).

BIBLIOGRAPHY

(67.1)   Emission, All Systems,
E. A. Shenyavskaya, A. A. Mal'tsev, and L. V. Gurvich,
Vestnik Moskov Univ. Ser 11: Khim 22, n°4, 104-5

(69.2)   Emission, A, B, C, D-X Systems,
E. A. Shenyavskaya, A. A. Mal'tsev, D. I. Kataev and
L. V. Gurvich,
Optika Spectrosk. S.S.S.R. 26, 937-44

ScF

Methods of Production and Experimental Technique

Absorption in a King furnace, Sc + AlF_3, T ~ 2000°C.
Emission from a hollow cathode discharge. Thermal emission.
Matrix isolation.

BAND SYSTEMS

System	Transition	Sources	Wavelength Limits	Degrading	Band Head, $v_{0,0}$	Remarks	Bibliography
I	$B^1\Pi \rightleftarrows X^1\Sigma^+$	Absorption	9500-8400	R	≤10661.2		(71.3, 67.10, 66.6)
II	$C^1\Sigma^+ \rightleftarrows X^1\Sigma^+$	Absorption	6600-5600	R	16092.0		(67.10, 66.6, 64.3)
III	$E^1\Pi \rightleftarrows X^1\Sigma^+$	Absorption	5300-4600	R	20326.8		(69.12, 67.10, 66.6, 64.3)
IV	$^1\Pi \rightleftarrows X^1\Sigma^+$	Absorption	3860-3650	R	26809.6		(67.10, 66.6, 64.3)
V	$^1\Pi \rightleftarrows X^1\Sigma^+$	Absorption, Hollow Cathode	2960-2800	R	34920.7		(67.10, 64.3, 63.1)
VI	$^1\Pi \rightleftarrows X^1\Sigma^+$	Absorption, Hollow Cathode	2640-2575	R	38806.1		(67.10)
VII	$^3\phi \rightleftarrows a^3\Delta$	Absorption	7100-6300	R	15234.4 15277.5 15317.6		(67.10)
VIII	$?^3\Pi \rightleftarrows a^3\Delta$	Absorption	~5450	R	18336.0 18361.4	Only 0,0 band	(69.12, 67.10)
IX	$?^3\Delta \rightleftarrows a^3\Delta$	Absorption	4560-4460	V	21963.9		(67.10)
X	$^3\phi \rightleftarrows a^3\Delta$	Absorption	~3680	R	27138.2 27171.1 27202.2	Only 0,0 band	(69.12, 67.10)

BAND SYSTEMS

System	Transition	Sources	Wavelength Limits	Degrading	Band Head, $\nu_{0,0}$	Remarks	Bibliography
XI	Triplet ?	Absorption	~3800	R	26300		(67.10)
XII	Triplet ?	Absorption	~2780	V	35942	Only 0,0 band	(69.12, 67.10, 63.1)

ScF

I. $B^1\Pi \rightleftarrows X^1\Sigma^+$ System

Band heads (R):

(v', v'')	(0, 0)	(1, 0)	(2, 0)
λ	9374.4	8889.85	8455.75

II. $C^1\Sigma^+ \rightleftarrows X^1\Sigma^+$ System

Band heads (R):

(v', v'')	(0, 1)	(0, 0)	(2, 0)	(3, 0)
λ	6506.88	6211.46	5792.84	5606.48

III. $E^1\Pi \rightleftarrows X^1\Sigma^+$ System

Band heads (R):

(v', v'')	(0, 1)	(0, 0)	(1, 0)
λ	5099.66	4917.18	4772.97

IV. $^1\Pi \rightleftarrows X^1\Sigma^+$ System

Band heads (R):

(v', v'')	(0, 1)	(0, 0)	(1, 0)
λ	3832.59	3728.53	3651.60

V. $^1\Pi \rightleftarrows X^1\Sigma^+$ System

Band heads (R):

(v', v'')	(2, 3)	(1, 2)	(0, 1)	(0, 0)	(3, 2)	(2, 1)	(1, 0)
λ	2948.5	2935.2	2922.3	2861.52	2842.5	2829.0	3651.6

VI. $^1\Pi \rightleftarrows X^1\Sigma^+$ System

Band heads (R):

(v', v'')	(0, 1)	(0, 0)
λ	2625.03	2575.78

VII. $^3\phi \rightleftarrows a^3\Delta$ System

Band heads (R), λ:

$^3\phi_2 - {}^3\Delta_1$

v', v''	0	1	2
0	6557.75	6845.23	
1		6590.85	6878.62

$^3\phi_3 - {}^3\Delta_2$

v', v''	0	1	2
0	6539.92	6826.20	
1	6307.14	6573.04	6859.60

$^3\phi_4 - {}^3\Delta_3$

v', v''	0	1	2
0	6523.21	6808.22	
1	6291.62	6556.48	6841.66

VIII. $?\,^3\Pi \rightleftarrows a^3\Delta$ System (5450Å)

Preliminary rotational analysis indicates the lower level to be $a^3\Delta$ with $\Delta\Lambda = 1$; however, the rotational structure is strongly perturbed and the assignment is therefore tentative.

Band heads of greatest intensity (Q); 5452.24 and 5444.70.

IX. $?\,^3\Delta \rightleftarrows a^3\Delta$ System (4560Å)

(0, 0) band sequence degrades V. Bands possess many closely packed heads. The transition is probably $^3\Delta \rightleftarrows a^3\Delta$ with the spin-orbit coupling of the two states very similar.

Band heads of greatest intensity:

(v', v'')	(0, 0)	(1, 1)	(2, 2)
λ	4559.22	4543.79	4528.80

X. $^3\phi \rightleftarrows a\,^3\Delta$ System

Resembles system VII, but much weaker. Only the (0, 0) band is intense.

Heads (R) of the (0, 0) band:

$$(^3\phi_2 - ^3\Delta_1) - 3582.95$$
$$(^3\phi_3 - ^3\Delta_2) - 3578.54$$
$$(^3\phi_4 - ^3\Delta_3) - 3574.38$$

XI. 3800Å System

Complex group of bands with many heads, degrading R.

Too complex for a singlet-singlet transition.

XII. 2780Å System

Most intense band of the group possesses a P head at 2781.69 and a Q head at 2781.44.

SPECTROSCOPIC CONSTANTS

State	T_o	$\Delta G_{1/2}$	$x_e\omega_e$	B_o	$\alpha_e \times 10^3$	$D_e \times 10^6$	r_o	Remarks	Bibliography
$^3\phi_4$	27202.2 + a_3	-	-	0.3463	-	-	1.910		(67.10)
$^3\phi_3$	27171.1 + a_2	-	-	0.3441	-	-	1.916		(67.10)
$^3\phi_2$	27138.2 + a_1	-	-	0.3413	-	-	1.924		(67.10)
$^3\Pi_2$	(18361.4 + a_3)[a]	-	-	(0.3677)[a]	-	-	(1.854)[a]		(67.10)
$^3\Pi_1$	(18336.0 + a_2)[a]	-	-	-	-	-	-		(67.10)
$^3\phi_4$	15317.6 + a_3	-	-	0.3530	3.10	-	1.891		(67.10)
$^3\phi_3$	15277.5 + a_2	564.5	2.96	0.3511	-	-	1.896		(67.10)
$^3\phi_2$	15234.4 + a_1	-	-	0.3490	-	-	1.902		(67.10)
$a\,^3\Delta_2$	a_3	643.05	-	0.3693	2.58	-	1.849		(67.10)
$a\,^3\Delta_2$	a_2	642.92	3.03	0.3652	2.54	-	1.860		(67.10)
$a\,^3\Delta_1$	a_1	642.85	-	0.3610	2.50	-	1.870		(67.10)
$^1\Pi$	38806.1	-	-	0.3671[b]	-	-	1.855		(67.10)

ScF

SPECTROSCOPIC CONSTANTS

State	T_o	$\Delta G_{1/2}$	$x_e\omega_e$	B_o	$\alpha_e \times 10^3$	$D_e \times 10^6$	r_o	Remarks	Bibliography
$^1\Pi$	34920.7	570	-	0.378^b	-	-	1.828		(67.10)
$^1\Pi$	26809.6	565.3	-	0.3449	2.5	-	1.914		(67.10)
$E^1\Pi$	20326.8	614.7	3.7	0.3615^b	2.96	-	1.869		(67.10)
$C^1\Sigma^+$	16092.0	584.3	2.64	0.3461	2.4	-	1.910		(67.10)
$B^1\Pi$	$\leq 10661.3^a$	≥ 581.9	2.28	0.3416^c	2.48	-	≤ 1.923		(67.10)
$X^1\Sigma^+$	0	728.0	3.8	0.3937	2.66		1.791		(67.10)

[a] analysis uncertain, upper state perturbed, [b] perturbed, [c] important Λ doubling.

Dissociation energy = 6.06 ± 0.14 eV, 139.8 kcal/mole, 48878 cm^{-1} (67.9).

Perturbations and General Information

The spin-orbit coupling constants of the triplet states are not known; however, the two $^3\phi$ and $^3\Delta$ states appear to be case a.

The energy difference from the $X^1\Sigma^+$ state to the $a^3\Delta$ state is not known. At $4°K$, absorption studies in a matrix indicate absorption from only the $X^1\Sigma^+$ state; however, at $2000°C$ absorption is seen to be equally intense from both the $X^1\Sigma^+$ and $a^3\Delta$ states implying the energy separation is <1388 cm^{-1}.

$$f_{abs}\left[\frac{^3\phi - ^3\Delta}{E^1\Pi - X^1\Sigma^+}\right] = 0.8 \pm 3 \ (69.12).$$

Radiative lifetime (69.12):

$$E^1\Pi \rightarrow X^1\Sigma^+ \ - \ \tau \approx 1.1 \times 10^{-4} \text{ sec}$$

$$^3\phi \rightarrow a^3\Delta \ - \ \tau \geq 10^{-4} \text{ sec}$$

BIBLIOGRAPHY

(63.1) System 5 in Emission, Vibrational Analysis,
L. V. Gurvich and E. A. Shenyavskaya,
Optics Spectrosc. 14, 161-2

(64.2) System 7 in Absorption, Rotational Analysis,
R. F. Barrow and W. J. M. Gissane,
Proc. Phys. Soc. 84, 615-6

(64.3) Four Singlet Systems in Absorption, Rotational Analysis,
R. F. Barrow, W. J. M. Gissane, R. C. Le Bargy,
G. V. M. Rose and P. A. Ross,
Proc. Phys. Soc. 83, 889-90

(64.4) R. F. Barrow, W. J. M. Gissane, R. C. Le Bargy,
G. V. M. Rose, and P. A. Ross,
"Rotational Analysis of Some Single Transitions in the Spectrum of Gaseous ScF,"
Proc. Phys. Soc. 83, 889-890

(66.5) Theoretical Calculations of the Ground State,
K. D. Carlson and C. Moser,
J. Chem. Phys. 44, 3259-65

ScF

(66. 6) Absorption and Fluorescent Spectra in Matrix Isolation,
D. MacLeod Jr. and W. Weltner Jr.,
J. Phys. Chem. 70, 3293-300

(67. 7) Analogies Between ScF and TiO,
C. J. Cheetham and R. F. Barrow,
Advances High Temp. Chem. 1, 7-41

(67. 8) LCAO-MO-SCF Calculations,
K. D. Carlson and C. Moser,
J. Chem. Phys. 46, 35-46

(67. 9) Dissociation Energy by Mass Spectra,
K. F. Zmbov and J. L. Margrave,
J. Chem. Phys. 47, 3122-5

(67. 10) Summary of Molecular Constants,
R. F. Barrow, M. W. Bastin, D. L. G. Moore, and C. J. Pott,
Nature 215, 1072-3

(68. 11) D. W. Green,
"Molecular Beam Studies of Scandium Monofluoride,"
U. C. Berkeley, Ph. D. Thesis Part 1, Jan 1968

(69. 12) Inspection of Most Levels with Molecular Beam,
L. Brewer and D. W. Green,
"The Low Lying Electronic States of ScF, TiO, ZrO,"
High Temp. Sci. 1, 26-45

(71. 13) R. F. Barrow and L. Pedersen,
"Rotational Analysis of the System $B^1\Pi - {}^1\Sigma^+$ of Gaseous ScF,"
J. Phys. B: Atom. Mol. Phys. 4, L11-L13

ScO

Methods of Production and Experimental Technique

Absorption in matrix isolation (67.15).

Emission of arc and flames + compounds of Sc; carbon flame (42.7).

In astrophysics; absorption in stellar atmospheres (47.8).

BAND SYSTEMS

	System	Transition	Sources	Wavelength Limits	Degrading	Band Head, $\nu_{0,0}$	Remarks	Bibliography
Orange	I	$A^2\Pi \rightleftarrows X^2\Sigma$	Carbon flame	7300-5740	R	16554.8 16440.6		(70.19, 67.15, 62.9, 42.7, 31.2, 31.3)
	II	$B^2\Sigma \rightleftarrows B^2\Sigma$	Carbon flame	5770-4500		20570.8		(68.18, 67.15, 62.9, 42.7, 31.2, 31.3)

ScO

I. $A^2\Pi \rightleftarrows X^2\Sigma$ System

Band heads of greatest intensity (70.19, 67.15, 62.9, 31.2, 31.3):

$A^2\Pi_{1/2} \rightarrow X^2\Sigma - Q_1$ heads

(v', v'')	(1, 2)	(0, 1)	(0, 0)	(1, 1)
λ(intensity)	6495.90(15)	6457.78(15)	6079.30(100)	6101.87(30)

$A^2\Pi_{3/2} \rightarrow X^2\Sigma - {}^RQ_{21}$ heads

(v', v'')	(1, 2)	(2, 2)	(0, 0)	(2, 1)
λ(intensity)	6446.24(30)	6109.93(40)	6036.17(200)	5772.74(15)

II. $B^2\Sigma \rightleftarrows X^2\Sigma$ System

Band heads (68.18, 67.15, 62.9):

v', v''		Intensity
0, 1	5096.73	20
0, 0	4858.09	30
	4857.79	20
2, 1	4706.97	10
1, 0	4673.10	10
	4672.62	10
4, 2	4571.30	10
3, 1	4536.59	8

All of the bands are subject to doubling because of nuclear hyperfine interaction in the ground state. Rotational level separations approach 0.25 cm^{-1} (68.18).

ScO

SPECTROSCOPIC CONSTANTS

State	T_o	ω_e	$x_e\omega_e$	B_o	$\alpha_e \times 10^3$	$D_e \times 10^6$	r_e	Remarks	Bibliography
$B^2\Sigma^+$	20570.79	825.47	4.21	0.48291	3.2		1.720	$\gamma = -0.0665$ cm^{-1}	(68.8, 62.9)
$A^2\Pi_{3/2}$	16554.8	874.6	4.99	0.5023	3.8		1.687	$p = -0.0748$ cm^{-1}	(70.19, 62.9, 32.4)
$A^2\Pi_{1/2}$	16440.6	875.0	4.98	0.5023	3.8		1.687		(70.19, 62.9, 32.4)
$X^2\Sigma^+$	0	964.95[a]	3.95	0.51342	3.3		1.668		(68.18, 62.9, 32.4)

[a] $\Delta G_{1/2}$.

Dissociation energy = 6.96 ± 0.8 eV, 160.4 kcal/mole, 56137 cm^{-1} (67.16, 67.17).

ScO

Perturbations and General Information

Nuclear hyperfine coupling constant for the $X^2\Sigma^+$ state:

$$= 0.063 \text{ cm}^{-1} - \text{gas phase (68.18)},$$
$$= 0.0670 \text{ cm}^{-1} - \text{Ar matrix (67.15)}.$$

Franck-Condon factors — Morse potential (64.11):

$A^2\Pi - X^2\Sigma$

v', v''	0	1	2	3	4	5	6	7
0	.93716	.05932	.00338	.00013	.00000	.00000	.00000	.00000
1	.06161	.80914	.11784	.01079	.00060	.00003	.00000	.00000
2	.00123	.12686	.67504	.17235	.02272	.00171	.00009	.00000
3	.00000	.00466	.19222	.54009	.21957	.03937	.00381	.00026
4	.00000	.00001	.01142	.25358	.41004	.25639	.06055	.00736
5	.00000	.00000	.00009	.02266	.30658	.29068	.28012	.08563
6	.00000	.00000	.00001	.00037	.03933	.34699	.18745	.28898
7	.00000	.00000	.00000	.00001	.00113	.06200	.37135	-
8	.00000	.00000	.00000	.00001	.00001	.00283	-	-

$B^2\Sigma - X^2\Sigma$

v', v''	0	1	2	3	4	5	6	7
0	.64301	.28458	.06254	.00894	.00092	.00007	.00000	.00000
1	.27940	.19311	.34921	.14291	.03062	.00428	.00043	.00003
2	.06521	.33440	.02097	.29217	.20978	.06413	.01173	.00147
3	.01080	.14462	.26984	.00682	.18873	.24545	.10494	.02459
4	.00141	.03595	.20407	.16474	.06003	.09008	.24463	.14625
5	.00015	.00640	.07264	.22736	.07083	.12007	.02438	.21199
6	.00001	.00090	.01704	.11361	.21214	.01431	.15430	.00027
7	.00000	.00010	.00301	.03442	.14964	.16794	.00038	-
8	.00000	.00001	.00043	.00753	.05798	.17223	-	-
9	.00000	.00000	.00005	.00129	.01553	-	-	-
10	.00000	.00000	.00000	.00018	-	-	-	-

BIBLIOGRAPHY

(27.1)   Generalities,
R. Mecke,
Z. Physik 42, 390-425

ScO

(31.2) Table of Visible Spectra,
W. F. Meggers and J. A. Wheeler,
Bur. Stand. J. Res. 6, 239-75

(31.3) A, B → X Systems, Interpretation,
L. W. Johnson and R. C. Johnson,
Proc. Roy. Soc. A 133, 207-19

(32.4) Molecular Constants,
W. Jevons,
"Report on Band-Spectra of Diatomic Molecules,"
Cambridge, G. B.

(33.5) Table of Visible Spectra,
G. Piccardi,
Gazz. Chim. Ital. 63, 127-38

(39.6) Generalities,
G. Piccardi,
Spectrochim. Acta 1, 249-60

(42.7) Spectral Reproduction,
A. Gatterer,
Ricerche. Spettroscop. Lab. Astrofis. Specola Vaticana 1, 153-79

(47.8) Absorption in Stellar Atmospheres,
D. N. Davis,
Astrophys. J. 106, 28-75

(62.9) A, B-X Systems, Rotational Analysis,
L. Akerlind,
Arkiv Fysik 22, 41-64

(64.10) Discussion of Hyperfine Structure and Ground State of ScO and LaO,
C. K. Jorgensen,
Molecular Phys. 7, 417-24

(64.11) Franck-Condon Factors, A, B-X Systems,
F. S. Ortenberg, V. B. Glasko, and A. I. Dimitriev,
"Vibrational Transition Probabilities for Band Systems of Some Diatomic Oxides II,"
Soviet Astronomy 8, 258-261

(65.12) Dissociation Energy by Mass Spectra,
A. Ben-Reuven,
J. Chem. Phys. 42, 2037-42

ScO

(65.13) Pure Precession. Relation Between $A^2\pi$ and $B^2\Sigma^+$,
R. A. Berg, L. Wharton, W. Klemperer, A. Büchler, and J. L. Stauffer,
J. Chem. Phys. 43, 2416-21

(65.14) Hartree-Fock Calculations and Ground State of ScO,
K. D. Carlson, E. Ludena, and C. Moser,
J. Chem. Phys. 43, 2408-15

(67.15) Optical Absorption Spectrum and Electron Spin Resonance in Ne at 4°K,
W. Weltner Jr., D. MacLeod, Jr., and P. H. Kasai,
J. Chem. Phys. 46, 3172-84

(67.16) Dissociation Energy by Mass Spectra,
L. L. Ames, P. N. Walsh, and D. White,
J. Phys. Chem. 71, 2707-18

(67.17) Dissociation Energy by Mass Spectra,
P. Coppens, S. Smoes, and J. Drowart,
Trans. Faraday Soc. 63, 2140-8

(68.18) $B^2\Sigma^+$-$X^2\Sigma^+$ System, Rotational Analysis, Nuclear Hyperfine Coupling Constant,
A. Adams, K. Klemperer and T. M. Dunn,
Can. J. Phys. 46, 2213-20

(70.19) A-X System,
C. Athenour, R. Bacis, J. L. Femenias, and R. Stringat,
"On the Orange System of ScO Excited with the aid of a Hollow Cathode Lamp,"
C. R. Acad. Sci. B 271, 567-70

(71.20) Estimation of $A'^2\Delta$ State Energy,
D. W. Green,
"Low Lying Electronic States of the Scandium Oxide, Yttrium Oxide and Lanthanum Oxide Molecules,"
J. Phys. Chem. 75, 3103-6

(72.L1) A-X System Rotational Analysis,
R. Stringat, C. Athenour, and J. L. Femenias,
"Rotational Analysis of the (0,0) Band of the Orange System of ScO,"
Can. J. Phys. 50, 395-403

ScS

Methods of Production and Experimental Technique

Neon matrix at $4°K$.

Band Systems

A system of bands, tentatively labled as $A^2\Pi \leftarrow X^2\Sigma^+$ has been observed between 8000 and 9000 Å:

Band	λ
(0, 0)	8836Å
	8681Å

Dissociation energy = 4.92 ± 0.11 eV, 113.4 kcal/mole, 39683 cm^{-1}.

BIBLIOGRAPHY

(67.1) P. Coppens, S. Smoes, and J. Drowart,
Trans. Faraday Soc. 63, 2140-8

(68.2) R. P. Steiger,
"Mass Spectrometric Investigation of the Vaporization, Thermodynamics and Dissociation Energies of LaS, ScS, YS, ZrS, and UO,"
Disc. Abstr. B 29, 2009

(72.L1) N. S. McIntyre, K. C. Lin, and W. Weltner, Jr.,
"ESR and Optical Spectra of the ScS and YS Molecules,"
J. Chem. Phys. 56, 5576-83

SeO

Methods of Production and Experimental Technique

Emission from flames: discharge.

High frequency discharge in SeO_2 (65.8).

Electronic resonance spectra observed (67.9).

BAND SYSTEMS

System	Transition	Sources	Wavelength Limits	Degrading	Band Head, $\nu_{0,0}$	Remarks	Bibliography
I	$b(^1\Sigma^+) \to X^3\Sigma^-$	Discharge	6150-5250	R	17349(0, 0)		(72.11)
II	$A^3\Sigma^- \to X^3\Sigma^-$	Discharge	5100-2920	R	34012-(F_1-F_1) 34082-(F_2-F_2)		(65.8, 63.6, 38.4)
III	$D \to b(^1\Sigma^+)$	Discharge	2435-2165	V	45992		(64.7)
IV	$F \to b(^1\Sigma^+)$	Discharge	2165-2080	V	47852		(64.7)
V	$C \to X^3\Sigma^-$	Discharge	2112-1917	V	50803		(64.7)
VI	$E \to X^3\Sigma^-$	Discharge	2015-1840	V	54131 54288		(64.7)

I. $b(^1\Sigma^+) \to X^3\Sigma^-$ System

Band heads (72.11):

(v', v'')	(0, 1)	(0, 0)	(1, 0)	(1, 1)	(2, 1)
λ	5487.6	5764.0	6081.2	5774.7	5501.5

II. $A^3\Sigma^- \to X^3\Sigma^-$ System

Transition is composed of three states: F_1, F_2, and F_3. F_2-F_2 and F_3-F_3 subsystems overlap, while F_1-F_1 subsystem is well separated from the others.

Band heads (65.8, 38.4):

F_1-F_1 Subsystem

(v', v'')	(0, 7)	(0, 6)	(0, 10)	(1, 10)
λ(intensity)	3578.7	3472.7	3932.4(8)	3854.6(4)

F_2-F_2 Subsystem

(v', v'')	(0, 7)	(0, 6)	(0, 10)	(1, 10)
λ(intensity)	3588.3	3481.8	3944.0(9)	3865.8(4)

III. $D \to b(^1\Sigma^+)$ System

Band heads (64.7):

(v', v'')	(1, 5)	(1, 4)
λ	2340.3	2295.3

IV. $F \to b(^1\Sigma^+)$ System

Band heads:

(v', v'')	(0, 2)	(0, 1)	(0, 0)
λ	2167.6	2127.8	2089.1

SeO

V. $\underline{C \rightarrow X^3\Sigma^-\text{ System}}$

Band heads:

(v', v'')	$(0, 1)$	$(0, 0)$
λ	1999.3	1963.9

VI. $\underline{E \rightarrow X^3\Sigma^-\text{ System}}$

Band heads:

(v', v'')	$(0, 4)$	$(0, 3)$
λ	1979.2	1944.6
	1973.3	1938.8

SeO

SPECTROSCOPIC CONSTANTS

State	T_e	ω_e	$x_e\omega_e$	B_e	$\alpha_e \times 10^3$	$D_e \times 10^6$	r_e	Remarks	Bibliography
F	65145.5	970	6.0	-	-	-	-		(64.7)
D	63361.5	942	3.5	-	-	-	-		(64.7)
E	54288[a] 54131[a]	-	-	-	-	-	-		(64.7)
C	50862	1034	6.0	-	-	-	-		(64.7)
$A^3\Sigma^-$ $\{F_3$	-	-	-	0.3443[b]	-	-	$\}$ 1.931[c]		(65.8, 63.6)
F_2	34012.2+x	521.0	4.0	0.3391[b]	-	-			(65.8, 63.6)
F_1	34081.75[a]	521.0	4.0	0.3332[b]	-	-			(65.8, 63.6)
$b(^1\Sigma^+)$	17338.5	885.2	5.85	-	-	-	-		(72.11)
$X^3\Sigma^-$ $\{F_3$	-	-	-	0.4762	3.4	-	$\}$ 1.640		(65.8, 63.6)
F_2	x	915.43	4.52	0.4704	3.2	-			(65.8, 63.6)
F_1	0	914.69	4.52	0.4655	3.2	-			(65.8, 63.6)

[a] T_o, [b] B_o, [c] r_o.

Dissociation energy = 4.3 ± 0.1 eV, 100 kcal/mole, 34683 cm^{-1}.

BIBLIOGRAPHY

(35. 1)   Preliminary Note to (38.4),
L. Bloch, E. Bloch and Choong Shin-Piaw,
C.R. Acad. Sci. 201, 654-5

(35. 2)   Preliminary Note to (38.4),
L. Bloch, E. Bloch and Choong Shin-Piaw,
C.R. Acad. Sci. 201, 824-5

(36. 3)   Flame Excitation,
R. K. Asundi, M. Jan-Kahn, and R. Samuel,
Proc. Roy. Soc. A 157, 28-49

(38. 4)   Excited Exchange. Vibrational Analysis,
Choong Shin-Piaw,
Ann. Physique 10, 173-290

(62. 5)   R. F. Barrow and E. W. Deutsch,
"Rotational Analysis of the $^3\Sigma^- - ^3\Sigma^-$ System of SeO,"
Proc. Phys. Soc. 80, 993-4

(63. 6)   A-X System Rotational Analysis. Dissociation Energy,
R. F. Barrow and E. W. Deutsch,
Proc. Phys. Soc. 82, 548-56

(64. 7)   C-X, E-X, D-b$^1\Sigma^+$, F-b$^1\Sigma^+$ Systems Vibrational Analysis,
P. B. V. Haranath,
J. Mol. Spectrosc. 13, 168-73

(65. 8)   A-X System, Vibrational Analysis,
P. B. V. Haranath,
"New Emission Bands of SeO in the Region 5100-5900 Å,"
Ind. J. Pure & App. Phys. 3, 75-76

(67. 9)   Observation of $X^3\Sigma^+$ and $^1\Delta$ in Electron Spin Resonance,
A. Carrington, G. N. Currie, P. N. Dyer, D. H. Levy, and T. A. Miller,
Chemical Comm. 641-2

(69. 10)   Electron Spin Resonance,
A. Carrington, G. N. Currie, D. H. Levy and T. A. Miller,
Molecular Phys. 17, 535-42

(72. 11)   b → X System. Vibrational Analysis,
V. S. Kushawaha and C. M. Pathak,
"A New Electronic Transition in SeO,"
Spectroscopy Letters 5, 393-99

SiBr

Methods of Production and Experimental Technique

Emission from a discharge into a stream of $SiBr_4$.
Absorption after flash photolysis of $SiBr_4$.

BAND SYSTEMS

System	Transition	Sources	Wavelength Limits	Degrading	Band Head, $\nu_{0,0}$	Remarks	Bibliography
I	$A^2\Sigma \rightarrow X^2\Pi_r$	High frequency discharge	6570-4330	R	20431.7 21275.5		(69.3)
II	$B'^2\Delta \rightarrow X^2\Pi$	High frequency discharge	4350-4150	-			(69.5)
III	$B(^2\Sigma) \rightarrow X^2\Pi$	High frequency discharge	3234-2875	V	33226.7 33644.7		(37.2, 35.1)
IV	$C^2\Pi \leftarrow X^2\Pi$	Absorption	2440-2340	V	41105		(71.6)
V	$D \leftarrow X^2\Pi$	Absorption	2239-2312	V	44088		(71.6)
VI	$E \leftarrow X^2\Pi$	Absorption	2264-2188	V	44585		(71.6)
VII	$F \leftarrow X^2\Pi$	Absorption	2221-2196	-	45017		(71.6)
VIII	?	Absorption	2185-2107	-	-		(71.6)

SiBr

I. $A^2\Sigma \to X^2\Pi_r$ System

Band heads (69.3):

$^2\Sigma \to {}^2\Pi_{1/2}$

(v', v'')	(0, 1)	(0, 2)	(1, 2)	(1, 3)	(2, 3)
λ	4894.0	4995.8	4934.7	5037.8	4975.7

$^2\Sigma \to {}^2\Pi_{3/2}$

(v', v'')	(0, 4)	(1, 4)	(1, 5)	(0, 5)
λ	5327.5	5257.7	5373.3	5446.3

III. $B(^2\Sigma) \to X^2\Pi$ System

Band heads, λ (Intensity) (37.2):

$^2\Sigma \to {}^2\Pi_{1/2}$

v', v''	0	1	2	3	4	5	6	7
0	2971.3 (6)	3009.2 (10)	3047.7 (9)	3086.8 (8)	3126.6 (7)	3167.5 (5)		
1	2922.0 (8)	2958.7 (7)		3033.7 (6)	3072.3 (7)	3111.6 (7)	3151.7 (6)	3192.4 (4)

$^2\Sigma \to {}^2\Pi_{3/2}$

v', v''	0	1	2
0	3008.8 (10)	3047.3 (9)	3086.8 (8)
1	2958.3 (7)	2995.7 (5)	
2	2875.0 (4)		

IV. $C^2\Pi \leftarrow X^2\Pi$ System

Band heads (71.6):

(v', v'')	(0, 0)	(1, 0)	(2, 0)	(3, 0)
λ	2432.06	2401.32	2395.06	2342.77

V. $D \leftarrow X^2\Pi$ System

Band heads (71.6):

(v', v'')	(0, 1)	(0, 0)	(1, 0)	(1, 1)
λ	2289.52	2267.47	2238.84	2260.23

VI. $E \leftarrow X^2\Pi$ System

Band heads (71.6):

(v', v'')	(0, 1)	(0, 0)	(1, 0)	(2, 0)
λ	2263.87	2242.21	2214.95	2188.41

VII. $F \leftarrow X^2\Pi$ System

Band heads (71.6):

(v', v'')	(0, 0)	(1, 0)
λ	2220.68	2196.04

SiBr

SPECTROSCOPIC CONSTANTS

State	T_e	ω_e	$x_e\omega_e$	B_o	$\alpha_e \times 10^3$	$D_e \times 10^6$	r_o	Remarks	Bibliography
E	44521	552	1.5					$A \sim 40$ cm^{-1}	(71.6)
D	44088[a]	565[b]							(71.6)
$C^2\Pi$	41051	531	2					$A = 12$ cm^{-1}	(71.6)
$B^2\Sigma^+$	33571.0	573.6	3.1	0.1771			2.15		(69.4, 35.1)
$B'^2\Delta$	23920								(69.5)
$A^2\Sigma^+$	20518.4	250.3	0.5						(69.3)
$X^2\Pi_{1/2}$	419.2	424.5	1.5						(69.3, 35.1)
$X^2\Pi_{3/2}$	0	424.2	1.5	0.1598			2.26		(69.3, 69.5, 35.1)

[a]T_o, [b]$\Delta G_{1/2}$.

Dissociation energy = 3.5 ± 0.5 eV, 81 kcal/mole, 28230 cm^{-1}.

Perturbations and General Information

B($^2\Sigma$) state predissociates for v' > 2 (35.1).

BIBLIOGRAPHY

(35.1)   B-X System, Vibrational Analysis,
E. Miescher,
Helv. Phys. Acta 8, 587-8

(37.2)   B-X System, Vibrational Analysis,
W. Jevons and L. A. Bashford,
Proc. Phys. Soc. 49, 554-67

(69.3)   A-X System, Vibrational Analysis,
K. B. Rao and P. B. V. Haranath,
"The Emission Band Spectrum of SiBr in the Visible Region,"
J. Phys. B 2, 1381-4

(69.4)   B-X System, Rotational Analysis,
L. A. Kuznetsova and Yu. Lizualpov,
"Rotational Analysis of the $^2\Sigma - ^2\Pi$, Band System of SiBr Molecule,"
Vestn. Mosk. Univ. Khim. 24, 103-4

(69.5)   B'-X System,
L. A. Kuznetsova and Yu. Ya. Kuzyakov,
Zhur. Pricklad. Spectroscopii 10, 413

(71.6)   C, D, E, F-X Systems,
B. A. Oldershaw and K. Robinson,
"Ultra Violet Spectrum of SiBr,"
Trans. Faraday Soc. 67, 1870-4

SiC

SiC

Dissociation energy = 4.62 ± 0.34 eV, 106.5 kcal/mole, 37250 cm^{-1} (64.2, 68.1).

BIBLIOGRAPHY

(58.1) J. Drowart, G. De Maria, and M. G. Inghram, J. Chem. Phys. 29, 1015-21

(64.2) JANAF Thermochemical Tables, Ed., Dow Chemical Co., Midland, Mich.

SiCl

Methods of Production and Experimental Technique

Absorption in a King furnace; pulsed discharge.
Emission from a discharge in $SiCl_4$ vapor mixed with argon.

BAND SYSTEMS

System	Transition	Sources	Wavelength Limits	Degrading	Band Head, $v_{0,0}$	Remarks	Bibliography
I	$A(^2\Sigma^+) \rightarrow X^2\Pi_r$	Discharge	5820-4550	R	22903 22696		(65.10)
II	$B^2\Sigma^+ \rightleftarrows X^2\Pi_r$	Discharge	3100-2850	V	34186.0 33978.1		(71.19, 69.15, 62.7, 60.6, 36.1)
III	$B'^2\Delta \rightleftarrows X^2\Pi_r$	Discharge	2830-2760	V	35617.9 35411.3		(66.12, 64.9, 63.8)
IV	$C^2\Pi \rightleftarrows X^2\Pi_r$	Discharge	2600-2350	V	41235 41040		(71.19, 36.1)
V	$D(^2\Pi) \rightleftarrows X^2\Pi_r$	Discharge	2340-2200	V	45006 44798		(71.19, 36.1)
VI	$E \leftarrow X^2\Pi_r$	Absorption	2229-2203	-	45169 45375		(71.19)
VII	$F \leftarrow X^2\Pi_r$	Absorption	2175-2164	-	45968 46179		(71.19)

SiCl

I. $A(^2\Sigma^+) \to X^2\Pi_r$ System

Only a single progression, v' = 0, is observed.
Band heads (65.10):

v', v''	0, 8	0, 7	0, 6	0, 5	0, 4
$\lambda(Q_2)$(Intensity)	5358.1(6)	5218.3(8)	5084.0(8)	4955.5(7)	4832.4(6)
$\lambda(R_1)$(Intensity)	5299.6(6)	5162.2(8)	5031.1(8)	4905.1(7)	4784.5(6)

II. $B^2\Sigma^+ \rightleftarrows X^2\Pi_r$ System

Band heads of the two subsystems (a) $^2\Sigma^+ - {}^2\Pi_{1/2}$; (b) $^2\Sigma^+ - {}^2\Pi_{3/2}$ (36.1):

v', v''	Sub-syst.	λ	Intensity	v', v''	Sub-syst.	λ	Intensity
0, 2	a	3036.6	4	1, 0	b	2865.8	8
1, 3	a	3020.5	6			2829.3	10
0, 2	b	3017.7	6	2, 0	a	2826.2	7
1, 3	b	3001.7	4			2823.6	6
0, 1	a	2988.8	8			2812.8	8
1, 2	a	2973.5	4	2, 0	b	2809.7	10
0, 1	b	2970.4	8			2807.2	10
0, 0	a	2942.2	8			2788.3	4
0, 0	b	2924.4	8			2783.6	6
1, 0	a	2882.9	8	3, 0	a	2772.0	7

III. $B'^2\Delta \rightleftarrows X^2\Pi_r$ System

Vibrational structure degrades R, rotational structure degrades V.
Band heads (P_2) (66.12, 64.9):

(v', v'')	(0, 0)	(1, 1)
λ	2823.38	2825.95

IV. $C^2\Pi \rightleftarrows X^2\Pi_r$ System

Band heads (P) of the two subsystems (a) $^2\Pi_{1/2} - ^2\Pi_{1/2}$; (b) $^2\Pi_{3/2} - ^2\Pi_{3/2}$ (36.1):

v', v''	Sub-syst.	λ	Intensity	v', v''	Sub-syst.	λ	Intensity
0, 3	a	2533.5	4	0, 0	a	2435.9	?
0, 3	b	2521.0	1	1, 1	a	2427.7	2
0, 2	a	2500.3	4	0, 0	b	2424.4	3
0, 2	b	2488.2	6	1, 1	b	2416.2	2
0, 1	a	2467.8	6	1, 0	a	2396.7	6
0, 1	b	2456.0	7	1, 0	b	2385.6	5

V. $D(^2\Pi) \rightleftarrows X^2\Pi_r$ System

Band heads of the two subsystems (a) $^2\Pi_{1/2} - ^2\Pi_{1/2}$; (b) $^2\Pi_{3/2} - ^2\Pi_{3/2}$ (71.19):

(v', v'')	(0, 1)	(0, 0)	(1, 0)	(2, 0)
λ (a)	2258.51	2231.63	2199.68	2168.88
λ (b)	2247.99	2221.37	2189.65	2159.11

VI. $E \leftarrow X^2\Pi_r$ System

Band heads of the two subsystems (a) $E - ^2\Pi_{1/2}$; (b) $E - ^2\Pi_{3/2}$:

(v', v'')	(0, 1)	(0, 0)
λ (a)	-	2213.22
λ (b)	2229.31	2203.20

VII. $F \leftarrow X^2\Pi_r$ System

Band heads of the two subsystems (a) $F - ^2\Pi_{1/2}$; (b) $F - ^2\Pi_{3/2}$:

(v', v'')	(0, 0)
λ (a)	2174.75
λ (b)	2164.82

SiCl

SPECTROSCOPIC CONSTANTS

State	T_e	ω_e	$x_e\omega_e$	B_e	$\alpha_e \times 10^3$	$D_e \times 10^6$	r_e	Remarks	Bibliography
F	46179[a]							$A \sim 0$ cm^{-1}	(71.19)
E	45375[a]							$A \sim 0$ cm^{-1}	(71.19)
D($^2\Pi$)	44943	659						$A \sim 0$ cm^{-1}	(71.17, 64.9, 36.1)
C$^2\Pi$	41165	674.2	2.20	0.2861[c]			1.94	$A = 10.5$ cm^{-1}	(71.17, 64.9, 64.11, 36.1)
B'$^2\Delta$	35617.1	511.1	5.6	0.2600[c]	2.5		2.035	$A = -2.73$ cm^{-1}	(66.12, 64.9, 63.8)
B$^2\Sigma^+$	34107	706.6	3.9	0.2794	2.5	0.135	1.971	$\gamma = 0.0098$ cm^{-1}	(69.15, 66.13, 62.7, 60.6)
A$^2\Sigma^+$	23010.4	296.4	(0.73)	0.1986	0.7		2.1[b]		(71.16, 65.10)
X$^2\Pi$	0	535.89	2.16	0.2559	1.56	0.216	2.058	$y_e\omega_e = 0.0053$ cm^{-1} $A = 207.21$ cm^{-1}	(71.16, 69.15, 66.13, 64.9, 62.7)

[a]T_o, [b]r_o, [c]B_o.

Dissociation energy = 4.5 ± 0.5 eV, 104 kcal/mole, 36300 cm^{-1}.

SiCl

Perturbations and General Information

Franck-Condon factors, r-centroids — Morse potential (71.18):

$B^2\Sigma - X^2\Pi$

v', v''	0	1	2	3	4
0	0.3595	0.3562	0.1827	0.0647	
	2.0130	1.9755	1.9380	1.9030	
	2924.4	2970.4	3017.6	3065.8	
1	0.3559		0.1719	0.1652	0.0645
	2.0605		1.9880	1.9520	1.9150
	2865.8		2955.0	3001.7	3049.3

Top = Franck-Condon factor
Middle = r-centroid
Bottom = Wavelength

Electronic transition moment: $R_e(r)$ = constant $(1 - 0.3856r)$ $1.9 \leq r \leq 2.1$ Å.

BIBLIOGRAPHY

(36.1)   B, C, D-X Systems, Vibrational Analysis,
W. Jevons,
Proc. Phys. Soc. 48, 563-73

(50.2)   B, B'-X Systems, Vibrational Analysis Possibly Incorrect,
S. N. Garg,
Proc. Nat. Inst. Sci. India A 19, 23-33

(52.3)   Absorption Spectrum,
K. Wieland and M. Heise,
Boll. Sci. Fac. Chim. Indisstr Bologna 10, 12-3

(54.4)   B → X System,
R. F. Barrow, G. Drummond, and S. Walker,
Proc. Phys. Soc. A 67, 186-7

(60.5)   B → X System,
B. A. Thrush,
Nature 186, 1044

SiCl

(60. 6)   B-X System, Rotational Analysis,
I. E. Ovcharenko, L. N. Tunitskii, and V. Yakutin,
Optics Spectr. 8, 393-5

(62. 7)   B-X System, Vibrational Analysis,
I. E. Ovcharenko and Yu. Ya. Kuzyakov,
Optics Spectr. 13, 362-5

(63. 8)   B'-X System, Rotational Analysis
I. E. Ovcharenko, Yu. Ya. Kuzyakov, and V. M. Tatevski,
Optika Spectrosk. Spornik Statei 2, 12-5

(64. 9)   B'-X System, Rotational Analysis,
R. D. Verma,
Can. J. Phys. 42, 2345-56

(65. 10)   A-X System, Vibrational Analysis,
N. Sanii and R. D. Verma,
Can. J. Phys. 43, 960-2

(65. 11)   C-X System, Rotational Analysis,
I. E. Ovcharenko and Y. Y. Kuzyakov,
"Fine Structure of Bands of the SiCl Molecule,"
Optika Spectrosk 18, 30-5

(66. 12)   B'-X System,
I. E. Ovcharenko, Yu. Ya. Kuzyakov and V. M. Tatevskii,
Optics Spectr. Suppl. 2 Mol. Spectroscopy 6-7

(66. 13)   B-X System, Rotational Analysis,
H. Cordes and F. Gehrke,
Z. Phys. Chem. 51, 281-9

(68. 14)   Dissociation Energy,
Y. Y. Kuzyakov,
"Determination of Energy of Dissociation of Certain
Monohalides of Carbon Subgroup,"
Vestn. Mosk. Univ. Khim. 3, 21-4

(69. 15)   B-X System, Rotational Analysis,
R. K. Mishra and B. N. Khanna
"On the B-X System of SiCl Molecule,"
Current Science 38, 361-2

(71.16) A-X System, Rotational Analysis,
S. R. Singhal and R. D. Verma,
"Rotational Analysis of the A-X System of the SiCl Molecule,"
Can. J. Physics 49, 407-411

(71.17) R. K. Pandey, K. N. Upadhya, and K. P. R. Nair,
"Rotational Structure of the C-X Band System of the SiCl Molecule,"
Ind. J. Pure and Appl. Physics 9, 36-38

(71.18) Franck-Condon Factors, r-Centroids, B-X,
J. Singh and P. S. Dube,
"On the Variation of Electronic Transition Moment with the Intermolecular Separation in SiCl and SnF Molecules,"
Ind. J. Pure Appl. Phys. 9, 164-5

(71.19) D, E, F → X Systems,
G. A. Oldershaw and K. Robinson,
"Ultraviolet Absorption Spectrum of Silicon Monochloride,"
J. Mol. Spectrosc. 38, 306-13

SiF

Methods of Production and Experimental Technique

Absorption in a King furnace.

Emission from a discharge across SiF_4. Discharge in a hollow cathode containing a mixture of Si powder and SiF_6 (He carrier).

BAND SYSTEMS

	System	Transition	Sources	Wavelength Limits	Degrading	Characteristic Bands, λ	Remarks	Bibliography
I	α	$A^2\Sigma^+ - X^2\Pi$	Discharge	4971-4234	R	4368.2(0,0)(R_1)		(58.6, 27.2, 11.1)
II	β	$B^2\Sigma^+ - X^2\Pi$	Discharge	3206-2685	V	2894.4(0,0)(P_2)		(58.6, 37.4, 36.3, 27.2, 11.1)
III	γ	$C^2\Delta - X^2\Pi$	Discharge	2654-2525	V	2539.2(0,0)(P_2)		(68.13, 27.2, 11.1)
IV		$C'^2\Pi - X^2\Pi$	Hollow cathode					(59.8)
V		$D'^2\Pi - X^2\Pi$	Hollow cathode					(59.8)
VI	η	$a^4\Sigma^- - X^2\Pi$	Discharge	3368-3340		3363(0,0) 3346(0,0)		(62.9, 27.2, 11.1)
VII		$D^2\Sigma^+ - X^2\Pi$	Hollow cathode	2178-1998	V	2108.7(0,0)(Q_2)		(58.6, 51.5)
VIII		$E - X^2\Pi$	Hollow cathode	1955-1948	V	1955(0,0)(P_2)		(58.6)
IX		$F - X^2\Pi$	Hollow cathode	1935-1928	V	1935(0,0)(P_2)		(58.6)
X		$G - X^2\Pi$	Hollow cathode	1951-1882	V	1925(0,0)(P_2)		(58.6)
XI		$H^2\Sigma^+ - X^2\Pi$	Hollow cathode	1951-1870	V	1919(0,0)(P_2)		(58.6)

BAND SYSTEMS

	System	Transition	Sources	Wavelength Limits	Degrading	Characteristic Bands, λ	Remarks	Bibliography
XII		$I - X^2\Pi$	Hollow cathode	1911-1904	V	1911(0,0)(P_2)		(58.6)
XIII		$B^2\Sigma^+ - A^2\Sigma^+$	Hollow cathode	8965-7068	V	8435.9(0,0)		(58.6)
XIV	ε	$C'^2\Pi - A^2\Sigma^+$	Hollow cathode	5644-4819	V	5193.6(0,0)		(59.8, 27.2, 11.1)
XV	ε	$D'^2\Pi - A^2\Sigma^+$	Hollow cathode	4230-3330	V	4183.3(0,0)		(59.8, 27.2, 11.1)
XVI		$D'^2\Pi - B^2\Sigma^+$	Hollow cathode					(59.8)
XVII	ξ	(a)	Discharge	6594-6270	V			(27.2)

[a] Bands have not been analyzed.

SiF

I. $\underline{A^2\Sigma^+ - X^2\Pi \text{ System } (\alpha)}$

Band heads ($^Q R_{12}$):

v', v''	3, 5	0, 1	4, 4	3, 3	2, 2
$\lambda(^Q R_{12})$	4850.5	4569.5	4531.6	4495.8	4462.0

v', v''	1, 1	0, 0	3, 2	2, 1	1, 0
$\lambda(^Q R_{12})$	4430.2	4400.5	4334.4	4301.3	4270.2

II. $\underline{B^2\Sigma^+ - X^2\Pi \text{ System } (\beta)}$

Band heads:

v', v''	λ(head)
0, 2	3042.4(P_2)
	3027.5(P_1)
0, 1	2967.1(P_2)
	2952.8(P_1)
0, 0	2894.4(P_2)
	2880.8(P_1)
1, 0	2813.0(P_2)
	2800.0(P_1)

III. $\underline{C^2\Delta - X^2\Pi \text{ System } (\gamma)}$

Band heads (P_2):

v', v''	0, 2	0, 1	1, 2	2, 3	3, 4
λ	2652.8	2595.1	2592.4	2589.9	2587.5

v', v''	4, 5	0, 0	1, 1	2, 2	3, 3
λ	2586.1	2539.2	2537.3	2535.4	2533.8

VI. $\underline{a^4\Sigma^- - X^2\Pi \text{ System } (\eta)}$

Band heads:

(v', v'')	(0, 0)	(1, 1)	(2, 2)
λ	3342	3354	3353

VII. $D^2\Sigma^+ - X^2\Pi$ System

Band heads:

(v', v'')	(0, 1)	(0, 0)	(1, 1)	(1, 0)
λ	2147.0	2108.7	2102.2	2065.5
	2139.6	2101.5	2095.2	2058.7

XIII. $B^2\Sigma^+ - A^2\Sigma^+$ System

Band heads:

(v', v'')	(0, 1)	(0, 0)	(1, 0)	(2, 1)	(2, 0)
λ	8964.5	8435.9	7778.5	7605.0	7221.2

XIV. $C'^2\Pi - A^2\Sigma^+$ System (ϵ)

Band head:

(v', v'')	(0, 0)
λ	5193.6

XV. $D'^2\Pi - A^2\Sigma^+$ System (ϵ)

Band heads:

(v', v'')	(?)	(0, 0)	(?)
λ	4229.7	4183.3	4011.8

XVII. ξ System

Bands:

λ | 6594 | 6492 | 6416 | 6397 | 6270

SiF

SPECTROSCOPIC CONSTANTS

State	T_e	ω_e	$x_e\omega_e$	B_e	$\alpha_e \times 10^3$	$D_e \times 10^6$	r_e	Remarks	Bibliography
I	52409[a]								(58.6)
$H^2\Sigma^+$	52179[a]	~1030							(58.6)
G	52020[a]	~1020							(58.6)
F	51770[a]	—							(58.6)
E	51224[a]	—							(58.6)
$D^2\Sigma^+$(b)	47418.6	1003.2	5.64	0.625	5		1.54		(58.6)
$D'^2\Pi$	46606.7	1032.9	5.28	0.6329	4.4		—		(59.8)
$C'^2\Pi$	41964.9	1031.9	4.45	0.6376	3.9		—		(59.8)
$C^2\Delta$	39438.4	892.41	7.01	0.60338	5.39		1.5714		(68.13)
$B^2\Sigma^+$(b)	34561.5	1011.2	4.825	0.62707	4.62		1.5714		(68.13, 58.6)
$a^4\Sigma^-$	29805.06	863.16	5.730	0.5786	5.02	1.05	1.604	$\gamma = 1.88 \times 10^{-3}$	(73.L1, 62.9)
$A^2\Sigma^+$	22858.4	718.5	10.167	0.57839	9.41		1.6049	$y_e\omega_e = 0.157$ cm^{-1}	(58.6)
$X^2\Pi$	0	857.20	4.735	0.58138	4.94	1.07	1.6008	$A = 161.9$ cm^{-1}	(73.L1, 68.13, 62.9, 58.6)

[a] T_0, [b] B and D states are considered Rydberg states where T_e is represented by $T_e = 58566 - \dfrac{109737}{(m-1.562)^2}$ with m = 4 and 5, m = 6 is identified with the H state.

Dissociation energy = 5.0 ± 0.5 eV, 115 kcal/mole, 40328 cm^{-1}.

Perturbations and General Information

Electronic transition moment:

$A^2\Sigma - X^2\Pi$, $R_e(r)$ = const. (70.18),

$B^2\Sigma - X^2\Pi$, $R_e(r)$ = const. exp. $[9.24 - 5.93\ r_{v',v''}]$ (70.19).

Potential energy curves — RKRV potential (66.12):

State	v	U(cm^{-1})	r_{min}(Å)	r_{max}(Å)
		SiF (Te = 0 cm^{-1})		
$X^2\Pi$	0	428.6	1.545	1.663
	1	1275.9	1.508	1.714
	2	2114.1	1.485	1.752
	3	2942.1	1.468	1.785
	4	3759.8	1.454	1.815
	5	4571.3	1.441	1.842
	6	5372.9	1.431	1.869
		SiF (Te = 22942.4 cm^{-1})		
$A^2\Sigma$	0	356.7	1.545	1.676
	1	1024.2	1.507	1.737
	2	1675.9	1.483	1.784
	3	2315.5	1.466	1.826
	4	2944.2	1.452	1.865
	5	3546.0	1.441	1.901
	6	4143.9	1.431	1.937
		SiF (Te = 34640.2 cm^{-1})		
$B^2\Sigma$	0	504.4	1.490	1.598
	1	1506.0	1.455	1.644
	2	2497.1	1.433	1.678
	3	3479.7	1.416	1.707
	4	4452.8	1.402	1.733
		SiF (Te = 39513.6 cm^{-1})		
$C^2\Sigma$	0	444.3	1.516	1.632
	1	1323.6	1.481	1.683
	2	2189.4	1.459	1.721
	3	3045.3	1.442	1.754
	4	3888.3	1.429	1.785

SiF

Franck-Condon factors, r-centroids — Morse potential:

$A^2\Sigma^+ - X^2\Pi$ (70.17)

v', v''	0	1	2	3
0	0.98710	0.0109	0.0016	0.0003
1	0.00886	0.9407	0.0466	0.0022
2	0.00393	0.0334	0.8397	0.1183
3	0.00007	0.0141	0.0744	0.6779
4		0.0006	0.0340	0.1232
5			0.0029	0.0668
6			0.0008	0.0095
7				0.0020
8				0.0003

Franck-Condon factors only.

$B^2\Sigma - X^2\Pi$ (69.15)

v', v''	0	1	2	3
0	0.618	0.306	0.083	-
	1.5830	1.5365	1.4865	1.4390
	2893.3	2965.8	3041.4	3119.9
1	0.318	0.145	0.328	-
	1.6430	1.5950	1.5360	1.4970
	2811.8	2866.8	2938.6	3025.4
2	0.002	0.362	0.000	-
	1.7005	1.6425	1.6030	1.5540
	2735.3	2801.4	-	2924.4
3	-	-	-	-
	1.7570	1.7060	1.6600	1.6100
	-	2725.1	2788.7	2843.2

Top   = Franck-Condon factor
Middle = r-centroid
Bottom = Wavelength

$C^2\Sigma - X^2\Pi$ (69.16)

v'', v'	0	1	2	3	4
0	0.8857 1.5883 2529.6	0.0996 1.4921 2585.1	0.0089 1.4199 2642.2	0.0008 1.3625 -	0.0000 1.3151 -
1	0.1072 1.7351 -	0.6902 1.5931 2527.7	0.1715 1.4967 2582.2	0.0238 1.4245 2638.7	0.0028 1.3670 -
2	0.0029 1.9997 -	0.1973 1.7400 -	0.5273 1.5979 2525.8	0.2185 1.5014 2579.7	0.0422 1.4291 2635.4
3	0.0000 3.7331 -	0.0084 2.0052 -	0.2719 1.7451 -	0.3904 1.6028 -	0.2474* 1.5062 2577.3
4	0.0000 - -	0.0000 3.7399 -	0.0163 2.0107 -	0.3325 1.7503 -	0.2850 1.6077 -

Top = Franck-Condon factor
Middle = r-centroid
Bottom = Wavelength

$B^2\Sigma^+ - A^2\Sigma^+$ (70.17)

v'', v'	0	1	2	3	4
0	0.5222(0) 1.575	0.3173(0) 1.635	0.1174(0) 1.682	0.3332(-1) 1.722	0.7874(-2) 1.759
1	0.3334(0) 1.533	0.4679(-1) 1.571	0.2544(0) 1.646	0.2140(0) 1.692	0.1025(0) 1.732
2	0.1124(0) 1.485	0.2849(0) 1.547	0.2597(-1) 1.642	0.7771(-1) 1.653	0.1988(0) 1.703
3	0.2640(-1) 1.430	0.2242(0) 1.507	0.9770(-1) 1.555	0.1294(0) 1.636	0.1606(-4) -
4	0.4785(-2) 1.363	0.9291(-1) 1.459	0.2274(0) 1.525	0.9543(-3) 1.379	0.1345(0) 1.645
5	0.7019(-3) 1.277	0.2672(-1) 1.404	0.1686(0) 1.485	0.1215(0) 1.539	0.4313(-1) 1.637
6	0.8505(-4) 1.160	0.5948(-2) 1.336	0.7551(-1) 1.439	0.1889(0) 1.508	0.1687(-1) 1.522

SiF

v'', v'	0	1	2	3	4
7	-	0.1082(-2)	0.2475(-1)	0.1354(0)	0.1231(0)
	-	1.249	1.383	1.469	1.525
8	-	0.1660(-3)	0.6488(-2)	0.6500(-1)	0.1594(0)
	-	1.135	1.317	1.424	1.495
9	-	0.2218(-4)	0.1429(-2)	0.2404(-1)	0.1154(0)
	-	0.981	1.234	1.371	1.458
10	-	-	0.2739(-3)	0.7362(-2)	0.6017(-1)
	-	-	1.128	1.308	1.415
11	-	-	-	0.1953(-2)	0.2523(-1)
	-	-	-	1.232	1.366

Top = Franck-Condon factor followed by factor of 10
Bottom = r-centroid

BIBLIOGRAPHY

(11.1) First Observations,
C. Porlazza,
Atti Accad. Lincol, Cl. Sci. Fis. Mat. Nat. Rend. 20, 486-90

(27.2) General Study,
R. C. Johnson and H. G. Jenkins,
Proc. Roy. Soc. A 116, 327-51

(36.3) β and γ Systems,
R. K. Asundi and R. Samuel,
Proc. Indian Acad. Sci. A 3, 346-59

(37.4) α and β Systems,
E. H. Eyster,
Phys. Rev. 51, 1078-86

(51.5) U. V. Bands,
W. H. Dovell and R. F. Barrow,
Proc. Phys. Soc. A 64, 98-9

(58.6) General Study,
J. W. C. Johns and R. F. Barrow,
Proc. Phys. Soc. 71, 476-84

(58.7) δ System,
J. W. C. Johns, G. W. Chantry, and R. F. Barrow,
Trans. Faraday Soc. 54, 1589-91

(59.8) ϵ System,
R. F. Barrow, D. Butler, J. W. C. Johns, and J. L. Powell,
Proc. Phys. Soc. 73, 317-20

(62.9) η System,
R. D. Verma,
Can. J. Phys. 40, 586-97

(64.10) Dissociation Energy,
T. C. Ehlert and J. L. Margrave,
J. Chem. Phys. 41, 1066-72

(66.11) Franck-Condon Factors, r-Centroids, C-X System,
S. Sankaranayanan and P. S. Narayanan,
"The Emission Spectrum of the γ System of SiF,"
Proc. Nat. Institute Sci. 32, 56-62

(66.12) R. B. Singh and D. K. Rai,
"Potential Curves for Some Diatomic Molecules P_2, PN, SiN, NBr, BaO, BeF, SiF and SnF,"
Indian J. Pure Appl. Phys. 4, 102-5

(68.13) γ System,
O. Appelblad, R. F. Barrow, and R. D. Verma,
J. Phys. B. Proc. Phys. Soc. 1, 274-82

(68.14) Y. Y. Kuzyakov,
"Determination of Energy of Dissociation of Certain Monohalides of Carbon Subgroup,"
Vestn. Mosk. Univ. Khim 3, 21-4

(69.15) B. S. Mohanty and O. N. Singh,
"Franck-Condon Factors and r-Centroids of Some Band Systems of the Monofluorides of Silicon, Calcium and Bismuth,"
Indian J. Pure Appl. Phys. 7, 109-111

(69.16) I. D. Singh and R. C. Maheshware,
"Franck-Condon Factors and r-Centroids for the C-X System of the SiF Molecule,"
Indian J. Pure Appl. Phys. 7, 708-9

SiF

(70.17) Franck-Condon Factors, r-Centroids, A,B-X, A-B Systems,
T. Wentink, Jr. and R. J. Spindler, Jr.,
"Franck-Condon Factors and r-Centroids for NO^+, CP, SiF, BF, BCl and BBr,"
J. Quant. Spectrosc. Radiative Transfer 10, 609-19

(70.18) Electronic Transition Moment, A-X System,
N. E. Kuz'menko, A. D. Smirnov, and Y. Y. Kuzyakov,
"Concerning Dependences of Electronic Transition Moment R_e on Internuclear Distance r for Bands of α-System of SiF Molecule,"
Vestn. Mosk. Univ. Khim 11, 357-9

(70.19) Electronic Transition Moment, B-X,
N. E. Kuz'menko, A. D. Smirnov, and Y. Y. Kuzyakov,
"The dependence of Electronic Transition Moment R_e on the Internuclear Separation $r_{v'v''}$ for B-X Band System of SiF Molecule,"
Vestn. Mosk. Univ. Khim 11, 478-480

(71.20) Hartree-Fock, SCF Calculations,
P. A. G. O'Hare and A. C. Wahl,
"Molecular Orbited Investigation of CF and SiF and their Positive and Negative Ions,"
J. Chem. Phys. 55 666-76

(73.L1) R. W. Martin and A. J. Merer,
"The $a^4\Sigma^- - X^2\pi$ Electronic Transition of SiF,"
Can. J. Phys. 51, 634-43

SiI

Methods of Production and Experimental Technique

Flash photolysis of SiI_4.
High frequency discharge.

BAND SYSTEMS

System	Transition	Sources	Wavelength Limits	Degrading	Band Head, $\nu_{0,0}$	Remarks	Bibliography
I	$A' \rightleftarrows X^2\Pi_{3/2}$	Microwave discharge	5500-4800	R	20247.04(0,0)		(72.L1, 70.4, 69.3)
II	$A^2\Sigma \leftarrow X^2\Pi_{1/2}$	Flash photolysis, absorption	4900-4200	R	21127.14(0,0)		(72.L1)
III	$B^2\Sigma^+ \leftarrow X^2\Pi_{1/2}$	Absorption	3190-2900	V	32434.4(0,0)	Possible first Rydberg state	(72.L1, 68.1)
IV	$C \leftarrow X^2\Pi$	Absorption	2400-2290	-	42772(0,0)		(72.5)
V	$D \leftarrow X^2\Pi$	Absorption	2350-2332	-	42859(0,0)		(72.5)
VI	$E \leftarrow X^2\Pi$	Absorption	2285-2267	-	44404(0,0)		(72.5)
VII	$F \leftarrow X^2\Pi$	Absorption	2257-2222	-	44995(0,0)		(72.5)
VIII	?	Absorption	2405-2296	-			(72.5)

SiI

I. $\underline{A' \rightleftarrows X^2\Pi_{3/2} \text{ System}}$

Band heads, λ (72.L1):

v', v''	0	1	2	3	4	5
0	4939.0	5027.7	5118.8			
1	4875.3	4961.7	5050.5	5142.0		
2		4899.9		5075.9	5167.6	5262.4
3			4927.3			5196.2

II. $\underline{A^2\Sigma \leftarrow X^2\Pi_{1/2} \text{ System}}$

Band heads, λ (72.L1):

(v', v'')	λ	(v', v'')	λ	(v', v'')	λ
(0, 1)	4815.8	(3, 0)	4600.6	(8, 0)	4406.0
(1, 1)	4768.6	(4, 0)	4559.2	(9, 0)	4371.0
(0, 0)	4733.2	(5, 0)	4519.0	(10, 0)	4337.3
(1, 0)	4687.8	(6, 0)	4480.0	(11, 0)	4305.0
(2, 0)	4643.5	(7, 0)	4442.3	(12, 0)	4274.8

III. $\underline{B^2\Sigma^+ \leftarrow X^2\Pi_{1/2} \text{ System}}$

Band heads, λ (72.L1):

v', v''	0	1	2
0	3083.1	3117.9	3153.2
1	3039.2	3072.8	
2	2996.6		
3	2955.4		
4	2915.4		

IV. $\underline{C \leftarrow X^2\Pi \text{ System}}$

Band heads, λ (70.4):

(v', v'')	(0, 1)	(0, 0)	(1, 0)	(2, 0)
λ	2357.03	2337.27	2311.38	2386.43

V. $\underline{D \leftarrow X^2\Pi \text{ System}}$

Band heads, λ (70.4):

(v', v'')	(0, 1)	(0, 0)
λ	2352.20	2332.50

VI. $\underline{E \leftarrow X^2\Pi \text{ System}}$

Band heads, λ (70.4):

(v', v'')	(0, 1)	(0, 0)
λ	2285.30	2266.66

VII. $\underline{F \leftarrow X^2\Pi \text{ System}}$

Band heads, λ (70.4):

(v', v'')	(0, 2)	(0, 1)	(0, 0)
λ	2257.3	2239.5	2221.8

SiI

SPECTROSCOPIC CONSTANTS

State	T_e	ω_e	$x_e\omega_e$	B_e	$\alpha_e \times 10^3$	$D_e \times 10^6$	r_e	Remarks	Bibliography
$B^2\Sigma^+$	32380.3	471.7	0.9						(72.L1, 68.1)
$A^2\Sigma$	21204.9	208.6	1.66	(0.085)				$\omega_e y_e = 0.079$ cm^{-1} $\omega_e z_e = 0.0055$ cm^{-1}	(72.L1, 70.4, 69.2)
A'	20289.7	275.7	5.6	(0.118)					(72.L1)
$X^2\Pi_{3/2}$	–	359.0	1.1				(2.451)		(72.L1, 70.4, 69.3, 68.1)
$X^2\Pi_{1/2}$	0	363.8	1.2	(0.123)					(72.L1, 70.4, 69.3, 68.1)

Dissociation energy ~24400 ± 1000 cm, 3.0 ± 0.1 eV.

Perturbations and General Information

It is possible that the A' state is the state which causes the perturbations in the (7, 0) and (8, 0) bands of the A-X system.

BIBLIOGRAPHY

(68. 1)   Flash Photolysis, B-X, Vibrational Analysis,
G. A. Oldershaw and K. Robinson,
Trans. Faraday Soc. 64, 616-9

(68. 2)   B-X System,
G. A. Oldershaw and K. Robinson,
"Ultra-violet Absorption Spectra of GeI and SiI,"
Trans. Faraday Soc. 64, 2256-9

(69. 3)   A-X System, Vibrational Analysis,
A. Lakshimnarayana and P. B. V. Haranath,
"The Band Spectrum of Silicon Monoiodide,"
Current Science 38, 136

(70. 4)   A-X System, Vibrational Analysis,
A. Lakshimnarayana and P. B. V. Haranath,
"The Emission Band Spectrum of Silicon Monoiodide,"
J. Phys. B 3, 576-8

(72. 5)   C, D, E, F ← X Systems,
G. A. Oldershaw and K. Robinson,
"Ultra-Violet Spectra of GeI and SiI: Lower Wavelength Bands,"
J. Mol. Spectrosc. 44, 602-4

(72. L1)   New A-X System,
J. Billingsley,
"The Absorption and Emission Spectrum of Silicon Monoiodide,"
J. Mol. Spectrosc. 43, 128-147

SiN

Methods of Production and Experimental Technique

Emission from gaseous $SiCl_4$ + active nitrogen. Noncondensing discharge across a capillary of quartz. Radio frequency discharge. Microwave discharge into $SiCl_4$ + active nitrogen.

BAND SYSTEMS

System	Transition	Sources	Wavelength Limits	Degrading	Characteristic Bands, λ	Remarks	Bibliography
I	$B^2\Sigma \rightleftharpoons X^2\Sigma$	Active N_2 + $SiCl_4$	5260-3780	R	4141.5(1,1) 4116.3(0,0)		(68.11, 65.8, 60.6, 13.1)
II	$C \rightarrow A^2\Pi$	Active N_2 + $SiCl_4$	5600-3200	R	3832.9(0,0)		(65.8, 43.5, 40.4)

I. $\underline{B\,^2\Sigma \rightleftarrows X\,^2\Sigma \text{ System}}$

Band heads (68.11):

(v', v'')	(5, 5)	(4, 4)	(3, 3)	(2, 2)	(1, 1)	(0, 0)
λ	4270.9	4234.3	4200.5	4168.6	4141.5	4116.3

SiN

SPECTROSCOPIC CONSTANTS

State	T_o	ω_e	$x_e\omega_e$	B_o	$\alpha_e \times 10^3$	$D_o \times 10^6$	r_o	Remarks	Bibliography
C	50253.08 + x	697.3	3.3	0.7106			1.595		(68.11, 65.8)
$B^2\Sigma$	24236.47	1031.01	16.743	0.7183	11.0	1.48	1.584	$y_e\omega_e = 0.11722$ cm^{-1}	(69.12, 65.8, 29.3)
$A^2\Pi$	x	1045.3	6.5						(65.8)
$X^2\Sigma$	0	1151.68	6.560	0.7282	6.3	1.19	1.575	$\gamma_o = 0.015$ cm^{-1}	(69.12, 65.8, 29.3)

Dissociation energy = 4.5 ± 0.4 eV, 104 kcal/mole, 36296 cm^{-1}.

Perturbations and General Information

Potential energy curves — RKRV potential (66.9):

State	v	U(cm^{-1})	r_{min}(Å)	r_{max}(Å)
		SiN ($T_e = 0$ cm^{-1})		
$X^2\Sigma^+$	0	574.2	1.519	1.631
	1	1712.7	1.483	1.679
	2	2838.2	1.460	1.714
	3	3950.5	1.443	1.744
	4	5049.7	1.428	1.771
	5	6135.8	1.415	1.797
	6	7208.7	1.404	1.821
	7	8268.4	1.394	1.844
	8	9314.9	1.385	1.866
		SiN ($T_e = 24299.4$ cm^{-1})		
$B^2\Sigma^+$	0	511.3	1.525	1.644
	1	1509.7	1.490	1.699
	2	2475.7	1.468	1.742
	3	3411.0	1.452	1.779
	4	4317.4	1.438	1.815
	5	5197.0	1.426	1.848
	6	6052.1	1.416	1.880
	7	6885.5	1.408	1.911

Franck-Condon factors, r-centroids — Morse potential (63.7):

$B^2\Sigma^+ - X^2\Sigma^+$

v', v''	0	1	2	3	4	5	6	7	8
0	0.978								
	1.596								
1	0.022	0.907							
	1.395	1.624							
2		0.073	0.772						
		1.404	1.651						
3			0.159	0.575	0.200				
			1.420	1.669	1.752				
4				0.267	0.346	0.253	0.095		
				1.446	1.688	1.764	1.819		

SiN

v', v''	0	1	2	3	4	5	6	7	8
5				0.022	0.361	0.139	0.255	0.144	
				1.383	1.499	1.706	1.776	1.828	
6					0.065	0.394	0.018	0.195	
					1.394	1.614	1.725	1.788	
7						0.139	0.330	0.010	0.095
						1.410	1.654	1.742	1.801
8							0.226	0.188	
							1.440	1.685	
9									0.048
									1.711

Top   = Franck-Condon factor
Bottom = r-centroid

B state displays vibrational inversion, with v' = 5 maximum from the reaction "active" N_2 + $SiCl_4$ (65.8).

Rotational perturbation in B-X system at v' = 4, K' = 13, 14, 19, and 21 (65.8).

BIBLIOGRAPHY

(13.1) W. Jevons,
Proc. Roy. Soc. A 89, 187-93

(25.2) R. S. Mulliken,
Phys. Rev. 26, 319-24

(29.3) F. A. Jenkins and H. de Laszlo,
Proc. Roy. Soc. A 122, 103-21

(40.4) R. Bernard,
Ann. Physique 13, 5-77

(43.5) C → A System,
L. H. Woods,
Phys. Rev. 63, 426-30

(60.6) B ← X System,
B. A. Thrush,
Nature 186, 1044

(63. 7) Intensity Distribution of B-X System,
A. E. Stevens and H. I. S. Ferguson,
Can. J. Phys. 41, 240-5

(65. 8) B-X, C-A Systems, Vibrational Analysis,
K. Schofield and H. P. Broida,
"Chemiluminescent Emission from the Reactions of Volatile Silicon Compounds and Active Nitrogen,"
Photochemistry and Photobiology 4, 989-1002

(66. 9) R. B. Singh and D. K. Rai,
"Potential Curves for some Diatomic Molecules P_2, PN, SiN, NBr, BaO, BeF, SiF, SnF,"
Indian J. Pure and Appl. Physics 4, 102-5

(68. 10) F. Jenc,
"Ground State Reduced Potential Curves (R.P.C.) of BeF, CS, SiN, P_2 and GeO,"
Spectrochim. Acta 24a, 259

(68. 11) B → X, C → A Systems, Rotational Analysis,
S. Nagaraj and R. D. Verma,
Can. J. Phys. 40, 1597-602

(69. 12) B-X System, Rotational Analysis,
T. M. Dunn, K. M. Rao, S. Nagaraj, and R. D. Verma,
Can. J. Phys. 47, 2128

SiO

Methods of Production and Experimental Technique

Absorption in SiO_2 at $T > 1500°C$ (45.7) where $Si + SiO_2$, $T > 1300°C$ (54.11) in a graphite furnace.

Emission from a carbon arc + compounds of Si. Flames and discharges into $SiCl_4$ stock tube methods (68.20).

Flash absorption - discharge into a stream of SiH_4+O_2 (71.29).

Radio frequency discharge into $SiCl_4$ vapor + argon (70.24).

Excitation of $SiO(Cl)_6$ (72.34).

Molecular beams electric resonance (70.28).

BAND SYSTEMS

System	Transition	Sources	Wavelength Limits	Degrading	Band Head, $\nu_{0,0}$	Remarks	Bibliography
I	$A^1\Pi \rightleftarrows X^1\Sigma^+$	Discharge	2930-2100	R	42640.1		(72.35, 69.21, 53.10, 32.5, 24.1)
II	$E^1\Sigma \leftarrow X^1\Sigma^+$	Absorption	2000-1715	R	52580		(71.29, 54.11)
III	$F \leftarrow X^1\Sigma^+$	Absorption	1525-1415	R	68497		(54.11)
IV	$G \leftarrow X^1\Sigma^+$	Absorption	1400-1330	R	72011		(54.11)
V	$H \leftarrow X^1\Sigma^+$	Absorption	1330-1270	R	76391		(54.11)
VI	$(I) \leftarrow X^1\Sigma^+$	Absorption	1256	R	(79605)		(54.11)
VII	$b^3\Sigma \rightarrow a^3\Pi_r$	Discharge	4300-4200	V, R	23642 23569 23498		(70.24, 61.12)
VIII	$c^3\Pi \rightarrow a^3\Pi_r$	Discharge	3950-3700	R	25620 26092 26562 27024		(72.32)
IX	$?^3\Sigma \rightarrow a^3\Pi_r$	Discharge	2950	-	33898.3		(72.32)

I. $\underline{A^1\Pi \rightleftarrows X^1\Sigma^+}$ System

Rotational structure is perturbed.

Band heads, λ(intensity) (72.35, 53.10, 32.5, 24.1):

v', v''	0	1	2	3	4	5	6	7
0	2344.3 (5)	2413.77 (7)	2486.80 (6)	2563.80 (5)	2644.79 (4)	2730.1 (2)	2820.04	
1	2298.9 (6)	2365.72 (6)	2436.3 (3)	2509.91 (4)	2587.15 (5)	2669.04 (9)	2755.01 (6)	2845.73 (2)
2	2255.9 (4)		2387.97 (5)	2459.04 (3)		2611.26 (4)	2693.7 (9)	2780.51 (7)
3	2215.4 (2)	2277.2 (1)	2342.45 (1)	2410.25	2481.9 (2)		2636.05	2718.8 (4)
4	2176.6 (1)	2236.3 (2)		2364.46				
5		2197.4 (0)						

II. $\underline{E^1\Sigma^+ \leftarrow X^1\Sigma^+}$

Band heads, λ (71.29, 54.11):

(v', v'')	(3, 1)	(1, 0)	(0, 0)	(1, 1)	(1, 2)	(0, 2)
λ	1875.22	1878.01	1901.85	1922.41	1968.48	1994.78

III. $\underline{F \leftarrow X^1\Sigma^+}$ System

Band heads, λ(intensity) (54.11):

v', v''	0	1	2	3
0	1459.91(10)	1486.58(6)		
1	1436.84(8)	1462.66(9)	1489.25(6)	
2	1414.67(4)	1439.72(8)	1465.43(6)	1491.84(6)
3		1417.78(4)		
4			1421.14(6)	

SiO

IV. $G \leftarrow X^1\Sigma^+$ System

Band heads, λ(intensity) (54.11):

v', v''	0	1
0	1388.68(5)	
1	1373.06(6)	1396.59(4)
2		1381.01(4)

V. $H \leftarrow X^1\Sigma^+$ System

Band heads, λ(intensity) (54.11):

(v', v'')	(0, 0)	(0, 1)
λ(Intensity)	1309.20(5)	1330.37(4)

VI. $(I) \leftarrow X^1\Sigma^+$ System

System is uncertain. Seen as a single band at 1256.2 Å (54.11).

VII. $b^3\Sigma \rightarrow a^3\Pi_r$ System

Band heads (R_{11}), λ (70.24) $\Delta v(^Q R_{12} - R_{11})$ ~13.0 cm^{-1}:

(v', v'')	(0, 0)	(1, 1)	(2, 2)	(3, 3)
λ	4228.6	4243.8	4262.6	4285.3

VIII. $c^3\Pi_i \rightarrow a^3\Pi_r$ System

Considerable overlapping with SiN bands.
Band heads (R_2) (72.32):

(v', v'')	(0, 0)	(2, 0)	(3, 0)
λ	3917.5	3778.7	3715.2

IX. $?^3\Sigma^+ \rightarrow a^3\Pi_r$ System

Considerable overlapping with SiCl bands.
Three bands observed (72.32):

| λ | 2948.8 | 2955.6 | 2962.3 |

SiO

SPECTROSCOPIC CONSTANTS

State	T_e	ω_e	$x_e\omega_e$	B_e	$\alpha_e \times 10^3$	$D_e \times 10^6$	r_e	Remarks	Bibliography
(I)	(79605)								(54.11)
H	76391	1138.3	9.0						(54.11)
G	72011	831.6	5.0						(54.11)
F	68497.2	1116.5	7.2						(54.11)
$?^3\Sigma$	(65757)[a]								(72.32)
$c^3\Pi_i$	~57620[a]	481.6	3.4	0.572				$A = -21.0 \text{ cm}^{-1}$	(72.32)
$b^3\Sigma$	~55570[a]			0.6841	7.9	1.7[b]		$\lambda = 0.298 \text{ cm}^{-1}$ $\gamma = -0.002 \text{ cm}^{-1}$	(70.24)
$E^1\Sigma^+$	52578.31[a]	675.52	4.204	0.54727		1.43	1.7399		(71.29, 54.11)
$g(\Pi\text{or}\Delta)$	≤44700							Perturber of $D^1\Pi$	(53.10)
$i(^1\Sigma^-)$	≤43300	~680						Perturber of $D^1\Pi$	(53.10)
$e^3\Sigma^-$	≤43068	≥684	~4					Perturber of $D^1\Pi$	(53.10)
$f(\Pi\text{or}\Delta)$	≤42800	~660						Perturber of $D^1\Pi$	(53.10)

SiO

SPECTROSCOPIC CONSTANTS

State	T_e	ω_e	$x_e\omega_e$	B_e	$\alpha_e \times 10^3$	$D_e \times 10^6$	r_e	Remarks	Bibliography
$A^1\Pi$	42640.4	852.7	6.44	0.6305	6.95		1.619		(72.35, 69.21, 53.10)
$a^3\Pi_r$	~32000			0.6789	4.4	1.48[b]	1.565[d]	$A = 72.55$ cm^{-1}	(72.32, 70.24, 61.13)
$X^1\Sigma^+$	0	1241.4	5.92	0.72675	5.04		1.5097		(71.29, 69.21, 68.19, 53.10)

[a]T_o, [b]D_o, [c]Measured from the (5, 7) band, [d]r_o.

Dissociation energy = 8.20 ± 0.08 eV, 189 kcal/mole, 66139 cm^{-1} (72.L3)

SiO

Perturbations and General Information

$A^1\Pi$ state is extensively perturbed (53.10).

$c^3\Pi_i$ state is extensively perturbed (72.32).

Ionization potential = 11.6 ± 0.2 eV, 93600 cm^{-1} (69.22).

Radiative lifetimes:

$A^1\Pi - X^1\Sigma^+ \quad - \quad \tau = 9.6 \pm 1.0$ µsec (72.34),

$a^3\Pi - X^1\Sigma^+ \quad - \quad \tau = 48$ msec (70.23) - estimated in Ar matrix at 20°K.

Franck-Condon factors, r-centroids — RKR potential (72.33):

$A^1\Pi - X^1\Sigma^+$

v", v'	0	1	2	3	4	5	6
0	0.1405	0.2546	0.2487	0.1745	0.9883-1	0.4819-1	0.2108-1
	1.562	1.539	1.516	1.493	1.471	1.448	1.425
1	0.2790	0.1300	0.6938-3	0.5841-1	0.1379	0.1487	0.1120
	1.596	1.574	1.563	1.527	1.505	1.483	1.461
2	0.2741		0.1220	0.1014	0.7805-2	0.2168-1	0.8659-1
	1.629		1.584	1.562	1.544	1.515	1.495
3	0.1773	0.9730-1	0.9524-1	0.4276-2	0.9398-1	0.7964-1	0.9770-2
	1.661	1.638	1.617	1.590	1.572	1.552	1.533
4	0.8463-1	0.1922	0.1937-3	0.1093	0.3817-1	0.1264-1	0.8153-1
	1.693	1.670	1.657	1.627	1.606	1.581	1.562
5	0.3168-1	0.1698	0.7221-1	0.6296-1	0.2514-1	0.8902-1	0.1218-1
	1.729	1.702	1.680	1.658	1.636	1.616	1.598
6	0.9651-2	0.9708-1	0.1614	0.7670-3	0.1021	0.4429-2	0.5342-1
	1.755	1.733	1.711	1.689	1.668	1.649	1.626
7	0.2542-3	0.4084-1	0.1521	0.7734-1	0.3281-1	0.5475-1	0.5266-1
	1.786	1.763	1.741	1.720	1.699	1.678	1.658
8	0.5281-3	0.1348-1	0.9057-1	0.1505	0.8183-2	0.8432-1	0.3600-2
	1.816	1.794	1.772	1.750	1.730	1.708	1.688
9		0.3621-2	0.3914-1	0.1350	0.9246-1	0.9879-2	0.7773-1
		1.825	1.802	1.781	1.760	1.737	1.718
10		0.8108-3	0.1314-1	0.7816-1	0.1447	0.2542-1	0.5580-1
		1.855	1.833	1.811	1.790	1.770	1.747

Top = Franck-Condon factor followed by factor of ten
Bottom = r-centroid

SiO

Vibrational transition probabilities (72.36).

Potential energy curves — RKRV potential (65.16):

State	v	U(cm^{-1})	r_{min}(Å)	r_{max}(Å)	Te+U(cm^{-1})
X$^1\Sigma^+$	0	619.5	1.461	1.564	619.5
	1	1848.5	1.428	1.608	1848.5
	2	3067.5	1.407	1.640	3067.5
	3	4273.5	1.390	1.667	4273.5
	4	5465.5	1.376	1.692	5465.5
	5	6648.5	1.364	1.714	6648.5
	6	7817.5	1.353	1.736	7817.5
	7	8974.5	1.344	1.757	8974.5
	8	10118.5	1.335	1.776	10118.5
	9	11252.5	1.327	1.796	11252.5
	10	12372.5	1.320	1.815	12372.5
A$^1\Pi$	0	424.2	1.562	1.687	43259.5
	1	1263.2	1.524	1.742	44098.5
	2	2089.2	1.499	1.782	44924.5
	3	2904.2	1.480	1.818	45739.5
	4	3709.2	1.465	1.850	46544.5

Dipole moment - ^{28}SiO - μ = 3.0982D (70.28).

BIBLIOGRAPHY

(24.1) $^1\Pi \to {}^1\Sigma^+$ System,
W. Jevons,
Proc. Roy. Soc. A 106, 174-94

(25.2) Generalities,
R. Mecke,
Phys. Z. 26, 217-37

(27.3) Generalities,
R. Mecke,
Z. Physik 42, 390-425

(32.4) Preliminary Note to (32.5),
P. G. Saper,
Phys. Rev. 40, 465

(32.5) $^1\Pi \rightarrow {}^1\Sigma^+$ System, Rotational Analysis,
P. G. Saper,
Phys. Rev. 42, 498-508

(44.6) Dissociation Energy,
R. F. Barrow,
Proc. Phys. Soc. 56, 204-10

(45.7) $^1\Pi \rightarrow {}^1\Sigma^+$ System,
D. Sharma,
Proc. Nat. Acad. Sci. A, India 14, 37-44

(46.8) Dissociation Energy,
E. E. Vago and R. F. Barrow,
Proc. Phys. Soc. 58, 538-44

(49.9) Thermochemistry,
L. Brewer and D. F. Mastick,
UCRL Report No. 571

(53.10) Perturbation Analysis,
A. Lagerqvist and U. Uhler,
Arkiv Fysik 6, 95-111

(54.11) U. V. Absorption Spectrum, Vibrational Analysis,
R. F. Barrow and H. C. Rowlinson,
Proc. Roy. Soc. A 224, 374-88

(61.12) A-X System, Franck-Condon Factors,
A. T. MacGregor, R. W. Nicholls, and W. R. Jarmain,
Can. J. Phys. 39, 1215-6

(61.13) $^1\Sigma \rightarrow {}^3\Pi$ System, Rotational Analysis,
R. D. Verma and R. S. Mulliken,
Can. J. Phys. 39, 908-16

(62.14) Franck-Condon Factors, A-X System,
R. W. Nicholls,
J. Res. Nat. Bur. Stand. 66, 227-31

(63.15) Intensity Distribution in the $^1\Sigma - {}^3\Pi$ System,
T. C. James,
J. Chem. Phys. 38, 1094-7

(65.16) K. P. R. Nair, R. B. Singh, and D. K. Rai,
"Potential Energy Curves and Dissociation Energies of Oxides and Sulfides of Group IV A Elements,"
J. Chem. Phys. 43, 3570-3574

SiO

(67.17) F. Jenc,
"Ground State Reduced Potential Curves of Diatomic Molecules,"
J. Mol. Spectrosc. 24, 284-9

(67.18) Dissociation Energy,
P. Coppens, S. Smoes, and J. Drowart,
Trans. Faraday Soc. 63, 2140-8

(68.19) Microwave Spectra,
T. Törring,
Z. Naturforsch. A 23, 777-8

(68.20) R. P. Main, A. L. Morsell, and W. S. Hooker,
"Measurement of the Oscillator Strength of the SiO ($A^1\Pi - X^1\Sigma^+$) Band System,"
J. Quant. Spectrosc. Radiative Transfer 8, 1527-32

(69.21) A-X System,
J. Sinch, K. N. Upadhya, and K. P. R. Nair,
"On the Ultraviolet Band System of Silicon Monoxide,"
Indian J. Phys. 43, 665-73

(69.22) Dissociation Energy, Ionization Potential,
D. L. Hildenbrand and E. Murad,
J. Chem. Phys. 51, 807-11

(70.23) Radiative Lifetime, $^3\Pi - ^1\Sigma^+$ in Matrix,
B. Meyers, J. J. Smith, and K. Spitzer,
"Phosphorescent Decay Time of Matrix - Isolated GeO, GeS, SnO, SnS and the Lifetime of the Cameron Bands of the CO-Type Diatomics,"
J. Chem. Phys. 53, 3616-20

(70.24) b-a System,
S. Nagaraj and R. D. Verma,
"$A^3\Sigma - ^3\Pi$ Transition of the SiO Molecule,"
Can. J. Phys. 48, 1436-40

(70.25) S. Szöke, Z. S. Vajna, and G. Jalsovszky,
"Potential and Electron Energy Curves of Isoelectronic Molecules,"
Acta Chim. Acad. Sci. 63, 59-66

(70.26) A. D. Ruskin,
"Determination of Transition Probability for the Band System $A^1\Pi - X^1\Sigma^+$ (0, 1) Bands of the SiO Molecule by the Explosion Method,"
Vestn. Mosk. Univ. Khim. 11, 526-31

(70.27) A. D. Ruskin,
"Explosion in a Spherical Bombasa Method of Excitation for SiO Spectra,"
Vestn. Mosk. Univ. Khim. 11, 397-401

(70.28) J. W. Raymonda, J. S. Muenter, and W. A. Klemperer,
"Electric Dipole Moment of SiO and GeO,"
J. Chem. Phys. 52, 3458-61

(71.29) E-X System, Vibrational, Rotational Analysis,
N. Elander and A. Lagerqvist,
"On the $E^1\Sigma^+ - X^1\Sigma^+$ System of SiO in the Vacuum Ultraviolet Region,"
Physica Scripta 3, 267-273

(71.30) D. L. Hildenbrand,
"First Ionization of the Molecules BF, SiO, and GeO,"
J. Mass Spec. Ion. Phys. 7, 255

(72.31) A. D. Ruskin,
"Matrix Elements of Dipole Moment of the Bands $A^1\Pi - X^1\Sigma^+$ System of SiO,"
Vestn. Mosk. Univ. Khim. 2, 196-200

(72.32) C, ?-a Systems,
R. Cornet and I. Dubois,
"New Electronic Transitions of SiO,"
Can. J. Phys. 50, 630-5

(72.33) Franck-Condon Factors, r-Centroids, A-X,
H. S. Liszt and W. H. Smith,
"RKR Franck-Condon Factors for Blue and Ultraviolet Transmissions of some Molecules of Astrophysical Interest and some comments on the Interstellar Abundance of CH, CH^+, and SiH^+,"
J. Quant. Spectrosc. Radiative Transfer 12, 947-958

(72.34) W. H. Smith and H. S. Liszt,
"Radiative Lifetimes and Absolute Oscillator Strength for the SiO $A^1\Pi - X^1\Sigma^+$ Transition,"
J. Quant. Spectrosc. Radiative Transfer 12, 505-9

(72.35) G. Bosser, J. Lebreton, and L. Marsigny,
"Nouvelles Bandes du Système $A^1\Pi - X^1\Sigma^+$ de SiO,"
C. R. Acad. Sci. C 275, 531-4

(72.36) J. Hedelund and D. L. Lambert,
"Transition Probabilities for the Vibration - Rotation Bands of Silicon Monoxide,"
Astro. Physical Letters 11, 71-5

SiO

(71.L1) R. W. Wilson, A. A. Penzias, K. B. Jefferts, M. Kutner, and P. Thaddeus,
"Discovery of Interstellar Silicon Monoxide,"
Astrophys. J. 167, L97-L100

(72.L2) T. G. Heil and H. F. Schaefer III,
"Potential Curves for the Valence - Excited States of Silicon Monoxide. A Theoretical Study,"
J. Chem. Phys. 56, 958-68

(72.L3) Dissociation Energy,
D. L. Hildenbrand,
High Temp. Sci. 4, 244

SiS

Methods of Production and Experimental Technique

Absorption to 800-1000°C.

Emission from a positive column: high frequency discharge.

Ground state studied by microwave techniques.

BAND SYSTEMS

System	Transition	Sources	Wavelength Limits	Degrading	Band Head, $\nu_{0,0}$	Remarks	Bibliography
I	$D^1\Pi \rightleftarrows X^1\Sigma^+$	Discharge, Absorption	4000-2600	R		Upper state is perturbed	(56.7, 52.5, 47.4, 46.2, 38.1)
II	$E^1\Sigma^+ \rightleftarrows X^1\Sigma^+$	Discharge, Absorption	2600-2000	R			(61.8, 54.6, 47.4, 46.3)

SiS

I. $D^1\Pi \rightleftarrows X^1\Sigma^+$ System

Band heads in emission, λ(intensity) (38.1):

v', v''	0	1	2	3	4	5
0	2863.66(8)[a]	2926.06(10)[a]	2990.78(10)	3057.95(9)	3127.67(8)	3200.04(
1	2822.66(9)[a]	2883.25(8)[a]		3010.92(3)	3078.83(7)	3149.05(
2	2783.18(9)[a]	2842.18(4)[a]?	2902.90(4)	2966.35(5)		
3	2745.28(8)[a]	2802.60(2)[a]?	2862.01(4)			3053.03(
4	2708.78(7)[a]	2764.66(6)[a]				
5	2673.78(6)[a]	2728.17(5)[a]				
6	2640.10(4)[a]	2693.06(2)[a]				
7	2607.15(2)	2658.77(3)				
8		2678.65(4)				

[a] Observed in absorption

II. $E^1\Sigma^+ \rightleftarrows X^1\Sigma^+$ System

Band heads, λ (47.4, 46.3):

v', v''	0	1	2	3	4	5
0				2528.70	2576.45	
1			2458.06	2503.37	2550.06	2598.02
2		2391.36	2434.48	2478.78	2524.27	2571.5
3		2369.01	2411.30	2454.75	2499.37	2545.55
4	2306.91	2347.24	2388.78	2431.46	2475.16	
5	2286.60	2326.11	2366.88	2408.85		
6	2266.66	2305.50	2345.50			
7	2247.36	2285.66	2325.01			
8	2228.58	2266.20				
9	2210.38	2247.35				

SPECTROSCOPIC CONSTANTS

State	T_e	ω_e	$x_e\omega_e$	B_e	$\alpha_e \times 10^3$	$D_e \times 10^6$	r_e	Remarks	Bibliography
$E^1\Sigma^+$	41743.8	406.8	1.95	0.2213	1.3				(61.8)
$l^1\Sigma^-$	37290	404	1.1	0.213	1.0			Observed as perturbation to $D^1\Pi$ state	(56.7)
$^1\Delta$	37114	439	3.9	0.226	2.0			Observed as perturbation to $D^1\Pi$ state	(56.7)
$e^3\Sigma^-$	≤35150	≥407	1.7	≥0.221	2.0			Observed as perturbation to $D^1\Pi$ state	(56.7)
$D^1\Pi$	34908.64	513.1	2.9	0.2664	2.1				(52.5)
$X^1\Sigma^+$	0	749.6	2.58	0.3035290	1.4736		1.9293		(65.9, 52.5)

Dissociation energy = 6.39 ± 0.12 eV, 147 kcal/mole, 51540 cm^{-1}.

SiS

Perturbations and General Information

Dipole moment $X^1\Sigma^+$, μ_0 = 1.73 ± 0.06D (69.12, 69.13).

Radiative lifetime — $a^3\Pi - X^1\Sigma^+$, τ = 29 msec (estimated in Ar matrix at 20°K) (70.14).

Potential energy curves — RKRV potential (65.10):

State	v	U(cm-1)	r_{min}(Å)	r_{max}(Å)	Te+U(cm^{-1})
$X^1\Sigma^+$	0	374.1	1.876	1.986	374.1
	1	1118.9	1.840	2.032	1118.9
	2	1856.6	1.817	2.064	1856.6
	3	2591.8	1.799	2.092	2591.8
	4	3323.1	1.784	2.117	3323.1
	5	4046.4	1.770	2.140	4046.4
	6	4763.5	1.758	2.162	4763.5
	7	5478.8	1.748	2.183	5478.8
	8	6184.4	1.738	2.202	6184.4
	9	6888.4	1.729	2.221	6888.4
	10	7590.8	1.721	2.239	7590.8
	11	8278.1	1.713	2.258	8278.1
	12	8966.9	1.705	2.275	8966.6
	13	9649.0	1.698	2.292	9649.0
	14	10327.4	1.692	2.309	10327.4
	15	10998.8	1.685	2.326	10998.8
	16	11666.9	1.679	2.343	11666.9
	17	12315.4	1.674	2.358	12315.4
	18	12982.5	1.668	2.374	12982.5
$D^1\Pi$	0	255.4	1.996	2.129	35284.2
	1	763.0	1.955	2.186	35791.8
	2	1264.2	1.928	2.228	36293.0
	3	1759.3	1.907	2.264	36788.1
	4	2247.5	1.890	2.297	37276.3
	5	2730.5	1.874	2.328	37759.3
	6	3208.1	1.861	2.357	38236.9
	7	3667.6	1.850	2.384	38696.4
	8	4136.8	1.839	2.411	39165.6
	9	4599.0	1.829	2.438	39627.8
	10	5049.1	1.820	2.463	40077.9

State	v	U(cm^{-1})	r_{min}(A)	r_{max}(A)	Te+U(cm^{-1})
E$^1\Sigma^+$	0	201.4	2.188	2.337	42124.9
	1	602.7	2.140	2.400	42526.2
	2	1000.0	2.108	2.446	42923.5
	3	1394.6	2.084	2.485	43318.1
	4	1785.5	2.064	2.520	43709.0
	5	2171.2	2.046	2.552	44094.7
	6	2556.2	2.031	2.583	44479.7
	7	2923.1	2.016	2.613	44856.6
	8	3308.7	2.003	2.641	45232.2
	9	3678.2	1.991	2.669	45601.7
	10	4043.1	1.980	2.696	45966.6
	11	4404.5	1.969	2.723	46328.0
	12	4759.4	1.959	2.749	46682.9
	13	5109.8	1.949	2.776	47033.3
	14	5453.3	1.940	2.802	47376.8
	15	5789.0	1.929	2.830	47712.5
	16	6115.6	1.920	2.856	48039.1
	17	6436.7	1.911	2.884	48360.2
	18	6751.6	1.902	2.910	48675.1
	19	7056.4	1.891	2.942	48979.9
	20	7346.3	1.882	2.971	49269.8
	21	7626.6	1.868	3.008	49550.1
	22	7885.3	1.857	3.041	49808.8
	23	8131.6	1.839	3.088	50055.1
	24	8363.9	1.825	3.128	50287.4
	25	8553.9	1.810	3.170	50477.4
	26	8758.6	1.794	3.215	50682.1

BIBLIOGRAPHY

(38.1) R. F. Barrow and W. Jevons,
Proc. Roy. Soc. A 169, 45-65

(46.2) R. F. Barrow,
Proc. Phys. Soc. 58, 606-15

(46.3) E. E. Vago and R. F. Barrow,
Proc. Phys. Soc. 58, 538-44

(47.4) D. M. Thomas,
Thesis, Oxford

(52. 5) D-X System, Rotational Analysis, Perburations,
A. Lagerqvist, G. Nilheden, and R. F. Barrow,
Proc. Phys. Soc. A 65, 419-33

(54. 6) E-X System, Dissociation Energy,
S. J. Q. Robinson and R. F. Barrow,
Proc. Phys. Soc. A 67, 95-6

(56. 7) D-X System, Perturbation Analysis,
G. Nilheden,
Arkiv Fysik 10, 19-36

(61. 8) E-X System, Rotational Analysis,
R. F. Barrow, J. L. Deutsch, A. Lagerqvist and
B. Westerlund,
Proc. Phys. Soc. 78, 1307-9

(65. 9) Microwave Spectra,
J. Hoeft,
Z. Naturforsch. A 20, 1327-9

(65. 10) K. P. R. Nair, R. B. Singh, and D. K. Rai,
"Potential Energy Curves and Dissociation Energies of Oxides
and Sulfides of Group IVA Elements,"
J. Chem. Phys. 43, 3570-4

(67. 11) Thermochemistry, Dissociation Energy,
P. Coppens, S. Smoes, and J. Drowart,
Trans. Faraday Soc. 63, 2140-8

(69. 12) Dipole Moment,
A. N. Murty and R. F. Curl Jr.,
J. Mol. Spectrosc. 30, 102-10

(69. 13) Stark Effect,
J. Hoeft, F. J. Lovas, E. Tiemann, and T. Törring,
Z. Naturforsch. A 24, 1422-3

(70. 14) Radiative Lifetime, $^3\Pi - {}^1\Sigma$ in Matrix Isolation,
B. Meyers, J. J. Smith, and K. Spitzer,
"Phosphorescent Decay Time of Matrix-Isolated GeO, GeS,
SnS, SnO and SnS and the Lifetime of the Cameron Bands
of CO-type Diatomics,"
J. Chem. Phys. 53, 3616-20

(70. 15) S. Szöke, Z. S. Vajna and G. Jalsovasyky,
"Potential and Electron Energy Curves of Isoelectronic
Molecules,"
Acta Chim Acad. Scien. 63, 59-66

(72.L1) E. Tiemann, E. Renwanz, J. Hoeft, and T. Törring, "Isotope Effect in the Rotation Spectrum of SiS," Z. Naturforsch 27a, 1566-70

SiSe

Methods of Production and Experimental Technique

Absorption at 800-1000°C.

Emission from a positive column.

Ground state studied by microwave technique.

BAND SYSTEMS

System	Transition	Sources	Wavelength Limits	Degrading	Band Head, $v_{0,0}$	Remarks	Bibliography
I	$D(^1\Pi) \rightleftarrows X^1\Sigma^+$	Discharge, Absorption	3700-2800	R			(46.2, 39.1)
II	$E(^1\Sigma^+) \leftarrow X^1\Sigma^+$	Absorption	2770-2450	R			(46.2)

I. $\underline{D(^1\Pi) \rightleftarrows X^1\Sigma^+ \text{ System}}$

Simple heads.

Band heads in emission, λ(intensity) (46.2, 39.1):

v', v''	0	1	2	3	4	5	6	7
0	3089.29 (1)[a]	3145.32 (9)[a]	3202.99 (10)[a]	3262.38 (10)[a]	3323.71 (8)	3387.10 (5)		
1	3051.78 (8)[a]	3106.41 (9)[a]	3162.80 (5)[a]		3280.27 (2)	3342.18 (9)	3405.96 (7)	3471.66 (4)
2	3015.77 (7)[a]	3069.11 (3)[a]	3123.95 (1)	3180.58 (4)	3238.88 (3)		3361.41 (2)	3425.23 (2)
3	2981.10 (6)[a]	3033.29 (1)[a]				3257.69 (1)		
4	2947.18 (4)[a]	2997.99 [a]						
5	2914.41 (1)[a]	2964.12 [a]						
6	2882.74 [a]	2931.60 (1)[a]						
7	2851.81 [a]	2899.72 [a]						
8	2821.51 [a]	2868.27 [a]						

[a] Observed in absorption

II. $\underline{E(^1\Sigma^+) \leftarrow X^1\Sigma^+ \text{ System}}$

Band heads, λ (46.2):

v', v''	0	1	2	3	4
0		2645.23	2685.83	2727.50	2769.55
1		2624.03	2663.98	2704.92	
2	2564.78	2603.43	2642.92	2683.00	
3	2545.48	2583.40	2622.21	2661.88	
4	2526.88	2564.13	2602.35	2641.42	2681.88
5	2508.74	2545.48	2583.02		
6	2490.75				
7	2473.20				
8	2456.38				

SiSe

SPECTROSCOPIC CONSTANTS

State	T_e	ω_e	$x_e\omega_e$	B_o	$\alpha_e \times 10^3$	$D_e \times 10^6$	r_e	Remarks	Bibliography
$E(^1\Sigma^+)$	38505.9	308.8	1.95					$y_e\omega_e = -0.032$ cm^{-1}	(46.2, 39.1)
$D(^1\Pi)$	32450.3	399.8	1.93						(46.2, 39.1)
$X^1\Sigma^+$	0	580.0	1.78	0.1916233	0.7767		2.0583		(65.3, 46.2, 39.1)

Dissociation energy = 5.45 ± 0.25 eV, 126 kcal/mole, 43958 cm^{-1}.

BIBLIOGRAPHY

(39. 1) D, E → X Systems,
R. F. Barrow,
Proc. Phys. Soc. 51, 267-73

(46. 2) D, E ← X Systems,
E. E. Vago and R. F. Barrow,
Proc. Phys. Soc. 58, 538-44

(65. 3) Microwave Spectra,
J. Hoeft,
Z. Naturforsch. A 20, 1122-4

SiTe

Methods of Production and Experimental Technique

Absorption at 800-1000°C.

Emission from a positive column.

BAND SYSTEMS

	System	Transition	Sources	Wavelength Limits	Degrading	Band Head, $\nu_{0,0}$	Remarks	Bibliography
	I	$D(^1\Pi) \rightleftarrows X^1\Sigma^+$	Discharge	3900-3225	R			(46.2, 39.1)
	II	$E(^1\Sigma^+) \leftarrow X^1\Sigma^+$	Absorption	3100-2800	R			(46.2)

I. $D(^1\Pi) \rightleftarrows X^1\Sigma^+$ System

Band heads in emission (46.2, 39.1):

v', v''	0	1	2	3	4
0	3496.58(4)[a]	3556.08(10)[a]	3617.28(10)[a]	3680.26(9)	3745.16(9)
1	3456.19(8)[a]	3514.32(10)[a]	3574.19(5)[a]		
2	3416.96(7)[a]	3473.73(6)[a]	3532.20(2)	3592.33(4)	3654.15(3)
3	3378.98(5)[a]	3434.63[a]	3491.54(3)	3550.23(1)	
4	3342.36(3)[a]	3396.74[a]	3452.59(3)		
5	3306.80(2)[a]	3360.11[a]			
6	3272.46[a]	3324.53(1)[a]			
7	3238.7[a]	3290.15(1)[a]			
8		3256.4[a]			

[a] Obtained in absorption

II $E(^1\Sigma^+) \leftarrow X^1\Sigma^+$ System

Band heads, λ (46.2):

v', v''	0	1	2	3	4
0		2993.52	3036.92	3081.42	
1		2972.76	3015.48	3059.08	3103.73
2	2912.04	2952.96	2994.99		
3	2893.28	2933.87			
4	2875.00	2914.97			
5	2856.97	2896.54			
6	2839.26	2878.34			

SiTe

SPECTROSCOPIC CONSTANTS

State	T_e	ω_e	$x_e\omega_e$	B_e	$\alpha_e \times 10^3$	$D_e \times 10^6$	r_e	Remarks	Bibliography
$E(^1\Sigma^+)$	33991	242	3.63					$y_e\omega_e = 0.13$ cm^{-1}	(46.2)
$D(^1\Pi)$	28661.8	338.6	1.70						(46.2, 39.1)
$X^1\Sigma^+$	0	481.2	1.30						(46.2, 39.1)

Dissociation energy = 5.2 ± 0.4 eV, 120 kcal/mole, 41940 cm^{-1}.

BIBLIOGRAPHY

(39.1) Emission Spectrum,
R. F. Barrow,
Proc. Phys. Soc. 51, 267-73

(46.2) Absorption Spectrum,
E. E. Vago and R. F. Barrow,
Proc. Phys. Soc. 58, 538-41

(67.3) Equilibrium Study,
G. Exsteen, J. Drowart, A. Van der Auwera-Mahieu, and R. Callaerts,
J. Phys. Chem. 71, 4130-1

(68.4) Dissociation Energy,
R. F. Brebrick,
J. Chem. Phys. 49, 2584-92

SmF

SmF

Dissociation energy = 5.47 ± 0.20 eV, 126 kcal/mole, 44100 cm^{-1} (67.1).

BIBLIOGRAPHY

(67.1) K. F. Zmbov and J. L. Margrave, J. Inorg. Nucl. Chem. 29, 59-63

SmO

Methods of Production and Experimental Technique

Emission from an arc or flame upon adding Sm salts.

Band Systems

Bands observed $7000 > \lambda > 4400$ Å, degrading R (57.1).

Characteristic bands:

| λ | 6556.8^a | 6533.0 | 6531.9^a | 6509.6^a | 6507.6 | 6484.3^a | 6071.0 | 5987.8 | 5698.3 | 5680.8 |

^{a} Most intense emission.

Dissociation energy = 6.16 ± 0.13 eV, 142 kcal/mole, 49700 cm^{-1} (67.2).

BIBLIOGRAPHY

(57.1) A. Gatterer, J. Junkes, E. W. Salpeter, and B. Rosen, "Molecular Spectra of Metallic Oxides," Ed. Specola Vaticana, 1957

(67.2) L. L. Ames, P. N. Walsh, and D. White, J. Phys. Chem. 71, 2707-18

SmS

SmS

Dissociation energy (estimated) = 4.01 ± 0.18 eV, 92.6 kcal/mole, 32400 cm^{-1} (69.1).

BIBLIOGRAPHY

(69.1)   S. Smoes, P. Coppens, C. Bergman, and J. Drowart, Trans. Faraday Soc. 65, 682-7

SnCl

Methods of Production and Experimental Technique

Absorption in $SnCl_2$ in a carbon furnace $T > 1200°C$ (42.6).

Emission from a discharge into $SnCl_4$ (26.1). Active nitrogen + $SnCl_4$ (28.2) Argon + $SnCl_4$ (67.9).

Flash photolysis of $SnCl_4$ (69.12).

BAND SYSTEMS

System	Transition	Sources	Wavelength Limits	Degrading	Band Head, $\nu_{0,0}$	Remarks	Bibliography
I	$A^2\Sigma \rightarrow X^2\Pi_{1/2, 3/2}$	Discharge	7800-4950	R	15957 18315		(67.9, 65.8, 64.7)
II	$B'^2\Delta \rightleftarrows X^2\Pi_r$	Discharge	3910-3730 3550-3480	R	26579.1 28665.3		(64.7, 28.2)
III	$B^2\Sigma \rightleftarrows X^2\Pi_r$	Discharge	3405-3065 3134-2830	V	31262.5 33622.6		(69.12, 26.1, 26.5)
IV	$C^1 \leftarrow X^2\Pi_{1/2}$	Absorption	2465-2352	V	41229		(69.12)
V	$C \leftarrow X^2\Pi_{1/2, 3/2}$	Absorption	2326-2228	V	41383.9 43676.1		(70.13, 69.12, 42.6)
VI	?	Discharge	4900-3950	-	max. ~23000	Continuum	(26.1)
VII	$E \leftarrow X^2\Pi_r$	Absorption	3500-3000	-	max. ~30000	Continuum	(42.6)
VIII	$F \leftarrow X^2\Pi_r$	Absorption	2950-1950	-	max. >40000	Continuum	(42.6)

SnCl

I. $A^2\Sigma \rightarrow X^2\Pi_r$ System

Band heads (64.7):

$^2\Sigma - {}^2\Pi_{3/2}$

(v', v'')	(0, 1)	(0, 0)	(1, 0)	(1, 1)	(2, 2)
λ	6405.3	6266.8	6177.4	6312.3	6357.7

$^2\Sigma - {}^2\Pi_{1/2}$

(v', v'')	(0, 1)	(0, 0)	(1, 0)	(1, 1)	(2, 2)
λ	5564.8	5460.0	5392.0	5494.2	5528.2

III. $B^2\Sigma \rightleftarrows X^2\Pi_r$ System

Band heads (26.1):

$^2\Sigma - {}^2\Pi_{3/2}$

(v', v'')	(0, 1)	(0, 0)	(1, 0)	(2, 0)	(3, 1)
λ(Intensity)	3234.4(7)	3197.8(7)	3154.3(7)	3112.4(5)	3105.4(4)

$^2\Sigma - {}^2\Pi_{1/2}$

(v', v'')	(0, 1)	(0, 0)	(1, 0)	(2, 0)	(3, 1)
λ(Intensity)	3004.5(6)	2973.4(8)	2935.8(10)	2899.4(7)	2893.0(5)

IV. $C^1 \leftarrow X^2\Pi_r$ System

(v', v'')	(0, 1)	(0, 0)	(1, 0)	(2, 0)	(3, 1)
λ	2443.65	2422.62	2398.66	2375.19	2371.98

V. $C \leftarrow X^2\Pi_r$ System

Band heads (69.12):

(v', v'')	(0, 1)	(0, 0)	(1, 0)	(2, 0)
λ	2307.40	2289.00	2268.34	2248.22

SPECTROSCOPIC CONSTANTS

State	T_e	ω_e	$x_e\omega_e$	B_e	$\alpha_e \times 10^3$	$D_e \times 10^6$	r_e	Remarks	Bibliography
C	43650.2	399.3	1.1						(70.13, 69.12, 42.6)
C^1	41229	431.8	1.7						(70.13, 69.12)
$B^2\Sigma$	33582.6	432.5	1.2						(26.1)
$B'^2\Delta$	28966 28692	301	4						(64.7, 28.2)
$A^2\Sigma$	18373	232.2	0.7						(67.9, 64.7)
$X^2\Pi_{3/2}$	2360.1	354.4	1.05						(70.13, 64.7, 42.6, 28.2)
$X^2\Pi_{1/2}$	0	351.1	1.06						(70.13, 64.7, 42.6, 28.2)

Dissociation energy = 3.2 ± 0.2 eV, 74 kcal/mole, 25810 cm^{-1}.

BIBLIOGRAPHY

(26. 1)   B → X System,
W. Jevons,
Proc. Roy. Soc. A 110, 365-90

(28. 2)   A → X System,
W. F. C. Ferguson,
Phys. Rev. 32, 607-10

(35. 3)   Absorption,
H. Trivedi,
Indian J. Phys. 9, 331-45

(35. 4)   H. G. Howell and G. D. Rochester,
Proc. Univ. Durham Phil. Soc. 9, 126-34

(36. 5)   Questionable Discussion of Terms,
H. Lessheim and R. Samuel,
Indian J. Phys. 10, 7-12

(42. 6)   C, E and F ← X Systems,
C. A. Fowler Jr.,
Phys. Rev. 62, 141-3

(64. 7)   A → $X^2\pi_{1/2, 3/2}$ Systems,
G. Pannetier, P. Deschamps, L. Marsigny, N. Luquent, and J. Guillaume,
"Vibrational Study of a New Band System of the Emission Spectra of SnCl,"
J. Chem. Phys. Physicochim. Biol. 61, 1142-6

(65. 8)   System I Vibrational Analysis,
P. R. K. Sarma and P. Venkateswarlu,
J. Mol. Spectrosc. 17, 252-64

(67. 9)   System I, Vibrational Analysis,
P. Deschamps,
Thesis, Paris

(68. 10) $A^2\Sigma^+ \to X^2\pi_{1/2}$. Vibrational Analysis,
G. Pannetier and P. Deschamps,
J. Chim. Phys. 65, 1164-70

(68. 11) Yu. Ya. Kuzyakov,
"Determination of Energy of Dissociation of Certain Monohalides of Carbon Subgroups,"
Vestn. Mosk. Univ. Khim. **3**, 21-4

(69. 12) B, C', C ← X Systems, Vibrational Analysis,
G. A. Oldershaw and K. Robinson,
"New Ultraviolet Bands of Tin Monochloride,"
J. Mol. Spectrosc. **32**, 469-74

(70. 13) C ← X System, Vibrational Analysis,
W. Richter,
"The U. V. Spectrum of the $^{120}Sn^{35}Cl$ in the Region Between 2800 and 3450 Å,"
Z. P. Chemie Neuefolge, **71**, 303-10

SnF

Methods of Production and Experimental Techniques

Absorption by (PbF_2 + Sn) in a carbon furnace T ≈ 1300-1800°C.
Emission from a high pressure discharge into SnF_4. Hollow cathode (PbF_2 + Sn).

BAND SYSTEMS

System	Transition	Sources	Wavelength Limits	Degrading	Band Head, $\nu_{0,0}$	Remarks	Bibliography
I	$A^2\Sigma^+ \rightarrow X^2\Pi_r$	Hollow cathode	6300-4600	R	17736.1 20055.6		(72.10, 62.5, 39.2)
II	$B^2\Sigma^+ \rightleftarrows X^2\Pi_r$	Hollow cathode	3260-2660	V	31837.0 34156.1		(62.5, 37.1)
III	$C^2\Delta \rightleftarrows X^2\Pi_r$	Hollow cathode	2700-2400	V	38524.7 40772.8		(68.7, 59.4, 37.1)
IV	$D \rightarrow X^2\Pi_r$	Hollow cathode	2700-2400	V	39044.5 41381.4		(68.7, 59.4, (27.1))
V	$E(^2\Pi) \rightarrow X^2\Pi_r$	Hollow cathode	2700-2400	V	39820.2 41856.1		(68.7, 59.4, 27.1)
VI	$F(^2\Sigma^+) \rightarrow X^2\Pi_r$	Hollow cathode	2400-2200	V	43235.7 45552.6		(68.7, 59.4, 27.1)
VII	$G \rightarrow X^2\Pi_r$	Hollow cathode	2350-2150	V	44121.9 46351.8		(68.7, 59.4, 27.1)
VIII	$B^2\Sigma^+ \rightarrow A^2\Sigma^+$	Hollow cathode	8300-6600	V	14099		(68.7, 59.4, 27.1)
IX	$F(^2\Sigma^+) \rightarrow A^2\Sigma^+$	Hollow cathode	4300-3950	V	25510		(68.7, 59.4, 27.1)

I. $A^2\Sigma^+ \rightarrow X^2\Pi_r$ System

Band heads, λ (62.5):

$^2\Sigma \rightarrow ^2\Pi_{3/2}$

v', v''	0, 2	0, 1	0, 0	1, 0	2, 0
λ(Intensity)	6032.6(6)	5827.5(8)	5635.9(8)	5506.8(7)	5384.8(4)

$^2\Sigma \rightarrow ^2\Pi_{1/2}$

v', v''	0, 2	0, 1	0, 0	1, 0
λ(Intensity)	5287.6(5)	5132.6(9)	4985.9(10)	4883.3(10)

v', v''	3, 1	2, 0	4, 1	3, 0
λ(Intensity)	4826.7(5)	4786.4(8)	4734.9(4)	4696.0(5)

II. $B^2\Sigma^+ \rightleftarrows X^2\Pi_r$ System

$B^2\Sigma^+ \leftarrow X^2\Pi_{3/2}$

v', v''	0	1	2	3
0	3141.2	3199.65	3259.8	
1	3076.3	3132.3	3190.0	3249.2
2	3014.6	3068.3		
3			3060.5	
4			3000.4	

$B^2\Sigma^+ \leftarrow X^2\Pi_{1/2}$

v', v''	0	1	2	3	4
0	2927.8	2978.1	3029.7		
1	2871.4	2919.8	2969.3	3020.1	
2	2817.6				
3	2766.3	2811.0	2856.9		
4		2760.1	2804.3		2895.6

SnF

III. $C^2\Delta \rightleftarrows X^2\Pi_r$ System

Band heads (P_1) (68.7):

(v', v'')	(0, 1)	(0, 0)	(1, 0)	(1, 1)	(1, 2)
λ	2488.2	2453.0	2417.7	2451.9	2486.7

IV.-VII. 2700-2100Å Region

Contains a large number of bands, degrading V, with four identified systems. G → X system is the most intense (59.4):

$G \rightarrow X^2\Pi_{3/2}$

(v', v'')	(0, 1)	(0, 0)	(1, 0)
λ	2296.1	2265.75	2234.9

$G \rightarrow X^2\Pi_{1/2}$

(v', v'')	(0, 1)	(0, 0)	(1, 0)
λ	2184.0	2156.7	2128.7

VIII. $B^2\Sigma^+ \rightarrow A^2\Sigma^+$ System

Intense system.
Band heads (59.4):

v', v''	0, 4	0, 3	0, 2	0, 1	1, 2	1, 1	1, 0	2, 1
λ	8022.6	7769.7	7532.2	7305.6	7168.9	6963.7	6767.7	6654.9

IX. $F(^2\Sigma^+) \rightarrow A^2\Sigma^+$ System

Band heads (59.4):

(v', v'')	(0, 4)	(0, 3)	(0, 2)	(0, 1)
λ	4188.0	4118.5	4050.2	3983.9

SnF

SPECTROSCOPIC CONSTANTS

State	T_e	ω_e	$x_e\omega_e$	B_e	$\alpha_e \times 10^3$	$D_e \times 10^6$	r_e	Remarks	Bibliography
G	46438.8[a] 46351.8[a]	609.9[b]	-	-	-		-	$A = 87$ cm^{-1}	(59.4)
$F(^2\Sigma^+)$	45552.6	688.2	4.6	-	-		-		(59.4)
$E(^2\Pi)$	42137.1 41856.1	677.0	3.0	-	-		-	$A = 281$ cm^{-1}	(59.4)
D	41361.4[a]	622[b]	-	-	-		-	$A \sim 0$ cm^{-1}	(59.4)
$C^2\Delta$	40834.3 40763.2	607.5 607.2	5.68 5.16	-	-		-	$A = 71.1$ cm^{-1}	(68.7, 59.4)
$B^2\Sigma^+$	34107.9	677.6	2.74	0.2896[c]	-		1.887[d]	$\gamma \sim 0$ cm^{-1}	(62.5, 59.4)
B^1	33039.8[a]	-	-	-	-		-		(59.4)
$A^2\Sigma^+$	20137.8	420	2.20	0.2431	2.6		2.042	$\gamma_0 = 0.084$ cm^{-1}	(72.10, 62.5, 59.4)
$X^2\Pi_{3/2}$	2316.9	588.8	2.82	0.2738	1.1		1.9387		(72.10, 64.7, 62.5, 59.4)
$X^2\Pi_{1/2}$	0	583.3	2.69	0.2727	1.1		1.9426	$p_0 = -0.059$ cm^{-1}	(72.10, 64.7, 62.5, 59.4)

[a] T_0, [b] $\Delta G_{1/2}$, [c] B_0, [d] r_0.

Dissociation energy = 4.78 ± 0.13 eV, 38554 cm^{-1} (72.11).

SnF

Perturbations and General Information

Ionization potential = 7.0 eV, 56800 cm^{-1} (58.3).

Electronic transition moment, $A^2\Sigma^+ - X^2\Pi_r$ (71.9).

$^2\Sigma - ^2\Pi_{3/2}$: Re(r) = constant (1-0.5208r), $1.9 \le r \le 2.2$ Å

$^2\Sigma - ^2\Pi_{1/2}$: Re(r) = constant (1-0.5307r), $1.9 \le r \le 2.2$ Å

Franck-Condon factors, r-centroids — Morse potential (71.9):

$^2\Sigma - ^2\Pi_{3/2}$

v', v''	0	1	2	3
0	0.1793 1.9930 5635.9	0.2027 2.0404 5827.5	0.1218 2.0840 6032.6	0.0509 2.1284 6242.8
1	0.2077 1.9560 5506.8	0.0057 2.0068 5688.6	- - -	0.0660 2.0524 6087.7
2	0.1227 1.9152 5384.8	0.0694 1.9708 5554.9	0.0772 2.0196 5745.8	- - -

$^2\Sigma - ^2\Pi_{1/2}$

v', v''	0	1	2	3	4
0	0.2841 1.9912 4985.6	0.3499 2.0372 5132.6	0.2282 2.0808 5287.5	0.1022 2.1232 5451.6	- - -
1	0.3559 1.9548 4883.3	0.0219 2.0044 5025.2	0.0801 2.0496 5174.3	0.1995 2.0924 5331.8	0.1321 2.1348 5497.5
2	0.2281 1.9164 4786.4	0.0827 1.9700 4922.3	0.0779 2.0172 5067.0	- - -	- - -

Potential energy curves — RKRV potential (66.6):

State	v	U(cm^{-1})	r_{min}(Å)	r_{max}(Å)
		(Te = 0 cm^{-1})		
X$^2\Pi$	0	290.8	1.885	2.004
	1	868.1	1.846	2.053
	2	1440.2	1.820	2.088
	3	2006.4	1.799	2.118
	4	2567.8	1.781	2.144
	5	3123.4	1.766	2.169
	6	3647.2	1.752	2.192
	7	4219.8	1.740	2.214
	8	4760.8	1.728	2.235
		(Te = 34108.4 cm^{-1})		
A$^2\Sigma$	0	337.7	1.990	2.100
	1	1008.8	1.958	2.150
	2	1675.3	1.938	2.187
	3	2336.0	1.924	2.219
	4	2992.1	1.913	2.249
	5	3641.9	1.903	2.276
	6	4287.2	1.896	2.302
	7	4926.1	1.889	2.327
	8	5559.0	1.884	2.352
	9	6187.6	1.879	2.376
	10	6810.6	1.875	2.399
	11	7428.0	1.871	2.422
	12	8040.9	1.868	2.444

BIBLIOGRAPHY

(37.1) Absorption,
F. A. Jenkins and G. D. Rochester,
Phys. Rev. 52, 1135-40

(39.2) Emission,
T. Yuasa,
Proc. Phys. Math. Soc. 21, 497-507

(58.3) Estimated Dissociation Energy and Ionization Potential,
J. W. C. Johns and R. F. Barrow,
Proc. Phys. Soc. 71, 476-84

SnF

(59.4) Emission, Vibrational Analysis,
R. F. Barrow, D. Butler, J. W. C. Johns, and J. L. Powell,
Proc. Phys. Soc. 73, 317-20

(62.5) A-X, B-X Systems, Rotational Analysis,
R. F. Barrow, I. Kopp, and A. J. Merer,
Proc. Phys. Soc. 79, 749-52

(66.6) R. B. Singh and D. K. Rai,
"Potential Curves for Some Diatomic Molecules,"
Ind. J. Pure & Appl. Phys. 4, 102-105

(68.7) C-X System, Vibrational-Rotational Analysis,
A. N. Uzikov and Y. Y. Kuzyakov,
"Vibrational and Rotational Analysis of SnF Bands in the Region 2500 Å,"
Vestn. Mosk. Univ. Khim. 5, 33-5

(68.8) Yu. Ya. Kuzyakov,
"Determination of Energy of Dissociation of Certain Monohalides of Carbon Subgroup,"
Vestn. Mosk. Univ. Khim. 3, 21-4

(71.9) Franck-Condon Factors, r-Centroids, Transition Moment, A-X,
J. Singh and P. S. Dube,
"On the Variation of Electron Transition Moment with the Internuclear Separation in SiCl and SiF Molecules,"
Ind. J. Pure Appl. Phys. 9, 164-165

(72.10) S. R. Rai and J. Singh,
"Rotational Structure and Isotopic Shift in the (1,0) Band of the $A^2\Sigma^+ - X^2\Pi_g$ System of SnF Molecule,"
Spectroscopy Letters 5, 155-167

(72.11) M. Shafi, C. L. Beckel, and R. Engelke,
"Diatomic Molecule Ground State Dissociation Energies,"
J. Mol. Spectrosc. 42, 578-81

SnO

Methods of Production and Experimental Technique

Absorption.

Absorption and fluorescence in Ar, Kr and Xe matrices at 20°K (68.18).

Emission from arcs and flames + compounds of Sn.

Ground state studied by microwave techniques.

BAND SYSTEMS

System	Transition	Sources	Wavelength Limits	Degrading	Band Head, $\nu_{0,0}$	Remarks	Bibliography
I	$A \rightleftarrows X^1\Sigma^+$	Flame	6480-5250	-	18889.0		(67.16)
II	$B \rightleftarrows X^1\Sigma^+$	Flame	5840-3930	-	24199.0		(67.16)
III	$C \rightleftarrows X^1\Sigma^+$	Flame	5010-3700	-	25318.4		(67.16)
IV	$D^1\Pi \rightleftarrows X^1\Sigma^+$	Flame	4490-3070	-	29505.1		(67.16, 38.4, 33.2)
V	$E(^1\Sigma) \rightleftarrows X^1\Sigma^+$	Arc, Absorption	3110-2265	R	36138.0		(49.8, 45.6, 34.3)
VI	?	Absorption	2132-2037	R		Possibly not SnO	(49.8)
VII	$F \leftarrow X^1\Sigma^+$	Absorption	1730-1660	R	58754		(54.11)
VIII	$p \leftarrow X^1\Sigma^+$	Absorption	1775-1725	R	57491		(54.11)
IX	$o \leftarrow X^1\Sigma^+$	Absorption	1937-1926		Continuum		(54.11)

SnO

I. $A \rightleftarrows X^1\Sigma^+$ System

Weak system.

Band heads (67.21):

(v', v'')	(0, 2)	(0, 1)	(0, 0)	(1, 1)	(1, 2)
λ(Intensity)	5789.6(8)	5532.4(3)	5292.0(1)	5399.8(2)	5643.9(5)

II. $B \rightleftarrows X^1\Sigma^+$ System

Weak system.

Band heads (67.21):

(v', v'')	(0, 2)	(0, 1)	(2, 2)	(2, 1)
λ(Intensity)	4428.3(4)	4275.8(0)	4212.3(2)	4074.3(0)

III. $C \rightleftarrows X^1\Sigma^+$ System

Intense system.

Band heads (67.21):

(v', v'')	(0, 2)	(0, 1)	(1, 2)	(2, 2)
λ(Intensity)	4218.8(8)	4079.9(10)	4121.9(7)	4026.9(3)

IV. $D^1\Pi \rightleftarrows X^1\Sigma^+$ System

Band heads, λ(intensity) (33.2):

v', v''	0	1	2	3	4	5	6	7
0	3388.26 (6)	3484.50 (8)	3585.40 (7)	3691.39 (5)	3802.70 (2)	3919.51 (1)		
1	3323.45 (7)	3415.84 (5)	3512.9 (1)	3614.75 (3)	3721.23 (3)	3833.24 (2)	3950.97 (2)	
2	3262.37 (6)	3351.4 (2)	3444.64 (3)	3542.39 (1)		3752.30 (1)		3983.91 (3)
3	3205.79 (4)	3291.80 (2)	3381.72 (2)					3899.31 (2)
4				3406.94 (1)				
5				3344.71 (1)				

V. $\underline{E(^1\Sigma^+) \rightleftarrows X^1\Sigma^+}$ System

Band heads in emission, λ(intensity) (34.3):

v', v''	0	1	2	3	4	5	6
1		2790.0(3)	2854.5(4)	2921.8(8)	2990.5(8)	3063.1(3)	
2		2752.9(4)	2814.8(8)	2880.4(6)	2947.7(6)		
3	2658.1(6)	2717.0(7)	2777.4(6)			2973.6(4)	3043.6(6)
4	2623.3(5)	2680.8(6)	2740.1(6)			2931.0(5)	
5	2590.9(4)	2647.0(6)				2890.8(3)	
6	2559.9(3)	2614.2(5)					
7	2529.3(3)					2877.4(4)	

VI. <u>2132-2037Å System</u>

Bands (49.8):

λ | 2132 | 2117 | 2102 | 2084 | 2073 | 2069 | 2054 | 2050 | 2037

VII. $\underline{F \leftarrow X^1\Sigma^+}$ System

Band heads (54.1):

(v', v'')	(0, 1)	(1, 1)	(0, 0)	(2, 1)	(1, 0)
λ(Intensity)	1726.10(4)	1705.90(8)	1702.00(10)	1687.50(5)	1682.54(6)

VIII. $\underline{p \leftarrow X^1\Sigma^+}$ System

Band heads (54.11):

(v', v'')	(1, 1)	(0, 0)	(1, 0)
λ(Intensity)	1750.02(6)	1739.40(10)	1725.4(4)

IX. $\underline{o \leftarrow X^1\Sigma^+}$ System

Weak absorption continuum 1937 > λ > 1926Å (54.11).

SnO

SPECTROSCOPIC CONSTANTS

State	T_o	ω_e	$x_e\omega_e$	B_e	$\alpha_e \times 10^3$	$D_e \times 10^6$	r_e	Remarks	Bibliography
F	58752	724	2.1						(54.11)
p	57491	460[b]	-						(54.11)
o	~51775[a]	-	-						(54.11)
$E(^1\Sigma^+)$	36138	508.0	2.9						(49.8)
$D^1\Pi$	29505.1	582.6	3.08		0.31455	0.25	1.950		(71.22, 59.12)
C	25318.4	561.0	1.20						(67.16)
B	24199.0	595.0	3.90						(67.16)
A	18889.0	453.0	4.0						(67.16)
$a^3\Pi$	~20900								(70.20)
$X^1\Sigma^+$	0	822.4	3.73	0.3557190	2.1429		1.8325		(71.22, 67.16, 67.17)

[a] Continuum center, [b] $G_{1/2}$.

Dissociation energy = 5.4 ± 0.07 eV, 124.5 kcal/mole, 43555 cm^{-1}.

SnO

Perturbations and General Information

B and C states exhibit perturbations in the v' = 2 level (67.16).

D state exhibits perturbations.

radiative lifetime, $a^3\Pi - X^1\Sigma^+$ (70.20):

$\tau \sim 240$ μsec - matrix isolation in Ar at 20°K,

~ 330 μsec - estimated in gas phase.

Electronic transition moment, $D^1\Pi - X^1\Sigma^+$ (71.22):

$R_e(r) = $ constant $(1 - 1.119r + 0.356r^2)$, $1.8 \leq r \leq 2.2$ Å.

Intensities, wavelengths, Franck-Condon factors, r-centroids, band strengths, Einstein coefficients and oscillator strengths for the $D^1\Pi - X^1\Sigma$ system (71.22):

v', v''	Wavelength (Å)	q	I	r(Å)	$p(a_0^2 e^2)$	$A(s^{-1})$	f
0, 0	3388.2	0.1806	75.0	1.895	4.50-3*	11.74+4	2.02-4
0, 1	3484.5	0.2408	100.0	1.931	6.68-3	16.00+4	2.91-4
0, 2	3585.4	0.2137	87.5	1.970	6.70-3	14.74+4	2.83-4
0, 3	3691.4	0.1475	62.0	2.012	5.30-3	10.70+4	2.18-4
0, 4	3862.7	0.0632	25.0	2.057	2.64-3	4.65+4	1.03-4
0, 5	3919.5	0.0301	12.5	2.106	1.48-3	2.50+3	5.75-5
1, 0	3323.4	0.2107	87.5	1.875	4.93-3	13.63+4	2.25-4
1, 1	3415.8	0.1535	62.5	1.904	3.92-3	10.00+4	1.75-4
1, 2	3512.9	0.0295	12.4	1.944	8.52-4	1.99+4	3.69-5
1, 3	3614.7	0.0903	37.5	1.980	2.92-3	6.28+4	1.23-4
1, 4	3721.2	0.0901	37.0	2.022	3.51-3	6.59+4	1.36-4
1, 5	3833.2	0.0598	25.0	2.069	2.60-3	4.69+4	1.03-4
1, 6	3950.9	0.0595	25.0	2.120	3.09-3	5.08+4	1.18-4
2, 0	3262.4	0.1836	75.0	1.860	4.14-3	12.10+4	1.93-4
2, 2	3444.6	0.0899	37.5	1.917	2.39-3	5.96+4	1.05-4
2, 3	3542.4	0.0300	12.5	1.955	8.83-4	2.01+4	3.79-5
2, 5	3752.3	0.0296	12.0	2.040	1.17-3	2.25+4	4.74-5
2, 6	3864.8	0.0918	37.0	2.082	4.17-4	7.30+3	1.63-5

*The positive or negative numbers following an entry indicate the power of ten by which the entry is multiplied.

Dipole moment - $X^1\Sigma^+$, $\mu = 4.32 \pm 0.10$ D (69.19).

SnO

Potential energy curves — RKRV potential (65.15):

State	v	U(cm^{-1})	r_{min}(Å)	r_{max}(Å)	Te+U(cm^{-1})
X$^1\Sigma^+$	0	410.3	1.789	1.895	410.3
	1	1224.9	1.758	1.946	1224.9
	2	2029.2	1.739	1.984	2029.2
	3	2811.0	1.726	2.016	2811.0
	4	3603.2	1.716	2.046	3603.2
	5	4388.2	1.708	2.074	4388.2
	6	5167.7	1.702	2.100	5167.7
	7	5940.7	1.696	2.126	5940.7
	8	6705.1	1.692	2.152	6705.1
D$^1\Pi$	0	290.5	1.905	2.033	29915.4
	1	865.6	1.869	2.093	30490.5
	2	1429.4	1.847	2.140	31054.3
	3	1971.1	1.834	2.178	31596.0
	4	2554.3	1.823	2.216	32179.2
	5	3095.5	1.813	2.252	32720.4
	6	3628.2	1.805	2.287	33253.1
	7	4145.5	1.798	2.322	33770.4
	8	4648.4	1.792	2.355	34273.3

BIBLIOGRAPHY

(31.1) D, B, C → X Systems,
P. C. Mahanti,
Z. Physik 68, 114-25

(33.2) D, B, C → X Systems, Vibrational Analysis,
F. C. Connelly,
Proc. Phys. Soc. 45, 780-91

(33.3) E → X System, Vibrational Analysis,
F. W. Loomis and T. F. Watson,
Phys. Rev. 45, 805-6

(38.4) D → X System, Rotational Analysis,
W. Jevons,
Proc. Phys. Soc. 50, 910-3

(44. 5) Dissociation Energy,
R. F. Barrow,
Proc. Phys. Soc. 56, 204-10

(45. 6) E ← X System,
R. C. Blake and T. Iredale,
Nature 157, 229

(46. 7) Dissociation Energy,
S. P. Sinha,
Proc. Phys. Soc. 59, 610-21

(49. 8) E ← X System,
B. Eisler and R. F. Barrow,
Proc. Phys. Soc. A 62, 740-1

(49. 9) Thermochemistry,
L. Brewer and D. F. Mastick,
U. C. R. L. Report #571

(52. 10) Dissociation Energy,
G. Drummond and R. F. Barrow,
Proc. Phys. Soc. A 65, 148-9

(54. 11) Schumann Region,
R. F. Barrow and H. C. Rowlinson,
Proc. Roy. Soc. A 224, 374-88

(59. 12) D-X System, Rotational Analysis,
A. Lagerqvist, N. E. L. Nilsson, and K. Wiqartz,
Arkiv Fysik 15, 521-30

(64. 13) A, B ← X Systems, Rotational Analysis,
E. W. Deutsch and R. F. Barrow,
Nature 201, 815

(65. 14) Thermochemistry. Dissociation Energy,
R. Colin, J. Drowart, and G. Verhaegen,
Trans. Faraday Soc. 61, 1364-71

(65. 15) K. P. R. Nair, R. B. Singh, and D. K. Rai,
"Potential Energy Curves and Dissociation Energies of Oxides and Sulfides of Group IVA Elements,"
J. Chem. Phys. 43, 3570-4

SnO

(67.16)   A, B, C, D → X Systems, Vibrational Analysis,
M. M. Joshi and R. Yamdagni,
"Flame Emission Spectrum of SnO Molecule in the Visible Region,"
Indian J. Phys. 16, 275-85

(67.17)   Microwave Spectra,
T. Törring,
Z. Naturforsch. A 22, 1234-6

(68.18)   Absorption and Fluorescence in Ar, Kr and Xe Matrices,
J. J. Smith and B. Meyer,
J. Mol. Spectrosc. 27, 304-12

(69.19)   Stark Effect,
J. Hoeft, F. J. Lovas, E. Tiemann, R. Tischer, and T. Törring,
Z. Naturforsch. A 24, 1222-6

(70.20)   Lifetime Measurements in Matrix Isolation,
B. Meyers, J. J. Smith, and K. Spitzer,
"Phosphorescent Decay Time of Matrix Isolated GeO, GeS, SnO and SnS and the Lifetime of the Cameron Bands of the CO-type,"
J. Chem. Phys. 53, 3616-20

(70.21)   J. S. Ogden and M. J. Ricks,
"Matrix Isolation Studies of Group IV Oxides. III. Infrared Spectra and Structures of SnO, Sn_2O_2, Sn_3O_3 and Sn_4O_4,"
J. Chem. Phys. 53, 896-903

(71.22)   Franck-Condon Factors, r-Centroids, Transition Probabilities, D-X System,
P. S. Dube and D. K. Rai,
"Electronic Transition Moment Variation in $D^1\Pi - X^1\Sigma$ System of SnO Molecule and Determination of the Effective Vibrational Temperatures,"
J. Phys. B: Atom. Mol. Phys. 4, 579-83

SnS

Methods of Production and Experimental Technique

Absorption at 1000-1200°C.

Absorption spectrum and fluorescence in Ar, Kr and Xe matrices at 20°K.

Emission in a positive column of a discharge in a high density stream.

Ground state studied by microwave techniques.

Heating of tin and sulfur in a vacuum graphite furnace to 1400°C (66.11).

BAND SYSTEMS

System	Transition	Sources	Wavelength Limits	Degrading	Characteristic Bands, λ	Remarks	Bibliography
I	$B \leftarrow X^1\Sigma^+$	Absorption	5050-4270	-	4410.2(0,0)		(66.1)
II	$C \leftarrow X^1\Sigma^+$	Absorption	4660-4000	-	4248.9(0,0)		(66.1)
III	$D^1\Pi \rightleftarrows X^1\Sigma^+$	Absorption	4020-3200	R	3728.20 3662.80 3599.30 3557.15 3496.60 3457.24 3418.81 3382.68		(61.6, 35.1, 35.2)
IV	$E^1\Sigma^+ \rightleftarrows X^1\Sigma^+$	Absorption	3325-2500	R	3163.80 3116.76 3089.21 3062.23 2992.24 2967.23 2902.01 2878.88		(61.6, 53.5, 45.4, 35.2)
V	$p \leftarrow X^1\Sigma^+$	Absorption	1933-1860	R	1886.5 1873.2		(53.5)
VI	$q \leftarrow X^1\Sigma^+$	Absorption	1820-1788	R	1803.5 1788.0		(53.5)
VII	$r \leftarrow X^1\Sigma^+$	Absorption	1771-1758		Continuum		(53.5)

SnS

I. $\underline{B \leftarrow X^1\Sigma^+ \text{ System}}$

Band heads, λ(intensity) (66.11):

(v', v'')	(0, 1)	(0, 0)	(1, 2)	(0, 2)
λ(Intensity)	4505.3(6)	4408.9(4)	4537.8(2)	4605.5(3)

II. $\underline{C \leftarrow X^1\Sigma^+ \text{ System}}$

Band heads, λ(intensity) (66.1):

(v', v'')	(0, 1)	(0, 0)	(1, 0)	(2, 0)
λ(Intensity)	4337.8(9)	4248.6(10)	4183.8(10)	4135.4(4)

III. $\underline{D^1\Pi \leftrightarrows X^1\Sigma^+ \text{ System}}$

Band heads in absorption, λ(intensity) (35.2):

v', v''	0	1	2	3	4	5	6	7
0	3537.55 (8)	3599.30 (10)	3662.80 (10)	3728.20 (10)	3795.58 (8)	3865.32 (10)	3937.14 (5)	4011.00 (3)
1	3496.60 (10)	3557.15 (10)	3619.30 (8)			3816.60 (6)	3886.78 (9)	3958.90 (7)
2	3457.24 (10)	3516.50 (8)		3639.45 (5)	3703.85 (4)	3769.87 (4)		
3	3418.81 (10)		3536.15 (6)	3597.03 (5)			3790.91 (6)	
4	3382.68 (10)							
5	3345.52 (8)	3400.82 (4)						
6	3310.57 (8)	3364.48 (9)						
7	3276.51 (6)	3329.30 (7)						
8	3243.29 (4)	3295.20 (7)						

IV. $E\,^1\Sigma^+ \rightleftarrows X\,^1\Sigma^+$ System

Band heads, λ (35.2):

v', v''	0	1	2	3	4
0			3126.74	3174.10	3212.96
1			3098.50	3144.95	3193.25
2			3070.88	3116.76	3163.80
3			3043.93	3089.21	3135.24
4		2974.39	3017.84	3062.23	3107.60
5	2907.85	2949.51	2992.24	3035.67	
6		2925.35	2967.23	3010.09	
7	2861.66	2902.01	2943.02		
8	2839.29	2878.88	2919.32		

V.-VII. $p, q, r \leftarrow X\,^1\Sigma^+$ Systems

These UV bands are attributed to two band systems and a continuum (53.5):

p ← X is an extremely weak system,

q ← X is an intense system displaying only a v' = 0 progression,

r ← X is a continuum.

SnS

SPECTROSCOPIC CONSTANTS

State	T_o	ω_e	$x_e\omega_e$	B_o	$\alpha_e \times 10^3$	$D_e \times 10^6$	r_e	Remarks	Bibliography
r	>56470	-	-	-	-		-	Continuum	(53.5)
q	55928	-	-	-	-		-		(53.5)
p	(52220)	(395)	-	-	-		-		(53.5)
$E^1\Sigma^+$	32940	295.05	1.09	-	-		-		(61.6, 53.5)
$D^1\Pi$	28258.67	331.34	1.265	0.11988	0.7		(2.357)	$y_e\omega_e = -0.012$ cm^{-1}	(61.6)
$C^1(\Omega=1)$	23707.0	-	-	0.1075	-		(2.493)		(63.8, 61.6)
$C(\Sigma=0)$	23531.94	324.0	1.0	0.1214	-		(2.346)		(66.11, 63.8, 61.6)
$B(\Omega=0^+)$	22674.96	325.0	2.50	0.1184	-		(2.375)		(66.11, 63.8, 61.6)
$a^3\Pi$	18000	-	-	-	-		-		(70.17)
$X^1\Sigma^+$	0	487.7	1.34	0.13661	0.5		2.209		(66.11, 63.8, 61.6)

Dissociation energy = 4.8 ± 0.1 eV, 111 kcal/mole, 38715 cm^{-1}.

Perturbations and General Information

B state perturbed in the v' = 2 level (66.11).

C state perturbed in the v' = 1 level (66.11).

Dipole moment — $X^1\Sigma^+$, μ = 3.18 ± 0.10D (69.16).

Radiative lifetime, $a^3\Pi - X^1\Sigma^+$ (70.17):

τ = 520 μsec - Ar matrix at 20°K,

τ = 550 μsec - estimated in gas phase.

Potential energy curves — RKRV potential (65.10):

State	v	U(cm^{-1})	r_{min}(Å)	r_{max}(Å)	Te+U(cm^{-1})
$X^1\Sigma^+$	0	242.6	2.162	2.267	242.6
	1	727.2	2.128	2.310	727.2
	2	1208.9	2.105	2.340	1208.9
	3	1688.0	2.087	2.366	1688.0
	4	2164.5	2.072	2.389	2164.5
	5	2638.2	2.059	2.411	2638.2
	6	3109.1	2.047	2.430	3109.1
	7	3577.1	2.036	2.449	3577.1
	8	4043.1	2.027	2.467	4043.1
	9	4505.8	2.018	2.484	4505.8
	10	4966.1	2.009	2.501	4966.1
$D^1\Pi$	0	164.4	2.296	2.423	28502.3
	1	493.2	2.255	2.477	28831.1
	2	819.4	2.229	2.516	29157.3
	3	1143.1	2.209	2.549	29481.0
	4	1464.6	2.192	2.578	29802.5
	5	1783.2	2.177	2.606	30121.1
	6	2099.2	2.165	2.632	30437.1
	7	2413.4	2.153	2.657	30751.3
	8	2725.3	2.143	2.681	31063.2
	9	3034.0	2.133	2.704	31371.9
	10	3341.0	2.124	2.726	31678.9
	11	3644.6	2.115	2.748	31982.5
	12	3946.1	2.108	2.770	32284.0

SnS

State	v	U(cm^{-1})	r_{min}(Å)	r_{max}(Å)	Te+U(cm^{-1})
E$^1\Sigma$	0	146.8	2.087	2.222	33181.8
	1	438.5	2.053	2.287	33473.5
	2	728.8	2.034	2.337	33763.8
	3	1016.2	2.021	2.381	34051.2
	4	1300.4	2.012	2.422	34335.4
	5	1584.6	2.006	2.461	34619.6
	6	1865.0	2.002	2.498	34900.0
	7	2141.0	1.999	2.535	35176.0
	8	2416.9	1.998	2.570	35451.9
	9	2689.4	1.998	2.604	35724.4
	10	2960.4	1.999	2.639	35995.4
	11	3231.5	2.001	2.673	36266.5

BIBLIOGRAPHY

(35.1) D ← X and Visible Systems,
E. N. Shawhan,
Phys. Rev. 48, 521-4

(35.2) D, E ← X and Visible Systems,
G. D. Rochester,
Proc. Roy. Soc. 150, 668-84

(36.3) Predissociation-Possibly Incorrect,
E. N. Shawhan,
Phys. Rev. 49, 810-2

(45.4) E ← X System,
D. Sharma,
Proc. Nat. Acad. Sci. A 14, 217-23

(53.5) E ← X and U.V. Systems,
R. F. Barrow, G. Drummond, and H. C. Rowlinson,
Proc. Phys. Soc. A 66, 885-8

(61.6) D, E ← X Systems, Rotational Analysis and Visible Systems,
A. E. Douglas, L. L. Howe, and J. R. Morton,
J. Mol. Spectrosc. 7, 161-93

(62.7) Thermochemistry,
R. Colin and J. Drowart,
J. Chem. Phys. 37, 1120-5

(63.8) Discussion of Visible Systems,
R. F. Barrow, P. W. Fry, and R. C. Le Bargy,
Proc. Phys. Soc. 81, 697-704

(65.9) Microwave Spectra,
J. Hoeft,
Z. Naturforsch. A 20, 313-6

(65.10) K. P. R. Nair, R. B. Singh, and D. K. Rai,
"Potential-Energy Curves and Dissociation Energies of Oxides and Sulfides of Group IVA Elements,"
J. Chem. Phys. 43, 3570-4

(66.11) C, B ← X Systems, Vibrational Analysis,
R. Yamdagni and M. M. Joshi,
"The Visible Absorption Spectrum of the SnS Molecule,"
Ind. J. Phys. 40, 495-500

(66.12) Dissociation Energy,
J. Drowart and P. Goldfinger,
Quarterly Rev. 20, 545-57

(67.13) Dissociation Energy,
P. Coppens, S. Smoes, and J. Drowart,
Trans. Faraday Soc. 63, 2140-8

(68.14) Matrix Isolation,
J. J. Smith and B. Meyer,
J. Mol. Spectrosc. 27, 304-12

(69.15) Dipole Moment,
A. N. Murty and R. F. Carl Jr.,
J. Mol. Spectrosc. 30, 102-10

(69.16) Dipole Moment,
J. Hoeft, F. J. Lovas, E. Tiemann, R. Tischer, and T. Törring,
Z. Naturforsch. A 24, 1222-6

(70.17) Radiative Lifetime, a-X in Matrix Isolation,
B. Meyers, J. J. Smith, and K. Spitzer,
"Phosphorescent Decay Time of Matrix Isolated GeO, GeS, SnO, and SnS and the Lifetime of the Cameron Bands of CO-type Diatomics,"
J. Chem. Phys. 53, 3616-20

SnSe

Methods of Production and Experimental Technique

Absorption at 900-1200°C.

Emission from a positive column of a high density discharge in a stream of SnSe.

Ground state studied by microwave techniques.

BAND SYSTEMS

System	Transition	Sources	Wavelength Limits	Degrading	Characteristic Bands, λ	Remarks	Bibliography
I	a ← $X^1\Sigma^+$	Absorption	6350-5600	R	6122.41 6078.50 6040.57		-
II	A ← $X^1\Sigma^+$	Absorption	5630-4880	R	5392.83 5362.44 5360.45 5329.67		(70.11, 38.1)
III	B ← $X^1\Sigma^+$	Absorption	4810-4390	R	4440.1		(70.11, 38.1)
IV	C ← $X^1\Sigma^+$	Absorption	4635-4190	R	4345.9		(70.11, 38.1)
V	D ⇌ $X^1\Sigma^+$	Absorption	4150-3400	R	3817.80 3779.20 3694.18		(46.5, 45.3)
VI	E ⇌ $X^1\Sigma^+$	Absorption	3440-2840	R	3338.98 3324.00 3318.16 3303.22		(46.5, 45.3)
VII	F ⇌ $X^1\Sigma^+$	Absorption	2200-2075	R	2148.9 2134.0 2119.3		(46.5)

II. $A \leftarrow X^1\Sigma^+$ System

Band heads, (70.11):

(v', v'')	(0, 1)	(0, 0)	(1, 0)	(2, 0)
λ(Intensity)	5459.5(2)	5362.7(1)	5300.0(2)	5238.5(3)

III. $B \leftarrow X^1\Sigma^+$ System

Band heads (70.11):

(v', v'')	(0, 1)	(1, 1)	(0, 0)	(1, 0)
λ(Intensity)	4504.6(7)	4460.8(6)	4438.8(5)	4396.4(9)

IV. $C \leftarrow X^1\Sigma^+$ System

Band heads (70.11):

(v', v'')	(0, 1)	(0, 0)	(1, 0)	(2, 0)
λ(Intensity)	4407.6(7)	4344.7(10)	4304.3(8)	4265.3(4)

V. $D \rightleftarrows X^1\Sigma^+$ System

Band heads, λ (46.5):

v', v''	0	1	2	3	4	5
0		3679.95	3724.87	3770.88	3817.80	3865.62
1		3649.72	3694.18	3739.20	3785.24	3832.28
2	3577.49	3620.17	3663.95	3708.47	3753.88	3799.83
3	3549.50	3591.37	3634.38	3678.30	3722.86	3768.42
4	3522.18	3563.49	3605.69	3648.70	3692.60	

VI. $E \rightleftarrows X^1\Sigma^+$ System

Band heads, λ (46.5):

v', v''	0	1	2	3	4
0				3367.83	3405.44
1		3273.72	3309.42	3345.55	3382.25
2		3252.97	3288.46	3324.00	3360.62
3	3198.89	3233.03	3267.95	3303.22	3338.98
4	3179.64	3213.51	3247.69	3282.42	3318.16

SrF

VII. $F \rightleftarrows X^1\Sigma^+$ System

Preliminary analysis (46.5):

v', v''	0,7	0,6	0,5	0,4	0,3	0,2
λ	2194.8	2179.3	2163.9	2148.9	2134.0	2119.3

v', v''	0,1	?	1,1	?	1,0
λ	2104.7	2096.6	2092.0	2080.5	2077.5

SPECTROSCOPIC CONSTANTS

State	T_e	ω_e	$x_e\omega_e$	B_e	$\alpha_e \times 10^3$	$D_e \times 10^6$	r_e	Remarks	Bibliography
F	(47850)	(290)	–						(46.5)
E	30738.9	196.6	0.77					$y_e\omega_e = -0.0016$ cm^{-1}	(46.5)
D	27549.6	225.1	0.69						(46.5)
C	23067.8	218.0	1.50						(70.11, 46.5)
B	22578.8	218.8	0.50						(70.11, 46.5)
A	18696.5	222.0	0.50						(70.11, 46.5)
a	16760[a]	226.5	1.0						(46.5)
$X^1\Sigma^+$	0	331.2	0.74	0.0649978	0.1705		2.3256		(70.11, 66.9, 46.5)

[a] T_0.

Dissociation energy = 4.1 ± 0.1 eV, 95 kcal/mole, 33069 cm^{-1}.

SnSe

Perturbations and General Information

Dipole moment $X^1\Sigma^+$, $\mu = 2.82 \pm 0.09$D (69.10).

BIBLIOGRAPHY

(38. 1)   A, B, C, D Systems,
J. W. Walker, J. W. Straley, and A. W. Smith,
Phys. Rev. 53, 140-5

(43. 2)   D, E → X Systems,
R. F. Barrow and E. E. Vago,
Proc. Phys. Soc. 55, 326-8

(45. 3)   D ← X System,
D. Sharma,
Proc. Nat. Acad. Sci. A 14, 224-7

(45. 4)   E ← X System,
D. Sharma,
Proc. Nat. Acad. Sci. A 14, 228-31

(46. 5)   a, A, B, C, D, E, F → X Systems,
E. E. Vago and R. F. Barrow,
Proc. Phys. Soc. 58, 707-17

(46. 6)   E ← X System,
D. Sharma,
Nature 157, 663

(48. 7)   Dissociation Energy,
E. E. Vago and R. F. Barrow,
J. Chim. Phys. 45, 9-10

(64. 8)   Thermochemistry,
R. Colin and J. Drowart,
Trans. Faraday Soc. 60, 673-83

(66. 9)   Microwave Spectrum,
J. Hoeft,
Z. Naturforsch. A 21, 437-40

(69.10)   Microwave Spectra,
            J. Hoeft, F. J. Lovas, E. Tiemann, and T. Törring,
            Z. Naturforsch. A 24, 1843-4

(70.11)   A, B, C ← X Systems, Vibrational Analysis,
            R. Yamdagni,
            "Absorption Spectra of the SnSe Molecule in the Visible Region,"
            J. Mol. Spectrosc. 33, 531-7

SnTe

Methods of Production and Experimental Technique

Absorption at 900-1000°C.

Emission from the positive column of a high density discharge.

Fundamental state studies by microwave technique.

Ground state studied.

BAND SYSTEMS

System	Transition	Sources	Wavelength Limits	Degrading	Characteristic Bands, λ	Remarks	Bibliography
I	$A \leftarrow X^1\Sigma$	Absorption	6400-5350	R	6071.59 6007.31 5916.04 5911.88 5854.80 5795.77 5710.07 5653.72		(44.2)
II	$B \leftarrow X^1\Sigma$	Absorption	5310-4475	R	5040.04 4975.64 4912.85 4859.17 4807.68 4757.77 4699.42 4652.94		(44.2)
III	$C \leftarrow X^1\Sigma$	Absorption	4800-4290	R	4738.79 4690.92 4681.59 4634.74 4589.59 4579.98		(44.2)
IV	$D \leftrightarrows X^1\Sigma$	Absorption	4250-3700	R	4058.86 4029.63 3988.25 3960.19 3947.64 3920.20 3893.28 3853.96		(40.1)

SnTe

BAND SYSTEMS

System	Transition	Sources	Wavelength Limits	Degrading	Band Head, $\nu_{0,0}$	Remarks	Bibliography
V	$E \rightleftarrows X^1\Sigma$ $F \rightleftarrows X^1\Sigma$ $G \rightleftarrows X^1\Sigma$ $H \leftarrow X^1\Sigma$ (a)	Absorption	3750-3150	R			(46.4, 46.5, 45.3, 40.1)
VI	$I \leftarrow X^1\Sigma$	Absorption	2400-2180	R	2325.4 2311.68 2298.07 2247.90 2236.87 2225.86		(46.4, 46.5, 45.3)
VII	$J \leftarrow X^1\Sigma$ (b)	Absorption	2200-2100	R	2175.5 2163.3 2151.2 2139.4		(46.4, 46.5, 45.3)

(a) Vibrational classification is uncertain, (b) Classification uncertain.

SnTe

I. $\underline{A \leftarrow X^1\Sigma \text{ System}}$

Band heads, λ (44.2):

v', v''	0	1	2	3	4
0		6042.44	6137.79	6236.18	6337.07
1		5978.57	6071.59	6167.60	6266.28
2	5827.30	5916.04	6007.31	6101.29	
3	5768.17	5854.80	5944.66		
4	5710.07	5795.77	5884.06	5974.42	6065.93
5	5653.72	5737.74		5911.88	6002.47
6	5598.76	5681.59			
7	5545.50	5626.39			
8		5572.23			

II. $\underline{B \leftarrow X^1\Sigma \text{ System}}$

Band heads, λ (44.2):

v', v''	0	1	2	3	4
0		4968.48	5032.96	5098.64	5165.70
1	4851.06	4912.85	4975.64	5040.04	5105.87
2	4798.55	4859.17	4920.78	4984.01	
3	4748.51	4807.68	4867.92	4929.25	4999.56
4	4699.42	4757.77		4876.80	4938.40
5	4652.94			4825.82	4886.62
6	4607.15			4177.00	
7	4562.62				
8	4519.64				

III. $\underline{C \leftarrow X^1\Sigma \text{ System}}$

Band heads, λ (44.2):

v', v''	0	1	2
0	4671.96	4729.36	4787.43
1	4625.56	4681.59	4738.79
2	4579.98	4634.74	4690.92
3	4535.79	4589.59	4644.44
4	4493.07	4546.10	
5	4451.11	4502.89	
6	4410.50		
7	4371.00		
8	4333.04		
9	4295.71		

IV. $D \leftarrow X^1\Sigma$ System

Band heads, λ (44.2):

v', v''	0	1	2	3	4
0	3935.22	3975.82	4016.86	4058.86	4101.35
1	3907.97	3947.64	3988.25	4029.63	4071.86
2	3880.86	3920.20	3960.19		4042.40
3	3853.96	3893.28	3932.66	3972.59	4013.52
4	3827.92	3866.73		3945.25	3985.45
5	3802.81	3840.57	3878.68	3918.15	3957.71
6	3778.34				3930.89
7	3753.12	3790.31			3904.50

V. $E, F, G, H \leftarrow X^1\Sigma$ Systems

Classification is uncertain because of the numerous overlapping systems.

VI. $I \leftarrow X^1\Sigma$ System

Band heads, λ (46.5):

v', v''	0	1	2
0		2284.54	2298.07
1	2259.33	2272.62	2286.09
2	2247.90	2261.10	
3	2236.87		
4	2225.86		2251.91

SnTe

SPECTROSCOPIC CONSTANTS

State	T_o	ω_e	$x_e\omega_e$	B_e	$\alpha_e \times 10^3$	$D_e \times 10^6$	r_e	Remarks	Bibliography
J	(47245)	(230)	–						(46.5)
I	44018.4	229.7	1.25					$y_e\omega_e = 0.003$ cm^{-1}	(46.5)
H	(30789.0)	(201.0)	(0.6)						(46.4)
G	(29042.5)	(200.8)	(0.3)						(46.4)
F	(28465)	(98.0)	(1.0)						(46.4)
E	~(28000)	(150)	–						(46.4)
D	25404.1	179.1	0.40						(46.5, 44.2)
C	21397.8	218.1	0.98						(44.2)
B	20380.0	230.3	1.53					$y_e\omega_e = 0.013$ cm^{-1}	(44.2)
A	16803.5	178.5	0.44						(44.2)
$X^1\Sigma$	0	259.5	0.50				2.5228		(68.7, 46.5, 44.2)

Dissociation energy = 3.27 ± 0.01 eV, 75 kcal/mole, 26375 cm^{-1}.

Perturbations and General Information

Dipole moment ^{120}Sn ^{130}Te - $X^1\Sigma$, $\mu = 2.19 \pm 0.9 D (69.8)$.

BIBLIOGRAPHY

(40.1)   D, E → X Systems,
R. F. Barrow,
Proc. Phys. Soc. 52, 380-7

(44.2)   A, B, C, D Systems,
R. F. Barrow and E. E. Vago,
Proc. Phys. Soc. 56, 78-85

(45.3)   E, F, G, H, I Systems,
D. Sharma,
Proc. Nat. Acad. Sci. A 14, 232-40

(46.4)   F, G, H, I, J Systems,
D. Sharma,
Nature 157, 663

(46.5)   E, F, G, H, I, J Systems,
E. E. Vago and R. F. Barrow,
Proc. Phys. Soc. 58, 707-17

(64.6)   Mass Spectra,
R. Colin and J. Drowart,
Trans. Faraday Soc. 60, 673-83

(68.7)   Microwave Spectra,
J. Hoeft and E. Tiemann,
Z. Naturforsch. A 23, 1034-9

(69.8)   Microwave Spectra,
J. Hoeft, F. J. Lovas, E. Tiemann, and T. Törring,
Z. Naturforsch. A 24, 1422-3

SrBr

Methods of Production and Experimental Technique

Absorption.

Emission from flames — arc.

High frequency discharge into $SrBr_2$ (71.6).

BAND SYSTEMS

System	Transition	Sources	Wavelength Limits	Degrading	Band Head, $\nu_{0,0}$	Remarks	Bibliography
I	$A_1{}^2\Pi_{1/2} \rightleftarrows X^2\Sigma^+$	Arc, Absorption (1100-1500°C)	6905-6445	V	14702.3		(50.4)
	$A_2{}^2\Pi_{3/2} \rightleftarrows X^2\Sigma^+$		6680-6290	V	15003.6		(42.3, 31.2, 28.1)
II	$B^2\Sigma \rightleftarrows X^2\Sigma^+$	Arc, Absorption (1100-1500°C)	6695-6310	V	15354.8		(50.4, 42.3, 31.2)
III	$C_1{}^2\Pi_{1/2} \rightleftarrows X^2\Sigma^+$	Absorption (1550°C)	4235-4040	R	24338.1		(50.4)
	$C_2{}^2\Pi_{3/2} \rightleftarrows X^2\Sigma^+$		4180-3990	R	24659.8		(42.3)
IV	$D^2\Sigma \rightleftarrows X^2\Sigma^+$	Absorption (1550°C)	3505-3320	V	28973.8		(66.5, 50.4, 42.3)
V	$E^2\Sigma \rightarrow X^2\Sigma^+$	Discharge	3200-3000	V	32068.1		(66.5)
VI	$F^2\Pi \rightarrow A^2\Pi$	Discharge	5600-5200	-	-		(71.6)
VII	$G^2\Delta \rightarrow A^2\Pi$	Discharge	5300-5050	-	-		(71.6)
VIII	$H^2\Sigma \rightarrow A^2\Pi$	Discharge	5200-5000	-	-		(71.6)
IX	$F^2\Pi \rightarrow B^2\Sigma$	Discharge	5800-5450	-	-		(71.6)

SrBr

I. $A^2\Pi \rightleftarrows X^2\Sigma^+$ System

Five well developed sequences.

Band heads (P_1) (50.4, 42.3):

v', v''		v', v''	
0, 1	6906.4	1, 0	6706.5
1, 2	6903.1	2, 1	6704.4
2, 3	6899.9	3, 2	6702.5
0, 0	6805.9	2, 0	6610.8
1, 1	6803.3	3, 1	6607.9
2, 2	6800.7	4, 2	6605.9

II. $B^2\Sigma \rightleftarrows X^2\Sigma^+$ System

P heads of double bands; caused by spin doubling, λ (50.4, 31.2):

v', v''	$P_K + 1/2$	$P_K - 1/2$	v', v''	$P_K + 1/2$	$P_K - 1/2$
0, 2	6698.4		6, 6	6497.7	6499.8
1, 3	6694.8		1, 0	6418.4	6420.5
0, 1	6603.5	6605.3	2, 1	6416.7	6418.7
1, 2	6600.6	6602.6	3, 2	6415.0	6417.1
2, 3	6597.9	6599.8	12, 11	6401.2	6403.3
11, 12	6574.7	6576.6	2, 0	6329.2	6331.1
0, 0	6510.8	6512.9	3, 1	6327.9	6330.0
1, 1	6508.6	6510.6	4, 2	6325.7	6328.6
2, 2	6506.3	6508.4	15, 13	6314.2	6317.5

III. $C^2\Pi \rightleftarrows X^2\Sigma^+$ System

Band heads (R_1), λ (50.4, 42.3):

v', v''		v', v''	
4, 7	4224.2	0, 0	4107.2
5, 8	4225.6	1, 1	4109.0
6, 9	4227.0	2, 2	4110.9
13, 16	4236.6	1, 0	4073.0
2, 4	4184.1	2, 1	4075.0
3, 5	4185.6	3, 2	4076.9
		4, 3	4078.9
11, 13	4198.1	4, 2	4043.8
0, 1	4143.8	5, 3	4045.9
1, 2	4145.5	6, 4	4047.9
2, 3	4147.3	7, 5	4050.0

SrBr

IV. $D^2\Sigma \rightleftarrows X^2\Sigma^+$ System

Band heads (P) of $Sr^{79}Br$ (50.4, 42.3):

v', v''	0, 2	1, 3	0, 1	1, 2	0, 0	1, 0	2, 0
λ	3502.1	3497.8	3476.3	3472.2	3450.4	3421.3	3392.7

v', v''	3, 1	4, 2	4, 1	5, 2	6, 2	7, 3	10, 6
λ	3389.4	3386.1	3361.7	3358.5	3331.6	3328.6	3320.0

V. $E^2\Sigma \rightarrow X^2\Sigma^+$ System

Band heads (66.5):

(v', v'')	(0, 1)	(0, 0)	(1, 0)	(1, 1)
λ (Intensity)	3139.49(9)	3118.36(10)	3094.55(8)	3115.41(3)

SrBr

SPECTROSCOPIC CONSTANTS

State	T_e	ω_e	$x_e\omega_e$	B_e	$\alpha_e \times 10^3$	$D_e \times 10^6$	r_e	Remarks	Bibliography
$H^2\Sigma$	34360[a]	267.7							(71.6)
$G^2\Delta$	34260[a]	256.6						$A = 26$ cm^{-1}	(71.6)
$F^2\Pi$	33133[a]	256.7						$A = 83$ cm^{-1}	(71.6)
$E^2\Sigma$	32068.1[a]	248.0	0.65						(66.5)
$D^2\Sigma$	28958.2	247.8	0.55						(66.5, 50.4, 42.3)
$C_2{}^2\Pi_{3/2}$	24665.8	204.92	0.48						(50.4, 42.3)
$C_1{}^2\Pi_{1/2}$	24343.7	205.56	0.49						(50.4, 42.3)
$B^2\Sigma$	15352.0	222.0	0.55						(50.4, 42.3)
$A_2{}^2\Pi_{3/2}$	15000.7	222.07	0.53						(50.4, 42.3)
$A_1{}^2\Pi_{1/2}$	14699.4	222.18	0.53						(50.4, 42.3)
$X^2\Sigma^+$	0	216.5	0.51						(50.4, 42.3)

[a] T_o.

Dissociation energy (estimated) = 3.47 ± 0.65 eV, 80 kcal/mole, 28000 cm^{-1}.

SrBr

Perturbations and General Information

Radiative lifetimes (n.p. 7):

$$A_1\,^2\Pi_{1/2} \to X^2\Sigma^+ \quad : \quad \tau = 34.3 \text{ nsec}$$

$$A_2\,^2\Pi_{3/2} \to X^2\Sigma^+ \quad : \quad \tau = 33.2 \text{ nsec}$$

$$B^2\Sigma^+ \to X^2\Sigma^+ \quad : \quad \tau = 42.2 \text{ nsec}$$

$$C_1\,^2\Pi_{1/2} \to X^2\Sigma^+ \quad : \quad \tau = 30.3 \text{ nsec}$$

$$C_2\,^2\Pi_{3/2} \to X^2\Sigma^+ \quad : \quad \tau = 28.1 \text{ nsec}$$

BIBLIOGRAPHY

(28.1) A → X and C ← X Systems,
O. H. Walters and S. Barratt,
Proc. Roy. Soc. A 118, 120-37

(31.2) A → X, B → X Systems,
K. Hedfeld,
Z. Physik 68, 610-31

(42.3) All Systems in Absorption: Vibrational Analysis,
R. E. Harrington,
Thesis, Berkeley

(50.4) Molecular Spectra,
G. Hertzberg,
"Molecular Spectra and Molecular Structure.I.,"
2nd Ed., New York, D. Van Norstrand Co.,

(66.5) E → X System, Vibrational Analysis,
Y. P. Reddy and P. T. Rao,
"A New Ultraviolet Band System of the SrBr Molecule,"
Ind. J. Pure Appl. Phys. 4, 251-3

(71.6) F, G, H → A and F → B Systems,
B. K. R. Reddy, Y. P. Reddy, Ms. Ashrafunnisa,
and P. T. Rao,
"New Electronic Transitions in Diatomic SrBr Molecule,"
Current Science 40, 317-8

(n.p. 7) P. J. Dagdigian, H. W. Cruse, and R. N. Zare
"Radiative Lifetimes of the Alkaline Earth Monohalides,"
(to be published)

SrCl

Methods of Production and Experimental Technique

Absorption.

Emission of an arc flames and discharges.

Heating of $SrCl_2$ + He, $T \sim 700°C$ (65.9).

BAND SYSTEMS

System	Transition	Sources	Wavelength Limits	Degrading	Band Head, $\nu_{0,0}$	Remarks	Bibliography
I	$A_1{}^2\Pi_{1/2} \rightleftarrows X^2\Sigma^+$	Arc, Absorption ~1100-1500°C	6895-6480	V	14821.9		(70.11, 65.9, 42.6)
	$A_2{}^2\Pi_{3/2} \rightleftarrows X^2\Sigma^+$		6880-6465	V	15115.7		
II	$B^2\Sigma \rightleftarrows X^2\Sigma^+$	Arc, Absorption ~1100-1500°C	6600-6230	V	15721.5		(70.11, 65.9, 42.6)
III	$C_1{}^2\Pi_{1/2} \rightleftarrows X^2\Sigma^+$	Arc, Absorption ~1100-1500°C	4055-3895	R	25235.5 25392.0		(65.9, 42.6)
	$C_2{}^2\Pi_{3/2} \rightleftarrows X^2\Sigma^+$						
IV	$D^2\Sigma \leftarrow X^2\Sigma^+$	Absorption ~1700°C	3570-3200	V	28844.2		(42.6)
V	$E^2\Sigma \leftarrow X^2\Sigma^+$	Absorption ~1750°C	3160-2925	V	32223.4		(42.6)
VI	$F^2\Sigma \leftarrow X^2\Sigma^+$	Absorption ~1750°C	2945-2785	V	34287.8		(42.6)
VII	$D^2\Sigma \rightarrow A^2\Pi$	Discharge	7450-7100 7450-7130	-	13727.0 14021.8		(65.9)
VIII	$E^2\Sigma \rightarrow A^2\Pi_{1/2}$	Discharge	5850-5630	-	17396.1		(65.9)
IX	$F^2\Pi \rightarrow A^2\Pi$	Discharge	5780-5355 5815-5280	-	17884.6 18109.9		(65.9)

SrCl

BAND SYSTEMS

System	Transition	Sources	Wavelength Limits	Degrades	Band Head, $\nu_{0,0}$	Remarks	Bibliography
X	$G^2\Delta \rightarrow A^2\Pi$	Discharge	5350-4780 5270-4970	-	18996.4 19264.8		(65.9)
XI	$D^2\Sigma \rightarrow B^2\Sigma$	Discharge	7800-7360	-	13123.7		(65.9)
XII	$F^2\Pi \rightarrow B^2\Sigma$	Discharge	5810-5690 5870-5670	-	17208.4 17277.5		(65.9)
XIII	$H^2\Sigma \rightarrow B^2\Sigma$	Discharge	5570-5170	-	18562.7		(65.9)

SrCl

I. $A^2\Pi \rightleftarrows X^2\Sigma^+$ System

Band heads (Q_2) λ:

v', v''		v', v''	
1, 3	6881.2	11, 11	6583.8
2, 4	6876.6		
3, 5	6872.0	1, 0	6481.9
		2, 1	6479.8
0, 1	6747.7	3, 2	6478.0
1, 2	6743.7		
		10, 9	6465.0
0, 0	6613.8		
1, 1	6610.8		
2, 2	6607.8		

II. $B^2\Sigma \rightleftarrows X^2\Sigma^+$ System

Double headed bands caused by spin doubling.

Band heads (P) in emission:

v', v''	$P_{K+1/2}$	$P_{K-1/2}$	v', v''	$P_{K+1/2}$	$P_{K-1/2}$
3, 5	6600.4		0, 0	6359.0	6362.3
4, 6	6597.3		1, 1	6357.3	6360.5
5, 7	6594.4		2, 2	6356.5	6358.8
0, 1	6482.8	6485.2	8, 8	6347.3	6349.6
1, 2	6480.3	6482.7	1, 0	6338.3	6342.3
2, 3	6477.8	6480.1	2, 1	6337.3	6341.4
19, 20	6484.6		3, 2	6336.5	6340.4
			20, 19	6328.9	

III. $C^2\Pi \rightleftarrows X^2\Sigma^+$ System

Band heads in emission, λ (35.3), intensities in absorption (int.$_a$) (28.1):

v', v'' (Int.$_a$)	Br.	λ
0, 1(2)	Q_1	4009.3
	R_1	4008.4
0, 2(2)	Q_2	3983.4
	R_2	3982.5
0, 0(10)	Q_1	3961.6
	R_1	3960.8
0, 0(10)	Q_2	3937.1
	R_2	3936.4
1, 0(0)	Q_1	3918.2
	R_1	3917.3
1, 0(0)	Q_2	3894.0
	R_2	3893.2

IV. $D^2\Sigma \leftarrow X^2\Sigma^+$ System

Band heads (P) of $Sr^{35}Cl$ (42.6):

v', v''	λ	v', v''	λ	v', v''	λ
0, 2	3539.4	2, 0	3385.7	8, 3	3263.6
1, 3	3533.6	3, 1	3381.3	9, 4	3260.3
2, 4	3527.9	4, 2	3377.0	10, 5	3256.9
0, 1	3502.4	3, 0	3347.3	12, 6	3220.1
1, 2	3497.0	4, 1	3343.3	13, 7	3216.9
2, 3	3491.7	5, 2	3339.3	14, 8	3214.2
0, 0	3465.9	5, 1	3306.3		
1, 0	3425.2	6, 2	3302.6		
2, 1	3420.6	7, 3	3299.0		
3, 2	3415.8				

SrCl

V. $\underline{E^2\Sigma \leftarrow X^2\Sigma^+ \text{ System}}$

Band heads (P) (42.6):

v', v''	λ	v', v''	λ	v', v''	λ
0, 2	3161.2	2, 0	3037.7	8, 4	2964.3
1, 3	3156.5	3, 1	3034.1	9, 5	2961.3
		4, 2	3030.5	10, 6	2958.4
0, 0	3102.4				
2, 2	3094.0	3, 0	3006.7	8, 3	2938.6
		4, 1	3003.4	9, 4	2935.8
1, 0	3069.6	5, 2	3000.0	10, 5	2933.1
2, 1	3065.7				

VI. $\underline{F^2\Sigma \leftarrow X^2\Sigma^+ \text{ System}}$

Weak band heads (P):

v', v''	0, 1	1, 2	0, 0	1, 0	2, 0
λ	2941.4	2935.8	2915.6	2885.1	2855.4

v', v''	3, 1	4, 1	5, 2	6, 2	7, 3
λ	2850.7	2822.1	2817.7	2790.1	2785.9

SPECTROSCOPIC CONSTANTS

State	T_o	ω_e	$x_e\omega_e$	B_e	$\alpha_e \times 10^3$	$D_e \times 10^6$	r_e	Remarks	Bibliography
$F^2\Sigma$	34256.7	364.6	1.08						(65.9, 42.6)
$E^2\Sigma$	32201.8	346.3	1.10						(65.9, 42.6)
$D^2\Sigma$	28822.9	344.8	1.04						(42.6)
$C^2\Pi$	25401.7 25245.5	279.3 280.4	0.80 0.56						(42.6)
$B^2\Sigma$	15779.5	306.35	0.98						(42.6)
$A^2\Pi$	15112.6 14818.5	309.13 309.6	0.97 0.99						(70.11, 65.1, 42.6)
$X^2\Sigma^+$	0	302.3	0.95						(70.11, 65.1, 42.6)

Dissociation energy = 4.16 eV, 95.9 kcal/mole, 33540 cm^{-1} (70.10).

SrCl

Perturbations and General Information

Radiative lifetimes (n. p. 12):

$$A_1{}^2\Pi_{1/2} \to X^2\Sigma^+ : \tau = 31.3 \text{ nsec}$$

$$A_2{}^2\Pi_{3/2} \to X^2\Sigma^+ : \tau = 30.4 \text{ nsec}$$

$$B^2\Sigma^+ \to X^2\Sigma^+ : \tau = 38.8 \text{ nsec}$$

$$C_1{}^2\Pi_{1/2} \to X^2\Sigma^+ : \tau = 26.1 \text{ nsec}$$

$$C_2{}^2\Pi_{3/2} \to X^2\Sigma^+ : \tau = 26.0 \text{ nsec}$$

BIBLIOGRAPHY

(28.1) A ← X, C ← X Systems,
O. H. Walters and S. Barratt,
Proc. Roy. Soc. A 118, 120-37

(31.2) A → X (Flame) System, Vibrational Analysis,
K. Hedfeld,
Z. Physik 68, 610-31

(35.3) A → X, C → X (Arc) Systems,
A. E. Parker,
Phys. Rev. 47, 349-58

(38.4) Interpretation,
K. R. More and S. D. Cornell,
Phys. Rev. 53, 806-11

(42.5) A → X System,
A. Gatterer,
Ricerche Spettroscop. Lab. Astrofis. Specola Vaticana 1, 153-79

(42.6) All Systems in Absorption, Vibrational Analysis,
R. E. Harrington,
Thesis, Berkeley

SrCl

(50. 7) Molecular Spectra,
G. Hertzberg,
"Molecular Spectra and Molecular Structure . I.,"
2nd Ed., New York, D. Van Norstrand Co.

(63. 8) V. F. Zhitkevich, A. I. Lyuty, N. A. Nesterko,
V. S. Rossikhin, and I. L. Tsikora,
"A Spectroscopic Study of Dissociation and Ionization
Processes in a Flame,"
Optic. Spect. 10, 17-9

(65. 9) A-X, D, E, F, G-A and D, F, H-B Systems,
M. M. Novikov and L. V. Gurvich,
"A New Study of the Emission Spectrum of the SrCl Molecule."
Optika Spectrosc. 19, 76-7

(70. 10) D. L. Hildenbrand,
"Dissociation Energies and Chemical Bonding in the Alkaline-
Earth Chlorides from Mass Spectrometric Studies,"
J. Chem. Phy. 52, 5751-9

(70. 11) A → X System,
J. Singh, K. P. R. Nair, K. N. Upadhya, and D. K. Rai,
"On the Red Bands of SrCl Molecule,"
Opt. Pasa. Apl. 3, 76-80

(71. 12) C. D. Jonah and R. N. Zare,
"Formation of Group IIA Halides by Two Body Radiative
Association,"
Chem. Phys. Letters 9, 65-67

(n. p. 12) P. J. Dagdigian, H. W. Cruse, and R. N. Zare,
"Radiative Lifetimes of the Alkaline Earth Monohalides,"
(to be published)

SrF

SrF

Methods of Production and Experimental Technique

Absorption at $1500 \leq T \leq 2000^\circ C$.

Emission from an arc.

BAND SYSTEMS

System	Transition	Sources	Wavelength Limits	Degrading	Band Head, $\nu_{0,0}$	Remarks	Bibliography
I	$A^2\Pi \rightleftarrows X^2\Sigma^+$	Arc	6870-6283	V	15072.7 15351.9		(31.4, 29.3, 28.2, 21.1)
II	$B^2\Sigma \rightleftarrows X^2\Sigma^+$	Arc	5785-5622	R	17297.9		(31.4, 29.3, 21.1)
III	$C^2\Pi \rightleftarrows X^2\Sigma^+$	Arc	3795-3600	R	26954.5 27017.9		(41.5, 29.3, 21.1)
IV	$D^2\Sigma \leftarrow X^2\Sigma^+$	Absorption (1500-2000°C)		V	28322.6		(41.5)
V	$E^2\Pi \leftarrow X^2\Sigma^+$	Absorption (1500-2000°C)		V	31560.6 31646.7		(41.5)
VI	$F^2\Sigma \leftarrow X^2\Sigma^+$	Absorption (1500-2000°C)		V	32869.2		(41.5)
VII	$G^2\Pi \leftarrow X^2\Sigma^+$	Absorption (1500-2000°C)		V	34796.3		(41.5)
VIII	$G' \leftarrow X^2\Sigma^+$	Absorption (1500-2000°C)	< 2500			Continuum	(41.5)

I. $A^2\Pi \rightleftarrows X^2\Sigma^+$ System

$\Delta V = +1$ series converges at $(v', v'') = (38, 37)$ to a "tail band".
Band heads (Q_2) in emission λ (intensity) (31.4, 29.3):

v', v''		v', v''	
0, 1	6729.4(4)	1, 0	6306.1(7)
1, 2	6724.0(3)	2, 1	6304.8(8)
2, 3	6718.6(3)	3, 2	6303.5(8)
0, 0	6512.0(10)		
1, 1	6508.7(9)		
2, 2	6505.5(9)	Tail	6283.1

II. $B^2\Sigma \rightleftarrows X^2\Sigma^+$ System

Double banded heads caused by spin doubling.
Band heads (R) in emission (29.3):

	λ (Int.)	
v', v''	$R_{K-1/2}$	$R_{K+1/2}$
2, 2	5784.8(8)	5777.6(8)
1, 1	5782.1(9)	5774.8(8)
0, 0	5779.4(10)	5772.0(8)
6, 5	5640.8(4)	
5, 4	5636.9(4)	
4, 3	5633.0(3)	
1, 0	5621.8(0)	

III. $C^2\Pi \rightleftarrows X^2\Sigma^+$ System

Weak emission.
R heads in absorption (41.5):

v', v''	2, 2	1, 1	0, 0	5, 4	4, 3	3, 2
λ	3720.5	3714.9	3708.9	3668.3	3663.9	3659.0

v', v''	2, 1	1, 0	5, 3	4, 2	3, 1	2, 0
λ	3653.6	3647.7	3604.5	3599.7	3594.4	3588.6

SrF

IV. $D^2\Sigma \leftarrow X^2\Sigma^+$ System

Band heads (P) (41.5):

v', v''	λ	Int.	v', v''	λ	Int.
0, 1	3592.8	2	3, 1	3394.0	2
0, 0	3529.8	9	4, 2	3389.1	3
1, 1	3523.5	10	5, 3	3384.2	4
2, 2	3517.2	8	6, 4	3379.4	4
3, 3	3510.8	6	7, 5	3374.5	4
4, 4	3504.6	3	8, 6	3369.6	3
			9, 7	3364.8	3
1, 0	3462.8	4	10, 8	3359.9	2
2, 1	3457.3	7	11, 9	3355.1	2
3, 2	3451.7	7	12, 10	3350.3	1
4, 3	3446.1	6	13, 11	3345.5	1
5, 4	3440.6	5			
6, 5	3435.0	3			
7, 6	3429.5	2			
8, 7	3424.0	1			

V. $E^2\Pi \leftarrow X^2\Sigma^+$ System

Triple bands (P_1, P_2, and Q_2 heads) (41.5):

v', v''	0, 1	1, 2	0, 0	1, 1	2, 2
λ	3218.3(5)	3211.4(2)	3167.6(10)	3167.5(3)	3155.5(1)

v', v''	1, 0	2, 1	3, 2	4, 3
λ	3112.6(7)	3107.3(6)	3102.1(3)	3097.2(1)

VI. $F^2\Sigma^+ \leftarrow X^2\Sigma^+$ System

Band heads (41.5):

v', v''	0, 1	1, 2	2, 3	0, 0	1, 1	1, 0
λ	3088.3	3078.9	3069.6	3041.6	3032.9	2987.8
Int.	6	4	2	8	1	10

v', v''	2, 1	3, 2	4, 3	2, 0	3, 1	4, 2	5, 3
λ	2979.9	2972.3	2965.0	2936.4	2929.4	2922.7	2916.1
Int.	7	3	1	5	7	6	2

VII. $G^2\Pi \leftarrow X^2\Sigma^+$ System

Triple and quadruple bands (41.5):

(v', v'')	(0, 1)	(0, 0)	(1, 0)	(2, 1)	(2, 0)	(3, 1)
λ(Int.)	2914.8(4)	2873.1(10)	2826.7(8)	2821.0(5)	2781.9(6)	2776.7(5)

SrF

SPECTROSCOPIC CONSTANTS

State	T_o	ω_e	$x_e\omega_e$	B_o	$\alpha_e \times 10^3$	$D_o \times 10^7$	r_e	Remarks	Bibliography
$G^2\Pi$	34759.2	573.9	1.28	-	-	-	-		(41.5)
$F^2\Sigma$	32871.96	598.5	3.42	0.26966	-	2.23	1.997		(67.10)
$E^2\Pi$	31614.8 31528.7	564.4	3.20	-	-	-	-		(41.5)
$D^2\Sigma$	28296.6	552.1	2.15	-	-	-	-		(41.5)
$C^2\Pi$	27358.8	448	1.72	-	-	-	-		(67.11)
$B^2\Sigma$	17303.4	488.9	1.86	-	-	-	-		(41.5, 31.4, 29.3, 28.2, 21.1)
$A^2\Pi$	15352.0 15071.6	505.1 505.7	2.18 2.20	-	-	-	-		(67.11)
$X^2\Sigma^+$	0	498.0	2.15	0.24971	-	2.46	2.0757		(67.10, 67.11, 67.12, 63.6)

Dissociation energy = 5.45 ± 0.1 eV, 126 kcal/mole, 43960 cm^{-1}.

Perturbations and General Information

Radiative lifetimes (n. p. 13):

$$A_1{}^2\Pi_{1/2} \to X^2\Sigma^+ : \tau = 24.1 \text{ nsec}$$

$$A_2{}^2\Pi_{3/2} \to X^2\Sigma^+ : \tau = 22.6 \text{ nsec}$$

Quantum yields (n. p. 14):

$$Sr + F_2 \to SrF\ (X^2\Sigma^+,\ A^2\Pi,\ B^2\Sigma^+,\ C^2\Pi,\ D^2\Sigma) + F$$

$$\frac{A^2\Pi}{SrF_{tot.}} = 8.6 \times 10^{-4},\ \frac{B^2\Sigma^+}{SrF_{tot.}} = 3.1 \times 10^{-4},\ \frac{C^2\Pi}{SrF_{tot.}} = 6.6 \times 10^{-5},$$

$$\frac{D^2\Sigma}{SrF_{tot.}} = 3.2 \times 10^{-5}.$$

BIBLIOGRAPHY

(21.1)   A, B, C-X Systems,
S. Datta,
Proc. Roy. Soc. A 99, 436-55

(28.2)   A ← X System,
O. H. Walters and S. Barratt,
Proc. Roy. Soc. A 118, 120-37

(29.3)   A, B, C-X Systems, Vibrational Analysis,
R. C. Johnson,
Proc. Roy. Soc. A 122, 161-88

(31.4)   A, B ⇌ X Systems,
A. Harvey,
Proc. Roy. Soc. A 133, 336-50

(41.5)   All Systems in Absorption,
C. A. Fowler, Jr.,
Phys. Rev. 59, 645-52

(63.6)   Dissociation Energy by Mass Spectra,
G. D. Blue, J. W. Green, R. G. Bautista, and
J. L. Margrave,
J. Phys. Chem. 67, 877-88

SrF

(63. 7)   G. D. Blue, J. W. Green, T. C. Ehlert, and J. L. Margrave,
"Dissociation Energies of the Alkaline Earth Monofluorides,"
Nature 199, 804-805

(64. 8)   V. G. Ryabova and L. V. Gurvich,
"Determination of Dissociation Energies of Metal Halides from Equilibrium Flames. Dissociation Energies of CaF, CaF_2, SrF and SrF_2,"
Teplo. Vgsa. Temperatur. 2, 834-5

(64. 9)   Dissociation Energy by Mass Spectra,
T. C. Ehlert, G. D. Blue, J. W. Green, and J. L. Margrave,
J. Chem. Phys. 41, 2250-5

(67. 10)   F-X Systems, Rotational Analysis,
R. F. Barrow and J. R. Beale,
Chemical Comm. 606

(67. 11)   C, A-X Systems, Vibrational Analysis,
M. M. Novikov and L. V. Gurvich,
Optics Spectra. 22, 395-9

(67. 12)   Dissociation Energy by Mass Spectra,
R. F. Barrow and J. R. Beale,
Proc. Phys. Soc. 91, 483-8

(68. 13)   Dissociation Energy by Mass Spectra,
D. L. Hildenbrand,
J. Chem. Phys. 48, 3657-65

(71. 14)   ESR Spectra in a Matrix at $4°K$,
L. B. Knight Jr., W. C. Easly, W. Weltner Jr., and M. Wilson,
"Hyperfine Interaction and Chemical Bonding in MgF, CaF, SrF and BaF Molecules,"
J. Chem. Phys. 54, 322-9

(n.p. 13)   P. J. Dagdigian, H. W. Cruse, and R. N. Zare,
"Radiative Lifetimes of the Alkaline Earth Monohalides,"
(to be published)

(n.p. 14)   Quantum Yields,
D. J. Eckstrom, S. A. Edelstein, and S. W. Benson,
(to be published)

SrI

Methods of Production and Experimental Technique

Absorption.

Emission from a discharge.

Radio frequency discharge into Sr + I_2 + Ar (71.5).

BAND SYSTEMS

System	Transition	Sources	Wavelength Limits	Degrading	Band Head, $\nu_{0,0}$	Remarks	Bibliography
I	$A^2\Pi \rightleftarrows X^2\Sigma$	High frequency discharge	6932-6683	-	14420 14744		(71.5, 70.3, 28.1)
II	$B^2\Sigma \rightleftarrows X^2\Sigma$	High frequency discharge	4485-4373	V	14635		(71.5)
III	$C^2\Pi \rightleftarrows X^2\Sigma$	High frequency discharge	4500-4200	R	22664.5(Q_1)		(71.4)

SrI

I. $A^2\Pi \rightleftarrows X^2\Sigma^+$ System

Band heads (Q_2) (70.3):

(v', v'')	(0, 1)	(0, 0)	(1, 1)	(2, 2)
λ	6858.05	6780.81	6778.32	6775.71

II. $B^2\Sigma \rightleftarrows X^2\Sigma^+$ System

Band heads (71.5):

(v', v'')	(0, 0)	(1, 0)	(2, 0)
λ	6778.74	6748.55	6670.22

III. $C^2\Pi \rightleftarrows X^2\Sigma^+$ System

Band heads (R_1) (71.4):

(v', v'')	(0, 1)	(1, 2)	(2, 3)	(3, 4)	(7, 8)
λ(Intensity)	4444.21(0)	4444.72(0)	4445.17(2)	4445.67(2)	4447.65(1)

SrI

SPECTROSCOPIC CONSTANTS

State	T_o	ω_e	$x_e\omega_e$	B_e	$\alpha_e \times 10^3$	$D_e \times 10^6$	r_e	Remarks	Bibliography
$C^2\Pi$	22666.1[a]	170.9	−0.36					$A = 557.3 \text{ cm}^{-1}$	(71.4)
$B^2\Sigma$	14642	176	0.5						(71.5)
$A^2\Pi$	14744.7 14420.2	174.0 190.0	0.4 0.5						(71.5, 70.3)
$X^2\Sigma^+$	0	169.44	0.64						(71.4, 71.5, 70.3)

[a] Q_1 head.

Dissociation energy (estimated) = 3.47 ± 0.65 eV, 80 kcal/mole, 28000 cm^{-1}

SrI

Perturbations and General Information

Radiative lifetimes (n. p. 6):

$$A_1\,^2\Pi_{1/2} \rightarrow X^2\Sigma^+ \;:\; \tau = 43.3 \text{ nsec}$$

$$A_2\,^2\Pi_{3/2} \rightarrow X^2\Sigma^+ \;:\; \tau = 41.9 \text{ nsec}$$

$$B^2\Sigma^+ \rightarrow X^2\Sigma^+ \;:\; \tau = 46.0 \text{ nsec}$$

$$C_2\,^2\Pi_{3/2} \rightarrow X^2\Sigma^+ \;:\; \tau = 36.0 \text{ nsec}$$

BIBLIOGRAPHY

(28. 1) All Systems in Absorption,
O. H. Walters and S. Barratt,
Proc. Roy. Soc. A 118, 120-37

(39. 2) B, C → X Systems, Vibrational Analysis,
P. Mesnage,
Ann. Physique 12, 5-87

(70. 3) A → X System, Vibrational Analysis,
M. M. Shukla,
"Analysis of the A-X Band System of the SrI Molecule,"
Ind. J. Pure and Appl. Phys. 8, 855-6

(71. 4) C → X System, Vibrational Analysis,
B. R. K. Reddy, Y. P. Reddy, and P. T. Rao,
"The Visible Emission Spectrum of Strontium Monoiodide,"
J. Phys. B: Atom Mol. Phys. 4, 574-8

(71. 5) A, B → X System, Vibrational Analysis,
B. R. K. Reddy, Y. P. Reddy, C. G. R. Rao, and P. T. Rao,
"The Spectrum of SrI in the Photographic Infrared,"
Current Science 40, 186-7

(n. p. 6) P. J. Dagdigian, H. W. Cruse, and R. N. Zare,
"Radiative Lifetimes of the Alkaline Earth Monohalides,"
(to be published)

SrO

Methods of Production and Experimental Technique

Stellar absorption (47.8).

Emission from a carbon arc + Sr salts. Acetylene + air flame; carbon flame.

Emission from molecular beam of Sr + O_2 (72.40).

BAND SYSTEMS

	System	Transition	Sources	Wavelength Limits	Degrading	Band Head, $\nu_{0,0}$	Remarks	Bibliography
	I	$A^1\Sigma \to X^1\Sigma^+$	Arc	10450-7500	R	10868.82		(68.35, 56.26, 50.11, 33.5)
Blue	II	$B^1\Pi \to X^1\Sigma^+$	Arc	5280-3750	R	24635.6		(53.17, 53.18, 52.15, 32.3)
U.V.	III	$C^1\Sigma \to X^1\Sigma^+$	Arc	3710-3240	R	28546.4		(54.19, 52.16, 32.3)
	IV		Flame C	6115-5900	V (part.)			(42.7)
	V	?	Arc	11000				(56.26)

SrO

I. $A^1\Sigma \to X^1\Sigma^+$ System

Band heads, λ (intensity) (50.11):

v', v''	0	1	2	3
0	9195.82	9776.15		
1	8700.02		10426.2	10437.1
2	8257.82	8722.5	9236.80	
3	7852.84	8272.18		
4	7500.63	7882.28		
5		7522.83	7901.96	
6			7541.60	

II. $B^1\Pi \to X^1\Sigma^+$ System

Band heads, λ (intensity) (32.3):

v', v''	0	1	2	3	4	5	6
0	4058.0 (3)	4167.2 (5)	4281.0 (5)	4399.6 (6)	4523.3 (4)		
1			4189.1 (4)	4302.7 (4)	4420.9 (5)	4544.1 (5)	4672.6 (4)
2							4564.8 (5)
3							4463.3 (4)

III. $C^1\Sigma \to X^1\Sigma^+$ System

Band heads (32.3):

(v', v'')	(0, 1)	(0, 0)	(1, 0)	(2, 0)
λ(Intensity)	3586.9(3)	3503.8(6)	3445.2(4)	3389.8(4)

V. IR Bands

λ | 11135.9 | 11115.9 | 11088.5 | 11055.8 | 11018.7 | 11000.9 | 10976.6 | 10930.2 | 10907.1

SPECTROSCOPIC CONSTANTS

SrO

State	T_e	ω_e	$x_e\omega_e$	B_e	$\alpha_e \times 10^3$	$D_e \times 10^6$	r_e	Remarks	Bibliography
$C^1\Sigma$	28632.7	480.2	2.6	0.2742	2.1	0.35	2.132		(54.19, 52.16)
$B^1\Pi$	24702.8	519.9	3.24	0.2936	1.5	~0.5	2.060		(53.17, 53.18, 52.15)
(Q)	~11260	444	–	0.242	2	–	–	(a)	(50.12)
(Y)	~11143	450	–	0.253	2	–	–	(a)	(50.12)
(X)	~10950	443	–	0.2513	2	–	–	(a)	(50.12)
(Z)	~10893	444	–	0.248	2	–	–	(a)	(50.12)
$A^1\Sigma$	10885.0	619.6	~0.9	0.3047	1.1	0.32	2.027		(56.26, 50.12)
$X^1\Sigma^+$	0	653.3	4.0	0.33798	2.1	0.34	1.9199		(65.31)

(a) Constants for these states are derived from perturbation analysis of the $A^1\Sigma$ state.

Dissociation energy $\geq$ 5.70 eV, 131.5 kcal/mole, 46000 cm^{-1} (72.39).

SrO

Perturbations and General Information

A state perturbed by X, Y, Z and Q states (50.12).

Reaction order and reactive cross-sections (72.39).

$$Sr + N_2O \rightarrow SrO(A^1\Sigma) + N_2 \text{ - First order reaction - } \sigma \leq 27 A^2$$

$$Sr + NO_2 \rightarrow SrO(A^1\Sigma) + NO \text{ - Second order reaction in } NO_2 \text{ - } \sigma = 127 A^2$$

Franck-Condon factors, r-centroids — RKR potential (71.37):

$B^1\Pi - X^1\Sigma^+$

v", v'	0	1	2	3	4	5	6	7
0	.1082 1.990	.2286 1.961	.2550 1.943	.1978 1.907	.1186 1.879	.5778-1 1.851	.2341-1 1.820	.7924-2 1.787
1	.2587 2.025	.1593 1.996	.7495-2 1.970	.4151-1 1.939	.1352 1.913	.1616 1.886	.1228 1.858	.6905-1 1.829
2	.2922 2.061	.2316-2 2.013	.1045 2.005	.1081 1.977	.1018-1 1.957	.2483-1 1.914	.1050 1.891	.1388 1.864
3	.2032 2.098	.8746-1 2.072	.1041 2.037	.2532-2 2.027	.9105-1 1.986	.7085-1 1.960	.2569-2 1.964	.3216-1 1.891
4	.9605-1 2.137	.2083 2.107	(.8385-4) (1.898)	.1120 2.047	.3008-1 2.014	.2172-1 1.996	.8436-1 1.968	.3761-1 1.948
5	.3224-1 2.179	.1838 2.145	.9591-1 2.118	.4926-1 2.076	.4223-1 2.059	.7813-1 2.024	.7107-3 1.971	.4793-1 1.977
6	.7824-2 2.224	.9248-1 2.187	.1944 2.115	.1324-1 2.140	.9830-1 2.088	.1279-2 2.106	.7712-1 2.034	.2602-1 2.002
7	.1375-2 2.275	.3008-1 2.233	.1512 2.196	.1408 2.165	.4144-2 2.088	.9264-1 2.098	.1279-1 2.053	.4171-1 2.045
8	.1726-3 2.334	.6619-2 2.284	.6580-1 2.241	.1825 2.205	.6965-1 2.178	.3898-1 2.123	.5277-1 2.110	.4534-1 2.069
9		.9930-3 2.343	.1797-1 2.293	.1073 2.250	.1784 2.215	.1822-1 2.200	.7549-1 2.137	.1525-1 2.130
10			.3190-2 2.352	.3621-1 2.302	.1448 2.260	.1463 2.225	(.8624-4) (2.587)	.8993-1 2.148

Top = Franck-Condon factor followed by factor of ten.
Bottom = r-centroid

$C^1\Sigma - X^1\Sigma$

v'', v'	0	1	2	3	4	5	6	7
0	.6446-2	.2682-1	.5986-1	.9534-1	.1217	.1324	.1278	.1123
	2.024	2.006	1.989	1.973	1.957	1.942	1.957	1.913
1	.3813-1	.1026	.1372	.1157	.6237-1	.1663-1	(.2815-4)	.1043-1
	2.048	2.029	2.011	1.994	1.977	1.958	1.833	1.939
2	.1057	.1555	.8489-1	.9894-2	.8121-2	.4764-1	.7047-1	.5822-1
	2.073	2.053	2.033	2.010	2.008	1.987	1.969	1.953
3	.1826	.1034	.1363-2	.3974-1	.7860-1	.4660-1	.5423-2	.5546-2
	2.099	2.077	2.037	2.044	2.024	2.005	1.981	1.985
4	.2200	.1432-1	.5023-1	.8389-1	.1698-1	.6820-2	.4736-1	.5387-1
	2.126	2.098	2.088	2.065	2.041	2.040	2.015	1.996
5	.1963	.1589-1	.1039	.1217-1	.2583-1	.6539-1	.2612-1	.1868-3
	2.155	2.141	2.111	2.082	2.077	2.054	2.032	2.088
6	.7199-1	.1024	.4410-1	.2652-1	.7238-1	.8835-2	.1711-1	.5236-1
	2.184	2.164	2.135	2.124	2.097	2.067	2.067	2.043
7	.3061-1	.1664	.8084-3	.9056-1	.9306-2	.3329-1	.5475-1	.6098-2
	2.215	2.192	2.216	2.145	2.111	2.108	2.083	2.053
8	.1039-1	.1531	.6639-1	.7129-3	.2974-1	.6039-1	.3976-3	.3647-1
	2.248	2.223	2.202	2.169	2.158	2.128	2.044	2.094
9	.2816-2	.9580-1	.1465	.6771-1	.8492-1	.1424-2	.4958-1	.3375-1
	2.283	2.255	2.231	2.268	2.180	2.117	2.139	2.111
10	.6068-3	.4376-1	.1493	.1456	.3024-1	.4661-1	.3970-1	.6329-2
	2.320	2.290	2.263	2.242	2.202	2.191	2.158	2.160

SrO

BIBLIOGRAPHY

(27. 1)   Generalities,
R. Mecke and M. Guillery,
Z. Physik. 28, 514-31

(27. 2)   Generalities,
R. Mecke,
Z. Physik. 42, 390-425

(32. 3)   Blue and Violet System,
P. C. Mahanti,
Phys. Rev. 42, 609-21

(33. 4) $^1\Sigma \rightarrow {}^1\Sigma$ System,
W. F. Meggers,
Bur. Stand. J. Res. 10, 669-84

(33. 5) $^1\Sigma \rightarrow {}^1\Sigma$ System,
K. Mahla,
Z. Physik 81, 625-46

(35. 6)   Generalities,
P. C. Mahanti,
Indian J. Phys. 9, 517-36

(42. 7)   Excitation in a C Flame,
A. Gatterer,
Ricerche Spettroscops. Lab. Astrofis. Specola Vaticana 1, 153-79

(47. 8)   D. N. Davis,
Astrophys. J. 106, 28-75

(48. 9)   Dissociation Energy,
L. Huldt,
Thesis, Stockholm

(49. 10) $^1\Sigma \rightarrow {}^1\Sigma$ System,
G. Almkvist and A. Lagerqvist,
Nature 164, 665

(50. 11) $^1\Sigma \rightarrow {}^1\Sigma$ System
G. Almkvist and A. Lagerqvist,
Arkiv Fysik 1, 477-94

(50. 12) Perturbation,
G. Almkvist and A. Lagerqvist,
Arkiv Fysik 2, 233-51

(50. 13) Dissociation Energy,
L. Huldt and A. Lagerqvist,
Arkiv Fysik 2, 333-6

(51. 14) Dissociation Energy,
G. Drummond and R. F. Barrow,
Trans. Faraday Soc. 47, 1275-7

(52. 15) Blue System,
I. Kovacs and A. Budo,
Acta Phys. Acad. Sci. 1, n° 4, 469-70

(52. 16) Prelimonary Note to (54. 19),
G. Almkvist and A. Lagerqvist,
Nature 170, 885-6

(53. 17) Blue System,
I. Deezsi, E. Koczkas, and T. Matrai,
Acta Phys. Acad. Sci. 3, 95-103

(53. 18) Blue System,
I. Kovacs and A. Budo,
Ann. Physik 12, 17-25

(54. 19) Ultraviolet System,
A. Lagerqvist and G. Almkvist,
Arkiv Fysik 8, 481-8

(54. 20) Dissociation Energy,
A. Lagerqvist and L. Huldt,
Arkiv Fysik 8, 427-32

(54. 21) Dissociation Energy,
A. Lagerqvist and L. Huldt,
Z. Naturforsch. A 9, 991-2

(55. 22) Formula of Suggested Terms,
L. Huldt and A. Lagerqvist,
Arkiv Fysik 9, 227-8

(55. 23) Dissociation Energy,
R. F. Porter, W. A. Chupka, and M. G. Inghram,
J. Chem. Phys. 23, 1347-8

SrO

(55.24) Dissociation Energy,
A. G. Gaydon,
Proc. Roy. Soc. A 231, 437-45

(56.25) Spectrum of Burning Sr,
L. Huldt and A. Lagerqvist,
Arkiv Fysik 11, 347-56

(56.26) A-X System,
A. Lagerqvist and L. E. Selin,
Arkiv Fysik 11, 323-8

(56.27) I. V. Veitz and L. V. Gurvich,
Optika Spectrosk. U.S.S.R. 1, 22-33

(57.28) Dissociation Energy,
I. V. Veits and L. V. Gurvich,
Optika Spektrosk. S.S.S.R. 2, 145-9

(57.29) Dissociation Energy,
I. V. Veitz and L. V. Gurvich,
Zh. Fiz. Khim. 31, 2306-11

(64.30) Dissociation Energy,
J. Drowart, G. Exsteen, and G. Verhaegen,
Trans. Faraday Soc. 60, 1920-33

(65.31) ESR in a Molecular Beam,
M. Kaufman, L. Wharton, and W. Klemperer,
J. Chem. Phys. 43, 943-52

(65.32) P. J. Kalff, T. V. Hollander, and C. Th. J. Alkemade,
"Flame-Photometric Determinator of the Dissociation
Energies of the Alkaline-Earth Oxides,"
J. Chem. Phys. 43, 2299-2307

(67.33) K. Schofield,
"The Bond Dissociation Energies of Group IIA Diatomic
Halides,"
Chem. Reviews 67, 707-15

(67.34) I. V. Veits and L. V. Gurnich,
"Absorption Spectra of the Molecules of Substances not
Readily Volatile and of Radicals in Shock Waves,"
Dokl. Akad. Nank. S.S.S.R. 173, 1325-7

8.35) A-X System,
L. Brewer and R. Hauge,
"Near Infrared Bands of Diatomic CaO, and SrO,"
J. Mol. Spect. 25, 330-339

8.36) Hartree-Fock Calculations for Ground State,
M. Yoshimine,
"Computed Ground State Properties of BeO, MgO, CaO and SrO in Molecular Orbital Approximation,"
J. Phys. Soc. 25, 1100-1119

1.37) Franck-Condon Factors, r-Centroids, B, C-X Systems,
H. S. Liszt and W. H. Smith,
"RKR Franck-Condon Factors for Blue and Ultraviolet Transitions of some Metal Oxides,"
J. Quant. Spectrosc. Radiative Transfer 11, 1043-62

2.38) J. P. Mon,
"Raman Diffusion Spectrum of Strantium Oxide,"
J. Phys. Chem. Solids 33, 1257-60

2.39) Chemiluminescence, Reaction Cross-Sections,
J. Jonah, R. N. Zare, and Ch. Ottinger,
"Crossed Beam Chemiluminescence Studies of some Group IIA Metal Oxides,"
J. Chem. Phys. 46, 263-274

2.40) C. Batalli-Cosmovici and K. W. Michel,
"Dissociation Energy of SrO from a Crossed Beam Experiment,"
Chem. Phys. Letters 16, 77-80

SrS

Methods of Production and Experimental Technique

Absorption in a King furnace at $T \sim 2000°C$.

Band Systems

A single system in the region $3900 > \lambda > 3600 Å$, degrading R, is observed.

SPECTROSCOPIC CONSTANTS

State	T_o	ω_e	$x_e\omega_e$	B_e	$\alpha_e \times 10^3$	$D_e \times 10^6$	r_e	Remarks	Bibliography
$B^1\Sigma^+$	39281.4	286.8	0.84	0.10566	0.32		2.609		(70.5)
$X^1\Sigma^+$	0	388.38	1.31	0.12072	0.44		2.4405		(70.5)

Dissociation energy = 3.48 ± 0.19 eV, 80.2 kcal/mole, 28050 cm^{-1} (67.4, 64.3, 63.2).

SrS

Perturbations and General Information

All the rotational levels of the upper state are perturbed for $J \leq 50$ and the R heads are weak. The most clearly defined heads correspond to the (0, 2) and (1, 0) bands at 3881.1 and 3728.3Å, respectively (70.5).

BIBLIOGRAPHY

(37. 1)   L. S. Mathur,
          Proc. Roy. Soc. A 162, 83-94

(63. 2)   Dissociation Energy,
          J. R. Marquart and J. Berkowitz,
          J. Chem. Phys. 39, 283-5

(64. 3)   Dissociation Energy,
          R. Colin, P. Goldfinger, and M. Jeunehomme,
          Trans. Faraday Soc. 60, 306-16

(67. 4)   Dissociation Energy,
          E. D. Cater and E. W. Johnson,
          J. Chem. Phys. 47, 5353-7

(70. 5)   B ← X System, Vibrational-Rotational Analysis,
          M. Marcano and R. F. Barrow,
          Trans. Faraday Soc. 66, 1917-9

TaO

Methods of Production and Experimental Technique

Absorption in matrix isolation.

Emission from an arc. Radio frequency discharge.

Band Systems

A large number of systems, mostly with the $X^2\Delta$ state as the lower level have been identified (67.8, 65.6). The band heads, degrading R, do not characterize the arc spectra.

System	v', v''	λ
O'-X_2	1, 2	4651.9
L-X_1	1, 2	4504.2
	0, 1	4478.9
	0, 0	4282.0
Q'-X_2	0, 0	4201.9
M-X_1	0, 0	4154.7
	1, 0	4006.1
P-X_1	0, 1	3896.4
	0, 0	3747.7
	1, 0	3625.8
	2, 0	3513.8

In addition, bands have also been observed in the region $10000 > \lambda > 9000$Å (34.1).

TaO

Spectroscopic Constants

States	Ω	Λ(a)	T_o	ω_e	$x_e\omega_e$	B_v	$\alpha_e \times 10^3$	$D_v \times 10^7$	r_o	Lower State
V	5/2	Δ	36785	-	-	0.375	-	3.3	1.748	X_2
U	5/2	Δ	36615	-	-	0.3715	-	3.3	1.753	X_2
T	5/2	Δ	35886.20	890.4(c)	-	0.37688	-	2.70	1.7430	X_2
S	5/2	Δ	35785.83	870.8(c)	-	0.37536	-	2.79	1.7480	X_2
R	3/2	Δ	32373.60	885.0(c)	-	0.38393	-	2.289	1.7284	X_1
Q'	5/2	Δ	27290.63	896.09	-	0.380739	2.15	2.744	1.7332(b)	X_2
P	3/2	Δ	26673.04	902.06	4.08	0.376595	1.78	2.573	1.7431(b)	X_1
O'	7/2	ϕ	26121.50	894.9(c)	-	0.381304	-	2.745	1.7343	X_2
N	3/2	Π	25593.13	888.2(c)	-	0.377207	-	2.649	1.7462	X_1, X_2
M	5/2	ϕ	24058.42	890.31(d)	-	0.376144	1.84	2.635	1.7440(b)	X_2
L(e)	1/2	Π	23341.74	887.70(d)	-	0.376448	1.95	2.706	1.7452(b)	X_2
K'	7/2	ϕ	22918.75	903.06	3.56	0.379848	1.92	2.756	1.7356(b)	X_2
E	5/2	ϕ	15880.62	841.2(c)	-	0.38618	-	3.26	1.7233	X_1
C	3/2	Δ	13569.27	942.0(c)	-	0.387547	-	2.624	1.7230	X_1
B	5/2	ϕ	12852.02	930.6(c)	-	0.386851	-	2.674	1.7218	X_1
A'	1/2	Π	11062	-	-	0.389	-	-	1.72	X_1
A''	3/2	Δ	10860.95	933.2(c)	-	0.387291	-	2.668	1.7209	X_1
X_2	5/2	Δ	3505.43	1030.81	3.59	0.402649	1.87	2.503	1.6858(b)	
X_1	3/2	Δ	0.00	1028.69	3.51	0.401930	1.82	2.450	1.6872	

(a) Value of Λ is given and assumed to be a product of both states - case a.; (b) r_e; (c) Calculated using the Kratzer relation; (d) $\Delta G_{1/2}$; (e) Doubling constant: $|p_0| = 0.0927$ cm^{-1}, $|p_1| = 0.919$ cm^{-1}.

Dissociation energy = 8.4 ± 0.5 eV, 195 kcal/mole, 67750 cm^{-1}.

Perturbations and General Information

Comparison between matrix isolation and arc spectra (67.8):

	Ne Matrix at 4°K				Arc Spectra	
State	$\nu_{0,0}$	$\Delta G_{1/2}$	Intensity	Ω	$\nu_{0,0}$	$\Delta G_{1/2}$
Q	29240	895	28	-	-	-
P	26752	899	79	3/2	26673	895
O	26284	904	14	-	-	-
N	25624	887	54	3/2	25593	888
M	24145	889	78	5/2	24058	890
L	23396	887	80	1/2	23342	888
K	22333	895	64	-	-	-
J	22128	886	5	-	-	-
I?	20884	944	16	-	-	-
H	20804	910	19	-	-	-
G	18007	-	3	-	-	-
F	16718	922	3	-	-	-
E	15985	925	19	5/2	15881	841
D	14395	(943)	7	-	~14370	-
C	13700	941	11	3/2	13569	942
B	12989	931	14	5/2	12852	931
A?	12145	-	8	-	~12090	-
X_1	0	1028	-	3/2	0	1022

BIBLIOGRAPHY

(34.1)   C. C. Kiess and E. Z. Stowell,
Bur. Stand. J. Res. 12, 459-69

(49.2)   A. Gatterer and J. Junkes,
"Atlas des Reslinen,"
Vol. III. Specola Vaticana

(57.3)   Dissociation Energy,
M. G. Inghram, W. A. Chupka, and J. Berkowitz,
J. Chem. Phys. 27, 569-71

(57.4)   Rotational Analysis. Preliminary Note,
D. Premaswarup and R. F. Barrow,
Nature 180, 602-3

TaO

(57.5) Wavelengths,
A. Gatterer, J. Junkes, E. W. Salpeter, and B. Rosen,
"Molecular Spectra of Metallic Oxides,"
Ed. Specola Vaticana

(65.6) Matrix Isolation. Absorption,
W. Weltner Jr. and D. MacLeod Jr.,
J. Chem. Phys. 42, 882-91

(65.7) O. H. Krikorian and J. H. Carpenter,
"Enthalpies of Formation of Gaseous Tantalum Oxide and Tantalum Dioxide,"
J. Phys. Chem. 69, 4399-4400

(67.8) Rotational Analysis. Comparison Between Arc Spectra and Matrix,
C. H. Cheetham and R. F. Barrow,
Trans. Faraday Soc. 63, 1835-45

TeO

Methods of Production and Experimental Technique

Absorption (TeO_2 at $T > 800°C$).

Emission from a microwave discharge.

Band Systems

The rotational analysis of the $A^3\Sigma^- - X^3\Sigma^-$ system has been made using the isotope effect and assuming a single system. In emission, however, the spin orbit coupling becomes important because the A-X system appears to arise from two separate subsystems, 0^+-0^+, 1-1, with a splitting of 395 cm^{-1}. The 1-1 bands have not been observed in absorption.

I. ### $A0^+ \rightleftarrows X0^+$ System

Observed in the region $4500 > \lambda > 3075$Å, degrading R.
Band origins of ^{128}TeO in emission (65.7):

(v', v'')	(0, 6)	(1, 6)	(0, 5)	(1, 5)
λ	4268.95	4189.07	4136.54	4061.58

Intense bands in absorption (38.3):

(v', v'')	(5, 1)	(6, 1)	(6, 0)	(7, 0)
λ	3422.2	3382.9	3294.7	3259.1

II. ### A1 → X1 System

Observed in the region $4500 > \lambda > 3880$Å, degrading R.
Band possesses four branches.
Band origins of ^{128}TeO:

(v', v'')	(0, 7)	(0, 6)	(0, 5)	(0, 4)
λ	4888.10	4343.49	4206.10	4075.98

TeO

SPECTROSCOPIC CONSTANTS

State	T_o	ω_e	$x_e\omega_e$	B_e	$\alpha_e \times 10^3$	$D_e \times 10^6$	r_e	Remarks	Bibliography
						^{128}TeO			
A1	~30391.83	~458[a]		(0.2777)			(2.066)		(65.7)
A0$^+$	28036.48	445.25[a]		0.2763	5.5		2.072		(65.7)
X1	~2750	798.70	4.00	0.35735	2.37		1.822		(65.7)
X0$^+$	0	797.69	4.00	0.3560	2.38		1.825		(65.7)
						^{130}TeO			
A1	~30391.95	~458[a]		0.2771			(2.067)		(65.7)
A0$^+$	28037.04	(444.95)[a]		0.2760	5.2		2.071		(65.7)
X1	~2750	798.06	4.00	0.3564	2.36		1.822		(65.7)
X0$^+$	0	797.11	4.00	0.3554	2.37		1.825		(65.7)

[a] $\Delta G_{1/2}$.

Dissociation energy = 3.9 ± 0.1 eV, 90 kcal/mole, 31460 cm^{-1}.

Perturbations and General Information (65.7).

$A0^+$ and $A1$ states weakly perturb each other.

Extremely strong, homogeneous perturbations of the $A0^+$ state. The vibrational levels cannot be described by a simple formula $3 \leq v \leq 10$.

Predissociation is observed for the $A0^+ - X0^+$ transition for $v' \geq 9$ and for $A1 - X1$ transitions with $v' \geq 3$.

$A0^+$ state appears to possess a maximum in its potential curve of ~ 1150 cm^{-1}.

BIBLIOGRAPHY

(35.1) Preliminary Note to (38.3),
Choong Shin Piaw,
C.R. Acad. Sci. 201, 1181-3

(36.2) Preliminary Note to (38.3),
Choong Shin Piaw,
C. R. Acad. Sci. 202, 127-8

(38.3) Emission and Absorption Studies,
Choong Shin-Piaw,
Ann. Physique 10, 173-290

(41.4) Emission and Absorption Studies,
R. Migeotte,
Thesis, Liege

(59.5) Band Emission,
R. L. Purbrick,
J. Chem. Phys. 30, 962-3

(59.6) Band Emission,
P. B. V. Haranath, P. T. Rao, and V. Sivaramamurty,
Z. Physik. 155, 507-17

(65.7) A-X System, Rotational Analysis,
G. G. Chandler, H. J. Hurst, and R. F. Barrow,
Proc. Phys. Soc. 86, 105-15

TeO

(66.8) S. Drowart and P. Goldfinger,
"The Dissociation Energies of the Group VI A Diatomic Molecules,"
Quarterly Rev. 20, 545-57

ThO

Methods of Production and Experimental Technique

Emission from a microwave discharge in an arc.

BAND SYSTEMS

System	Transition	Sources	Wavelength Limits	Degrading	Characteristic Bands, λ	Remarks	Bibliography
I	$A^1\Sigma \to X^1\Sigma$	Microwave discharge	10500-8730	R	9420.75(0,0)		(65.5)
II	$B^1\Pi \to X^1\Sigma$	Microwave discharge	10000-8345	R	8971.73(0,0)		(65.5)
III	$C^1\Pi \to X^1\Sigma$	Microwave discharge	7550-6894	R	6894.42(0,0)		(65.5, 57.2)
IV	$D^1\Pi \to X^1\Sigma$	Microwave discharge	6350-6265	R	6265.58(0,0)	(0,0) sequence only	(65.5, 57.2)
V	$E^1\Sigma \to X^1\Sigma$	Microwave discharge	6600-5832	R	6121.43(0,0)		(65.5, 57.2)
VI	$F^1\Sigma \to X^1\Sigma$	Microwave discharge	5550-5449	R	5448.79(0,0)	(0,0) sequence only	(65.5, 57.2)
VII	$I^1\Pi \to X^1\Sigma$	Microwave discharge	5250-5107	R	5116.55(0,0) (Q)	(0,0) sequence only	(69.7, 57.2)
VIII	$M^1\Pi \to X^1\Sigma$	Microwave discharge	4650-4424	R	4596.43(0,0)		
IX	$K^1\Pi \to X^1\Sigma$	Microwave discharge	4650-4264	R	4414.91(0,0)		
X	$N(^1\Pi) \to X^1\Sigma$	Microwave discharge	4190-4110	R	4115.6(0,0)		(72.9)
XI	$G^1\Delta \to H^1\phi$	Microwave discharge	8100-7868	R	7868.23(0,0)	(0,0) sequence only	(69.7, 65.5, 57.2)

ThO

I. $A^1\Sigma \to X^1\Sigma$ System

Band heads, λ:

(v', v'')	(0, 1)	(1, 1)	(0, 0)	(1, 0)
λ	10282.34	9464.97	9420.75	8729.65

II. $B^1\Pi \to X^1\Sigma$ System

Band heads, λ:

(v', v'')	(0, 1)	(1, 1)	(0, 0)	(1, 0)
λ	9748.84	9014.30	8971.73	8345.33

III. $C^1\Pi \to X^1\Sigma$ System

Band heads, λ:

(v', v'')	(0, 1)	(1, 1)	(0, 0)	(1, 0)
λ	7344.98	6923.39	6894.42	6521.49

IV. $D^1\Pi \to X^1\Sigma$ System

Band heads, λ:

(v', v'')	(1, 1)	(0, 0)
λ	6287.96	6265.58

V. $E^1\Sigma \to X^1\Sigma$ System

Band heads, λ:

(v', v'')	(0, 1)	(1, 1)	(0, 0)	(1, 0)
λ	6473.95	6146.43	6121.43	5827.67

VI. $F^1\Sigma \to X^1\Sigma$ System

Band heads, λ:

(v', v'')	(1, 1)	(0, 0)
λ	5487.03	5448.79

VII. $I^1\Pi \to X^1\Sigma$ System

Band heads, λ:

(v', v'')	(1, 1)	(0, 0)
λ	5141.05(Q)	5116.55(Q)

VIII. $M^1\Pi \to X^1\Sigma$ System

Band heads, λ:

(v', v'')	(1, 1)	(0, 0)	(1, 0)
λ	4605.27	4596.43	4424.17

IX. $K^1\Pi \to X^1\Sigma$ System

Band heads, λ:

(v', v'')	(0, 1)	(1, 1)	(0, 0)	(1, 0)
λ	4595.57	4432.68	4414.91	4264.37

X. $N(^1\Pi) \to X^1\Sigma$ System

Band heads, λ:

(v', v'')	(0, 0)	(1, 1)	(2, 2)	(3, 3)
λ	4115.6	4129.6	4144.0	4158.8

XI. $G^1\Delta \to H^1\phi$ System

Band heads, λ:

(v', v'')	(1, 1)	(0, 0)
λ	7898.01	7868.23

ThO

SPECTROSCOPIC CONSTANTS

State	T_o	ω_e	$x_e\omega_e$	B_e	$\alpha_e \times 10^3$	$D_o \times 10^7$	r_e	Remarks	Bibliography
$N(^1\Pi)$	24290.9	815.19	3.27	–	–	–	–		(72.9)
$K^1\Pi$	22635.65	795.5[b]	–	0.3180[a,c] 0.3180[a,d]	–	2.00[c] 2.02[d]	–		(70.8)
$M^1\Pi$	21734.32	850.87[b]	–	0.325162[a,c] 0.325055[a,d]	–	2.059[c] 2.051[d]	–		(70.8)
J?	21406.8	~840	–	–	–	–	–		(72.9)
$I^1\Pi$	19539.06	800.85	1.47	0.3296[a,c] 0.3280[a,d]	–	2.40[c] 2.23[d]	1.85		(65.5)
$F^1\Sigma$	18337.56	–	–	0.321397[a]	–	2.042	–		(65.5)
$G^1\Delta$	17998(±5)	816(±2)	2.4(±4)	0.318192	1.28	1.936	1.881		(65.5)
$E^1\Sigma$	16320.37	829.26	2.30	0.323090	1.303	1.990	1.867		(65.5)
$D^1\Pi$	15946.22	839.2	2.50	0.321550[c] 0.325691[d]	1.30[c] 1.357[d]	1.850[c] 1.997[d]	1.866		(65.5)
$C^1\Pi$	14490.02	835.1	2.39	0.322455[c] 0.321618[d]	1.280[c] 1.29[d]	1.931[c] 1.873[d]	1.870		(65.5)
$B^1\Pi$	11129.14	842.80	2.18	0.324973[c] 0.32364[d]	1.299[c] 1.29[d]	1.942[c] 1.882[d]	1.864		(65.5)
$A^1\Sigma$	10600.82	846.4	2.44	0.323044	1.294	1.866	1.867		(65.5)

ThO

SPECTROSCOPIC CONSTANTS

State	T_o	ω_e	$x_e\omega_e$	B_e	$\alpha_e \times 10^3$	$D_o \times 10^7$	r_e	Remarks	Bibliography
$H^1\phi$	5305($\pm$5)	864($\pm$2)	2.4($\pm$4)	0.326427	1.26	1.864	1.858		(65.5)
$C^1\Sigma$	0.00	895.77	2.39	0.332644	1.302	1.833	1.840		(65.5)

$^a B_o$, $^b \Delta G_{1/2}$, ^{c} c level, ^{d} d level.

Dissociation energy = 8.59 $\pm$ 0.21 eV, 198 kcal/mole, 69250 cm^{-1} (66.6, 63.4, 60.3).

ThO

Perturbations and General Information

$I^1\Pi$ state in v = 0-2 perturbed by v = 2-4 levels of $G^1\Delta$ state (69.7).

$K^1\Pi$ state, v = 0-1 perturbed by v = 1-2 levels of $M^1\Pi$ state (70.8).

Franck-Condon factors, r-centroids — Morse potential (72.9):

$A^1\Sigma - X^1\Sigma$

v″, v′	0	1	2	3	4	5
0	0.8665-0 1.856	0.1226-0 1.764	0.1022-1 1.674	0.6596-3 1.585	0.37-4	0.18-5
1	0.1255-0 1.954	0.6339-0 1.862	0.2101-0 1.770	0.2780-1 1.680	0.2473-2 1.592	0.1758-3 1.503
2	0.7772-2 2.054	0.2205-0 1.960	0.4475-0 1.868	0.2677-0 1.776	0.5020-1 1.686	0.5776-2 1.598
3	0.2688-3 2.157	0.2185-1 2.059	0.2888-0 1.965	0.3014-0 1.874	0.3005-0 1.782	0.7520-1 1.693
4	0.57-5	0.1045-2 2.162	0.4085-1 2.064	0.3338-0 1.970	0.1906-0 1.880	0.3129-0 1.788
5	0.78-7	0.28-4	0.2538-2 2.166	0.6346-1 2.069	0.3590-0 1.976	0.1099-0 1.887
6	0.67-9	0.47-6	0.85-4	0.4920-2 2.171	0.8846-1 2.074	0.3676-0 1.981
7	0.37-11	0.49-8	0.17-5	0.1980-3 2.279	0.8333-2 2.175	0.1148-0 2.079
8			0.20-7	0.46-5	0.3941-3 2.284	0.1288-1 2.180
9			0.15-9	0.63-7	0.10-4	0.7052-3 2.288
10						0.21-4

Top = Franck-Condon factor followed by factor of ten
Bottom = r-centroid

$B^1\Pi - X^1\Sigma$

v'', v'	0	1	2	3	4	5	6
0	0.8990-0 1.854	0.9541-1 1.747	0.5414-2 1.631				
1	0.9603-1 1.968	0.7177-0 1.860	0.1704-0 1.753	0.1515-1 1.638			
2	0.4858-2 2.070	0.1728-0 1.975	0.5648-0 1.865	0.2275-0 1.759	0.2821-1 1.645	0.1819-2 1.503	
3		0.1365-1 2.076	0.2327-2 1.981	0.4371-0 1.870	0.2692-0 1.765	0.4369-1 1.652	0.3500-2 1.513
4			0.2555-1 2.082	0.2779-0 1.987	0.3315-0 1.875	0.2976-0 1.770	0.6078-1 1.659
5			0.1193-2 2.201	0.3983-1 2.088	0.3106-0 1.994	0.2453-0 1.880	0.3147-0 1.776
6				0.2270-2 2.208	0.5588-1 2.094	0.3325-0 2.000	0.1761-0 1.884
7					0.3779-2 2.214	0.7313-1 2.100	0.3452-0 2.007
8						0.5750-2 2.221	0.9110-1 2.106
9							0.8203-2 2.227

$C^1\Pi - X^1\Sigma$

v'', v'	0	1	2	3	4	5
0	0.8400-0 1.858	0.1450-0 1.774	0.1397-1 1.689			
1	0.1476-0 1.947	0.5705-0 1.863	0.2406-0 1.780	0.3734-1 1.696	0.3645-2 1.608	
2	0.1183-1 2.034	0.2499-0 1.953	0.3668-0 1.869	0.2961-0 1.786	0.6615-1 1.703	0.8435-2 1.616
3		0.3239-1 2.039	0.3144-0 1.959	0.2185-0 1.874	0.3197-0 1.793	0.9708-1 1.710
4		0.2117-2 2.128	0.5893-1 2.045	0.3480-0 1.964	0.1159-0 1.880	0.3190-0 1.799
5			0.5027-2 2.133	0.8900-1 2.050	0.3571-0 1.970	0.5051-1 1.884

Top = Franck-Condon factor followed by factor of ten
Bottom = r-centroid

ThO

$C^1\Pi - X^1\Sigma$ (Continued)

v'', v'	0	1	2	3	4	5
6				0.9528-2 2.138	0.1205-0 2.055	0.3473-0 1.976
7					0.1577-1 2.143	0.1516-0 2.061
8					0.1071-2 2.234	0.2381-1 2.148
9						0.1877-2 2.239

$D^1\Pi - X^1\Sigma$

v'', v'	0	1	2	3	4	5
0	0.8806-0 1.856	0.1113-0 1.758	0.7739-2 1.654			
1	0.1124-0 1.960	0.6679-0 1.861	0.1963-0 1.764	0.2176-1 1.662	0.1514-2 1.551	
2	0.6725-2 2.056	0.2006-0 1.966	0.4906-0 1.867	0.2575-0 1.771	0.4060-1 1.670	0.3682-2 1.561
3		0.1917-1 2.061	0.2666-0 1.971	0.3460-0 1.872	0.2975-0 1.778	0.6280-1 1.678
4			0.3630-1 2.067	0.3124-0 1.977	0.2313-0 1.877	0.3189-0 1.785
5			0.2375-2 2.160	0.5706-1 2.072	0.3402-0 1.982	0.1435-0 1.881
6				0.4668-2 2.165	0.8041-1 2.077	0.3523-0 1.988
7					0.8011-2 2.169	0.1053-0 2.082
8						0.1254-1 2.174

Top = Franck-Condon factor followed by factor of ten
Bottom = r-centroid

ThO

$E^1\Sigma - X^1\Sigma$

v", v'	0	1	2	3	4	5
0	0.8688-0 1.856	0.1220-0 1.763	0.8790-2 1.660			
1	0.1222-0 1.956	0.6397-0 1.862	0.2120-0 1.770	0.2445-1 1.668	0.1558-2 1.541	
2	0.8648-2 2.043	0.2128-0 1.963	0.4554-0 1.867	0.2741-0 1.777	0.4515-1 1.676	0.3779-2 1.553
3		0.2401-1 2.049	0.2760-0 1.969	0.3104-0 1.871	0.3124-0 1.783	0.6915-1 1.685
4		0.1416-2 2.138	0.4431-1 2.054	0.3159-0 1.975	0.1996-0 1.875	0.3307-0 1.789
5			0.3382-2 2.144	0.6796-1 2.060	0.3363-0 1.982	0.1182-0 1.877
6				0.6449-2 2.149	0.9352-1 2.066	0.3408-0 1.988
7					0.1074-1 2.155	0.1198-0 2.072
8						0.1634-1 2.161
9						0.1163-2 2.251

$G^1\Delta - H^1\phi$

v", v'	0	1	2	3	4	5	6
0	0.8988-0 1.872	0.9517-1 1.763	0.5731-2 1.650				
1	0.9641-1 1.988	0.7157-0 1.878	0.1707-0 1.770	0.1617-1 1.658			
2	0.4626-2 2.097	0.1754-0 1.993	0.5587-0 1.884	0.2284-0 1.776	0.3034-1 1.665	0.2374-2 1.545	
3		0.1327-1 2.102	0.2383-0 1.999	0.4260-0 1.890	0.2702-0 1.783	0.4726-1 1.673	0.4593-2 1.554
4			0.2532-1 2.108	0.2865-0 2.005	0.3155-0 1.895	0.2977-0 1.789	0.6605-1 1.680

Top = Franck-Condon factor followed by factor of ten
Bottom = r-centroid

ThO

$G^1\Delta - H^1\phi$ (Continued)

v'', v'	0	1	2	3	4	5	6
5			0.1201-2 2.222	0.4020-1 2.113	0.3215-0 2.011	0.2252-0 1.901	0.3129-0 1.796
6				0.2359-2 2.227	0.5732-1 2.119	0.3446-0 2.017	0.1532-0 1.906
7					0.4050-2 2.232	0.7614-1 2.124	0.3572-0 2.023
8						0.6350-2 2.237	0.9613-1 2.130
9							0.9323-2 2.243

$I^1\Pi - X^1\Sigma$

v'', v'	0	1	2	3	4	5
0	0.9822-0 1.847	0.1737-1 1.582				
1	0.1607-1 2.143	0.9492-0 1.852	0.3339-1 1.590			
2	0.1726-2 1.978	0.2853-1 2.171	0.9191-0 1.857	0.4783-1 1.598	0.1647-2 2.273	0.1126-2 1.785
3		0.4895-2 1.979	0.3793-1 2.201	0.8917-0 1.861	0.6054-1 1.604	0.2685-2 2.259
4			0.9262-2 1.979	0.4478-1 2.232	0.8669-0 1.865	0.7138-1 1.609
5				0.1462-1 1.979	0.4952-1 2.266	0.8445-0 1.869
6					0.2077-1 1.978	0.5254-1 2.302
7						0.2755-1 1.977

Top = Franck-Condon factor followed by factor of ten
Bottom = r-centroid

BIBLIOGRAPHY

(45.1) Spectral Reproduction,
A. Gatterer, J. Junkes, and V. Frodl,
"Spektren der Seltenen Erden,
Specola Vaticana 347 pp

(57.2) Wavelengths of Bands,
A. Gatterer, J. Junkes, E. W. Salpeter, and B. Rosen,
"Molecular Spectra of Metallic Oxides,"
Ed. Specola Vaticana

(60.3) Dissociation Energy,
A. J. Darnell, W. A. MacCollum, and T. A. Milne,
J. Phys. Chem. 64, 341-6

(63.4) Dissociation Energy,
R. J. Ackermann, E. G. Rauh, and R. J. Thorn,
J. Phys. Chem. 67, 762-9

(65.5) All Systems, Rotational Analysis,
G. Edvisson, L. E. Selin, and N. Aslund,
Arkiv Fysik 30, 283-319

(66.6) Dissociation Energy,
"Proc. Symp. on Thermodynamics with Emphasis on Nuclear
Materials and Atomic Transport in Solids,"
Vienne 22--27 Juillet 1965 Publ. IAEA, Vienne 1966
R. J. Ackermann and R. J. Thorn,
1, 243-69

(69.7) Perturbations, I, G States,
G. Edvinsson, A. Von Bornstedt, and P. Nylen,
Arkiv Fysik 38, 193-218

(70.8) K, M-X States,
A. Von Bornstedt and G. Edvinsson,
Physica Scripta 2, 205

(72.9) J, N States, Franck-Condon Factors, r-Centroids,
T. Wentrink Jr.,
"The Isoelectronic Series of ScF Through ThO I. Notes on the
Band Spectra of TiO, HfO, and ThO,"
J. Quant. Spectrosc. Radiative Transfer 12, 1569-90

ThP

ThP

Dissociation energy = 3.8 ± 0.5 eV, 30650 cm^{-1}.

BIBLIOGRAPHY

(72.1)   M. Shafi, C. L. Beckel, and R. Ergelke,
"Diatomic Molecule Ground State Dissociation Energies,"
J. Mol. Spectrosc. 42, 578-81

TiBr

Methods of Production and Experimental Technique

Absorption in flash photolysis (TiBr$_4$ + He).
Radio frequency discharge into Ti + Br$_2$.

BAND SYSTEMS

System	Transition	Sources	Wavelength Limits	Degrading	Band Head, $\nu_{0,0}$	Remarks	Bibliography
I	$A^4\Pi \to X^4\Sigma$	High frequency discharge	4265-4185	R and V	23434.8		(70.2)
II	$C^4\Pi \to X^4\Sigma$	High frequency discharge	3850-3790	-	26053.0		(70.2)

TiBr

I. $A^4\Pi \to X^4\Sigma$ System

Band heads (70.2):

(v', v'')	(0, 0)	(1, 1)	(2, 2)
λ(Intensity)	4265.97(10)	4264.50(6)	4263.24(8)
	4244.84(10)	4242.70(5)	4241.12(5)
	4227.96(10)	4226.57(4)	4224.44(4)

II. $C^4\Pi \to X^4\Sigma$ System

Band heads (70.2):

(v', v'')	(0, 0)	(1, 1)	(2, 2)
λ(Intensity)	3837.25(7)	3838.25(7)	3839.24(7)
	3824.22(4)	3825.10(4)	3825.77(4)
	3808.38(4)	3809.57(3)	3810.32(2)
	3791.76(3)	3792.97(3)	-

BIBLIOGRAPHY

(69.1) Excitation. Spectral Reproduction,
A. Chatalic, P. Deschamps, and G. Pannetier,
C. R. Acad. Sci. C <u>268</u>, 1111-3

(70.2) A, C → X Systems,
C. Sivaji and P. T. Rao,
"The Band Spectrum of Titanium Monobromide,"
J. Phys. B: Atom Mol. Phys. <u>3</u>, 720-4

TiC

Dissociation energy ≤ 5.50 eV, 127 kcal/mole, 44361 cm^{-1} (58.1).

BIBLIOGRAPHY

(58.1) W. A. Chupka, J. Berkowitz, C. F. Giese, and M. G. Inghram, J. Phys. Chem., U.S.A. 62, 611-4

TiCl

Methods of Production and Experimental Technique

Absorption in flash photolysis.

Emission from discharges across $TiCl_4$ vapor.

Band Systems

Two systems have been observed in flash photolysis. Both systems contain intense (0, 0) sequences, degrading V and with multiple heads (69.4).

I. $^4\Pi - X^4\Sigma$ System (4192Å)

Band heads of the (0, 0) sequence, λ(intensity) (62.3):

Head	$^4\Pi_{-1/2}$	$^4\Pi_{1/2}$	$^4\Pi_{3/2}$	$^4\Pi_{5/2}$
Q	4192.8(10)	4188.1(9)	4183.1(9)	4179.4(6)
P	4197.2(0)	4192.8(10)	4188.1(9)	4184.5(8)
O	-	4207.5(4)	4202.6(1)	4199.5(2)
N	-	-	4234.1(0)	4227.3(0)

II. $^4\Sigma - X^4\Sigma$ System (3868Å)

(No additional information.)

TiCl

SPECTROSCOPIC CONSTANTS

State	T_o	ω_e	$x_e\omega_e$	B_e	$\alpha_e \times 10^3$	$D_e \times 10^6$	r_e	Remarks	Bibliography
$^4\Sigma$	~25850	–	–						(62.3)
$^4\Pi$	23883	438.1	1.98					$A \sim 25$ cm^{-1}	(62.3)
$X\,^4\Sigma$	0	379.7	3.41						(62.3)

Dissociation energy ≈ 4.3 ± 0.8 eV, 99 kcal/mole, 34680 cm^{-1}.

TiCl

BIBLIOGRAPHY

(37. 1) Partial Vibrational Analysis,
K. R. More and A. H. Parker,
Phys. Rev. 52, 1150-2

(49. 2) New Vibrational Analysis,
V. R. Rao,
Indian J. Phys. 23, 535-46

(62. 3) System I. Detailed Vibrational Analysis,
E. A. Shenyavskaya, Yu. Ya. Kuzyakov, and V. M. Tatevskii,
Optics Spectr. U.S.S.R. 12, 197-9

(69. 4) Observation in Flash Photolysis,
A. Chatalic, P. Deschamps, and G. Pannetier,
C. R. Acad. Sci. C 268, 1111-3

(69. 5) Absorption in Flash-Heating,
R. L. Diebner and J. G. Kay,
J. Chem. Phys. 51, 3547-54

TiF

Methods of Production and Experimental Technique

King furnace (Ti + AlF$_3$) in absorption at T $\sim$ 1650°C, in emission at T $\sim$ 2000°C.

Absorption in flash heating (69.2).

Emission from collision of He with TiF$_4$.

Band Systems

Intense bands in the region 4090 > λ > 3930Å, degrading V, have been observed. Structure of the (0, 0) sequence is complex, but has been identified as a $^4\Pi$ - $^4\Sigma^-$ transition (70.3, 69.2).

Band heads (Q), λ(intensity) (70.3):

(v', v")	$^4\Pi_{-1/2} - {}^4\Sigma$	$^4\Pi_{1/2} - {}^4\Sigma$	$^4\Pi_{3/2} - {}^4\Sigma$	$^4\Pi_{5/2} - {}^4\Sigma$
(0, 0)	4077.2(10)	4069.7(10)	4062.0(10)	4054.4(6)
(1, 1)	4066.6(3)	4058.5(2)	4051.4(4)	4043.7(1)
(2, 2)	4055.8(1)	4048.2(3)	4040.6(2)	4031.9(2)
(1, 0)	3968.6(2)	3961.0(2)	3954.3(0)	-

Two other systems have also been observed in the regions 3906 > λ > 3855Å and 3775 > λ > 3650Å.

Intense heads, (69.2):

λ(degrades) | 3894.93(V) | 3892.45(V) | 3696.49(R)

Spectroscopic Constants

Dissociation energy = 5.90 ± 0.35 eV, 136 kcal/mole, 47600 cm^{-1} (67.1).

Perturbations and General Information

Vibrational lasing action has been observed between 5 $\leq \lambda \leq$ 24 μm by exploding a Ti wire in (a) F$_2$, and (b) NF$_3$ (73.4).

TiF

BIBLIOGRAPHY

(67. 1)   Dissociation Energy,
K. F. Zmbov and J. L. Margrave,
J. Phys. Chem. 71, 2893-5

(69. 2)   Absorption in Flash Heating,
R. L. Diebner and J. G. Kay,
J. Chem. Phys. 51, 3547-54

(70. 3)   A-X System,
A. Chatalic, P. Deschamps, and G. Pannetier,
"Emission Spectra of Titanium Monofluoride,"
C. R. Acad. Sci. C 270, 146-9

(73. 4)   Vibrational Lasing,
W. W. Rice and W. H. Beattie,
"Metal Atom Oxidation Lasers,"
Chem. Phys. Letters 19, 82-5

TiH

Methods of Production and Experimental Technique

Emission from shock in $Ti + H_2 + Ar$.

Band Systems

Bands have been observed in the regions $4800 > \lambda > 4700$A, $5110 > \lambda > 5030$A, and $5500 > \lambda > 5210$A. Isotope shifts indicate transitions are due to (0, 0) sequences. Bands appear slightly predissociated (71.1).

Spectroscopic Constants

Dissociation energy is unknown.

BIBLIOGRAPHY

(71.1)   R. E. Smith and A. G. Gaydon,
"The Spectrum of Titanium Hydride, TiH and TiD,"
J. Phys. B: Atom. Mol. Phys. 4, 797-9

TiI

TiI

Methods of Production and Experimental Technique

High frequency discharge in Ti + I_2 mixtures.

BAND SYSTEMS

System	Transition	Sources	Wavelength Limits	Degrading	Band Head, $\nu_{0,0}$	Remarks	Bibliography
I	$A^4\Pi \to X^4\Sigma$	High frequency discharge	4500-4430	-	22220.9 22380.3 Q(0,0) 22526.6		(70.1)
II	$B^4\Sigma \to X^4\Sigma$	High frequency discharge	4050-4035	-	24705.4		(70.1)

I. $A^4\Pi \rightarrow X^4\Sigma$ System

Band heads (Q), (70.1):

(v', v'')	(0, 0)	(1, 1)	(2, 2)
λ(Intensity)	4499.02(4)	4495.74(3)	4493.05(2)
	4466.97(10)	4464.42(9)	4461.57(9)
	4437.95(10)	4435.17(9)	4432.51(7)

II. $B^4\Sigma \rightarrow X^4\Sigma$ System

Band heads, (70.1):

(v', v'')	(0, 0)	(1, 1)	(2, 2)	(3, 3)
λ(Intensity)	4046.55(10)	4045.27(9)	4043.77(9)	4042.31(5)

Spectroscopic Constants

Dissociation energy is unknown.

BIBLIOGRAPHY

(70.1) A, B-X Systems,
C. Sivaji, D. V. K. Rao, P. T. Rao,
"The Band Spectrum of Titanium Monoiodide,"
Current Sci. <u>39</u>, 153-4

TiN

Methods of Production and Experimental Technique

Absorption.

Microwave discharge in a mixture of $TiCl_4$, He, N_2.

Emission from a Knudsen cell.

BAND SYSTEMS

System	Transition	Sources	Wavelength Limits	Degrading	Band Head, $\nu_{0,0}$	Remarks	Bibliography
I	$(A)^2\Pi_r \rightarrow (X)^2\Sigma$	Microwave discharge	6250-6100	R	16286.16 $\|$ (0,0) 16126.50		(70.4)
II	?	Flash heating	3140-3014	V	-		(63.1)

I. $(A)^2\Pi_r \to (X)^2\Sigma$ System

Band heads, (70.4):

$^2\Pi_{1/2} - {}^2\Sigma$

(v', v'')	(0, 0)	(1, 1)
λ	6200.97	6196.39

$^2\Pi_{3/2} - {}^2\Sigma$

(v', v'')	(0, 0)	(1, 1)
λ	6140.18	6174.88

II. ? System

Complex bands, (63.1):

| λ | 3140 | 3117 | 3043 | 3014 |

TiN

SPECTROSCOPIC CONSTANTS

State	T_o	ω_e	$x_e\omega_e$	B_o	$\alpha_e \times 10^3$	$D_e \times 10^6$	r_o	Remarks	Bibliography
$(A)^2\Pi_{3/2}$	16273.95			0.6149			—		(70.4)
$(A)^2\Pi_{1/2}$	16119.56			0.6070			1.596	$p = 0.0384$ cm^{-1} $q = 0.0011$ cm^{-1}	(70.4)
$(X)^2\Sigma$	0	~1080		0.6225			1.581	$\gamma = -0.002$ cm^{-1}	(70.4, 67.2)

Dissociation energy = 4.82 ± 0.35 eV, 111.2 kcal/mole, 38880 cm^{-1} (70.5).

BIBLIOGRAPHY

(63.1) Shock Heating,
W. H. Parkinson and E. M. Reeves,
Can. J. Phys. 41, 702-4

(67.2) Theoretical Calculations,
K. D. Carlson, C. R. Claydon, and C. Moser,
J. Chem. Phys. 46, 4963-9

(68.3) Dissociation Energy,
K. A. Gingerich,
J. Chem. Phys. 49, 19-24

(70.4) A-X System, Rotational Analysis,
T. M. Dunn, L. K. Hanson, and K. A. Rubinson,
"Rotational Analysis of the Red Electronic Emission System of Titanium Nitride,"
Can. J. Phys. 48, 1657-63

(70.5) C. A. Stevens and F. J. Kohl,
"The Dissociation Energy of Gaseous Titanium Mononitride,"
NASA TN D-5027

TiO

Methods of Production and Experimental Technique

Isolation in a matrix. Emission from an arc and discharge. Carbon furnace. Exploding wire. In astrophysics: absorption in Stellar atmosphere (47.18).

Emission from a shock tube of $TiCl_4$, O_2, and Ar. (71.57).

Microwave discharge into He, $TiCl_4$ and O_2 (72.58).

Hydrogen flame in arc (71.52).

Absorption in a King furnace T $\sim$ 2300°C in argon (72.59).

Shock tube using He + TiO_2 + O_2 + Ar   (70.50).

Heating of TiO_2 in Knudsen cell 2200 < T < 2400°K.

BAND SYSTEMS

	System	Transition	Sources	Wavelength Limits	Degrading	Band Head, $\nu_{0,0}$	Remarks	Bibliography
γ	I	$A^3\phi \rightleftarrows X^3\Delta$	Arc, Absorption	8650-5700	R	14019.7 14095.9 14163.0		(65.35, 51.22, 29.2)
γ'	II	$B^3\Pi \rightleftarrows X^3\Delta$	Arc, Absorption	6700-5800	R	~16150		(71.53, 69.48, 65.35, 57.28)
α	III	$C^3\Delta \rightleftarrows X^3\Delta$	Arc, Absorption	6300-4050	R	19338.6 19347.4 19349.3		(70.49, 65.35, 65.36, 29.3)
	IV	$D \rightarrow X^3\Delta$	Flame	3260-3000	-	31767.2 31829.9 32081.1 32116.2		(70.51)
δ	V	$b^1\Pi \rightarrow a^1\Delta$	Arc	9100-8850	R	11272.77		(50.21)
β	VI	$c^1\phi \rightarrow a^1\Delta$	Arc	5800-4900	R	17840.60		(69.45)
	VII	$f^1\Delta \rightarrow a^1\Delta$	Microwave discharge	5240	-	19082.14		(72.58)
	VIII	$(g) \rightarrow a^1\Delta$	Flame	3650-3286	R	-		(70.51)
ϕ	IX	$b^1\Pi \rightarrow d^1\Sigma^+$	Arc	11050-10000	R	9054.0		(62.31, 59.29)
	X	$e^1\Sigma \rightarrow d^1\Sigma$	Discharge	4113-3847	-	24303.2		(72.59, 71.52)
	XI	$(g) \rightarrow d^1\Sigma$	Flame	3545-3488	R	-		(70.51)

TiO

I. $\underline{A^3\phi \rightleftarrows X^3\Delta}$ System (γ)

Band heads, λ (n. p. 63, 65.35, 51.22, 29.2):

v', v''	0	1	2	3
0	7132.96	7680.72	8313.62	
	7094.47	7636.01	8261.05	
	7060.65	7597.08	8215.46	
1	6720.54	7204.85	7758.63	8398.32
	6686.43	7165.55	7713.20	8345.14
	6656.67	7131.35	7673.62	8298.74
2	6356.43	6788.00		7837.28
	6326.03	6753.31		7791.20
	6299.49	6723.12		7750.92

II. $\underline{B^3\Pi \rightleftarrows X^3\Delta}$ System (γ')

Band heads (57.28):

v', v''	λ
0, 1	6629.0
	6596.2
	6569.3
1, 1	6275.7
	6239.7
0, 0	6216.8
	6185.6
	6161.6
2, 1	5923.6
	5904.7
1, 0	5849.8

TiO

III. $C^3\Delta \rightleftarrows X^3\Delta$ System (α)

Band heads, λ (n. p. 63):

v', v''	0	1	2	3	4
0	5170.18 5169.65 5172.23	5452.11 5451.52 5454.37	5763.47 5762.84 5766.03	6109.09 6108.31 6111.94	
1	4957.78 4957.23 4959.73		5500.76 5500.03 5503.12	5814.73 5813.92 5817.46	
2	4764.13 4763.56 4766.94	5002.32 5001.80 5005.52		5549.88 5549.36 5553.97	5866.76 5865.91 5870.92
3	4586.64 4586.29 4589.24	4807.12 4806.76 4809.94			
4	4423.72 4423.58 4425.98	4628.46 4628.34 4630.93	4850.96 4850.79 4853.64		

IV. $D \rightarrow X^3\Delta$ System

Band heads (70.51):

(v', v'')	(0, 1)	(0, 0)	(1, 0)
λ(Intensity)	3250.5 3243.9(7) 3217.9 3214.5	3147.9 3141.7(10) 3117.1 3113.7	3048.9 3042.5(9) 3019.4 3014.7

V. $b^1\Pi \rightarrow a^1\Delta$ System (δ)

Band heads:

v', v''	Head	λ(Intensity)
3, 3	R	9094.5(2)
2, 2	R	9014.6(2)
1, 1	Q	8949.8(3)
	R	8937.38(15)
0, 0	Q	8868.49(10)
	R	8859.64(50)

VI. $\underline{c^1\phi \to a^1\Delta}$ System (β)

Band heads:

v', v''	Head	λ(Intensity)
3, 3	Q	5700.61(2)
	R	5694.42(4)
2, 2	Q	5667.59(2)
	R	5661.55(7)
1, 1	Q	5635.27(2)
	R	5629.28(7)
0, 0	Q	5603.64(3)
	R	5597.68(8)

VIII. $\underline{(g) \to a^1\Delta}$ System

Band heads (70.51):

(v', v'')	(0, 3)	(0, 2)	(0, 1)	(0, 0)
λ(Intensity)	3642.4(1)	3519.2(3)	3402.4(4)	3292.1(5)
	3634.9	3511.8	3395.6	3286.2

IX. $\underline{b^1\Pi \to d^1\Sigma^+}$ System (ϕ)

Band heads (R):

(v', v'')	(0, 0)	(2, 1)	(1, 0)
λ	11032	10137	10025

X. $\underline{e^1\Sigma \to d^1\Sigma^+}$ System

Band heads (72.59):

(v', v'')	(0, 0)	(1, 0)	(2, 0)
λ	4113.6	3975.5	3847.7

XI. $\underline{(g) \to d^1\Sigma^+}$ System

Band heads (70.51):

(v', v'')	(3, 3)	(2, 2)	(1, 1)	(0, 0)
λ(Intensity)	3511.9(7)	3503.2(8)	3495.5(10)	3488.4(10)

TiO

SPECTROSCOPIC CONSTANTS

State	T_e	ω_e	$x_e\omega_e$	B_e	$\alpha_e \times 10^3$	$D_e \times 10^6$	r_e	Remarks	Bibliography
(g)	~(30955)[a]	–	–	–	–	–	–		(70.51)
$e^1\Sigma$	~27181.7	853.9	4.7	0.4892	2.5	–	–		(n. p. 63, 72.59, 71.52)
$f^1\Delta$	~19648.95	–	–	0.50223^b	–	0.630^c	1.67273^d		(n. p. 63, 72.58)
$c^1\phi$	~18470.09	917.55	4.42	0.52115^b	–	–	$(1.642)^d$		(n. p. 63, 69.45, 50.21)
$b^1\Pi$	~11899.6	918.7	3.75	0.5133	2.8	–	1.654	$q_0 = 0.00016$ cm^{-1}	(n. p. 63, 50.21)
$d^1\Sigma$	~2802.3	1023.8	4.60	0.5490	3.37	0.057	1.600		(n. p. 63, 69.31, 59.29)
$a^1\Delta$	~580	1016.3	3.93	0.5362^b	–	0.0594	1.619^d		(n. p. 63, 69.45, 50.21)
D	~31767.2	1040^e	–	–	–	–	–		(70.51)
$C^3\Delta_3$	19531.6							$\Delta F_{2,3}(0) =$ 94.2 cm^{-1}	(71.53, 65.36, 36.9, 29.3)
$C^3\Delta_2$	19437.4	837.86	4.54	0.4884^b	2.9	–	1.695	$\Delta F_{1,2}(0) =$ 95.7 cm^{-1}	
$C^3\Delta_1$	19341.7								

TiO

SPECTROSCOPIC CONSTANTS

State	T_e	ω_e	$x_e\omega_e$	B_e	$\alpha_e \times 10^3$	$D_e \times 10^6$	r_e	Remarks	Bibliography
$B^3\Pi_2$	16264.2							$\Delta F_{2,1}(0) =$ 16.2 cm^{-1} $\Delta F_{0,1}(0) =$ 21.6 cm^{-1}	(n.p. 63, 71.53, 69.48, 57.28)
$B^3\Pi_1$	16248.0	~865	—	0.50613[b]	—	0.6790[c]			
$B^3\Pi_0$	16226.4								
$A^3\phi_4$	14360.7							$\Delta F_{3,4}(0) =$ 168.4 cm^{-1} $\Delta F_{2,3}(0) =$ 172.9 cm^{-1}	(n.p. 63, 71.53, 65.36, 51.22)
$A^3\phi_3$	14192.3	867.71	3.94	0.5057[b]	3.1	0.6918[c]	1.664		
$A^3\phi_2$	14019.4								
$X^3\Delta_3$	197.5								(n.p. 63, 71.53, 65.36, 51.22)
$X^3\Delta_2$	96.4	1008.2	4.13	0.5338[b]	3.0	0.6059[c]	1.620		
$X^3\Delta_1$	0								

[a]T_o, [b]B_o, [c]D_o, [d]r_o, [e]$\Delta G_{1/2}$.

Dissociation energy = 6.97 ± 0.20 eV, 160.8 kcal/mole, 56200 cm^{-1} (71.10).

TiO

Perturbations and General Information

$e^1\Sigma$ state perturbed in v' = 0 levels for J < 9 and J = 25, 53, and 59 and in the v' = 1 level for J = 18 and 54 (72.59).

Franck-Condon factors, r-centroids — Morse potential.

$c^1\phi - a^1\Delta$ (69.45):

v', v''	0	1	2	3	4	5	6
0	0.9239 1.633	0.0724 1.775	0.0043 1.842	0.0002 1.941			
1	0.0746 1.505	0.7750 1.640	0.1371 1.783	0.0129 1.850	0.0007 1.946		
2	0.0022 1.291	0.1456 1.518	0.6323 1.646	0.1926 1.791	0.0258 1.857	0.0018 1.952	
3		0.0073 1.322	0.2103 1.531	0.4989 1.652	0.2372 1.798	0.0427 1.864	0.0037 1.957
4			0.0159 1.350	0.2664 1.543	0.3778 1.656	0.2698 1.806	0.0631 1.871
5			0.0003 0.813	0.0286 1.376	0.3116 1.556	0.2712 1.658	0.2901 1.814
6				0.0007 0.941	0.0457 1.400	0.3441 1.567	0.1816 1.656

$b^1\Pi - a^1\Delta$ (69.45):

v', v''	0	1	2	3	4	5	6
0	0.8089 1.640	0.1732 1.727	0.0175 1.807	0.0010 1.891			
1	0.1704 1.561	0.4985 1.646	0.2806 1.734	0.0467 1.814	0.0039 1.898	0.0002 1.986	
2	0.0196 1.478	0.2709 1.569	0.2811 1.653	0.3364 1.742	0.0826 1.821	0.0090 1.905	0.0005 1.993
3	0.0016 1.386	0.0513 1.487	0.3175 1.577	0.1380 1.659	0.3527 1.750	0.1211 1.829	0.0167 1.912
4		0.0058 1.397	0.0886 1.497	0.3242 1.584	0.0527 1.662	0.3403 1.757	0.1589 1.836
5		0.0005 1.286	0.0134 1.408	0.1263 1.506	0.3030 1.592	0.0108 1.657	0.3086 1.765
6			0.0013 1.301	0.0244 1.419	0.1604 1.514	0.2642 1.599	

Top = Franck-Condon factor
Bottom = r-centroid

$b^1\Pi - d^1\Sigma$ (69.45):

v', v''	0	1	2	3	4	5	6
0	0.6073	0.3153	0.0692	0.0082	0.0005		
	1.630	1.687	1.748	1.816	1.896		
1	0.2914	0.1542	0.3738	0.1524	0.0263	0.0023	0.0001
	1.579	1.639	1.695	1.755	1.823	1.904	2.009
2	0.0812	0.3105	0.0113	0.3165	0.2220	0.0527	0.0058
	1.531	1.585	1.662	1.702	1.762	1.831	1.912
3	0.0171	0.1578	0.2285	0.0098	0.2228	0.2676	0.0848
	1.485	1.537	1.592	1.623	1.711	1.770	1.838
4	0.0030	0.0486	0.1996	0.1316	0.0572	0.1329	0.2877
	1.439	1.491	1.543	1.599	1.643	1.722	1.777
5	0.0005	0.0113	0.0854	0.2040	0.0568	0.1072	0.0645
	1.392	1.445	1.497	1.549	1.608	1.653	1.735
6		0.0022	0.0253	0.1185	0.1807	0.0140	0.1406
		1.399	1.452	1.503	1.555	1.622	1.661

Top   = Franck-Condon factor
Bottom = r-centroid

$C^3\Delta - X^3\Delta$ (62.32):

v', v''	0	1	2	3	4	5	6
0	0.391	0.382	0.172	0.047	0.008	0.002	0.006
1	0.349	0.001	0.228	0.262	0.137	0.000	-
2	0.173	0.175	0.081	0.069	0.063	-	-
3	0.063	0.221	0.020	0.423	-	-	-
4	0.019	0.126	0.041	-	-	-	-
5	0.004	0.020	-	-	-	-	-

$A^3\phi - X^3\Delta$ (62.32):

v', v''	0	1	2	3	4	5	6
0	0.721	0.236	0.040	0.004	0.000	0.000	0.000
1	0.236	0.320	0.332	0.097	0.015	0.001	-
2	0.040	0.330	0.104	0.335	0.146	-	-
3	0.004	0.097	0.330	0.012	-	-	-
4	0.000	0.016	0.153	-	-	-	-
5	0.000	0.003	-	-	-	-	-
6	0.000	-	-	-	-	-	-

Franck-Condon factors only.

TiO

Band strengths (S), Einstein coefficients (A), and the oscillator strengths (f) for the $c^1\phi - a^1\Delta$ system of TiO molecule (72.L1):

v', v''	S ($a_0^2 e^2$)	A	f
1, 0	0.0307	4.12×10^5	1.75×10^{-3}
2, 1	0.0605	7.98×10^5	3.43×10^{-3}
3, 2	0.0882	11.42×10^5	4.97×10^{-3}
4, 3	0.1127	14.33×10^5	6.31×10^{-3}
0, 0	0.4146	4.79×10^6	2.25×10^{-2}
1, 1	0.3492	3.96×10^6	1.88×10^{-2}
2, 2	0.2861	3.19×10^6	1.57×10^{-2}
3, 3	0.2266	2.48×10^6	1.20×10^{-2}
4, 4	0.1720	18.56×10^5	9.13×10^{-3}

Vibrational lasing action has been observed between $10.5 \leq \lambda \leq 24$ μm by exploding a Ti wire in O_2 (73.62).

BIBLIOGRAPHY

(27.1) Generalities,
R. Mecke,
Z. Physik 42, 390-425

(29.2) γ System,
A. Christy,
Astrophys. J. 70, 1-10

(29.3) α System
A. Christy,
Phys. Rev. 33, 701-29

(29.4) γ System,
F. Lowater,
Proc. Phys. Soc. 41, 557-68

(29.5) Generalities,
R. Wildt,
Z. Physik 54, 856-79

(31.6) Astrophysics,
R. S. Richardson,
Astrophys. J. 73, 216-49

(33.7) Astrophysics,
R. S. Richardson,
Astrophys. J. 78, 354-71

(34.8) Astrophysics,
N. T. Bobrovnikoff,
Astrophys. J. 79, 483-91

(36.9) α System,
A. Bude,
Z. Physik 98, 437-44

(37.10) $^1\pi$-$^1\Delta$ System,
P. P. Dobronravin,
C. R. Acad. Sci. U.R.S.S. 17, 399-403

(37.11) Infrared System,
K. Wurm and H. J. Meister,
Z. Astrophys. 13, 199-204

(39.12) Astrophysics,
N. T. Bobrovnikoff,
Astrophys. J. 89, 301-10

(39.13) Astrophysics,
T. Tanaka, S. Nagasawa, and K. Saito,
Proc. Phys. Math. Soc. 21, 431-55

(41.14) Astrophysics,
A. H. Joy and M. L. Humason,
Publ. Astron. Soc. Pacific 53, 296

(42.15) Generalities,
H. Grouiller,
C. R. Acad. Sci. 214, 256-8

(42.16) Astrophysics,
P. Swings,
Publ. Astron. Soc. Pacific 54, 232-6

(43.17) β and γ' System,
F. P. Coheur,
Bull. Soc. Roy. Sci. Liege 12, 98-102

TiO

(47. 18) Stellar Spectrum,
D. N. Davis,
Astrophys. J. 106, 28-75

(48. 19) Infrared System,
C. C. Kiess,
Publ. Astron. Soc. Pacific 60, 252-3

(49. 20) L. Brewer and D. F. Mastick,
UCRL Report #571

(50. 21) $^1\pi$, $^1\phi$-$^1\Lambda$ Systems,
J. G. Phillips,
Astrophys. J. 111, 314-27

(51. 22) γ System, Rotational Analysis,
J. G. Phillips,
Astrophys. J. 114, 152-62

(52. 23) $a^1\Delta - X^3\Delta$ Separation,
J. G. Phillips,
Astrophys. J. 115, 567-8

(57. 24) Classification of Triplet States,
U. Uhler,
Thesis, Stockholm

(54. 25) Thermochemical Dissociation Energy,
Q. D. Wheatley,
Thesis, Univ. Kansas

(55. 26) Thermochemical Dissociation Energy,
W. O. Graves, M. Hocher, and H. L. Johnston,
J. Phys. Chem. 59, 127-31

(57. 27) Dissociation Energy by Mass Spectra,
J. Berkowitz, W. A. Chupka, and M. G. Inghram,
J. Phys. Chem. 61, 1569-72

(57. 28) γ' System. Wavelengths,
A. Gatterer, J. Junkes, E. W. Salpeter, and B. Rosen,
"Molecular Spectra of Metallic Oxides,"
Ed. Specola Vaticana, 1957

(59. 29) γ System. Rotational Analysis,
A. V. Pettersson,
Arkiv Fysik 16, 185-90

(61. 30) b-d System,
A. V. Pettersson and B. Lindgren,
"The 10025 Å TiO Band,"
Naturwissenschaften 49, 128-9

(62. 31) φ System. Rotational Analysis,
A. V. Pettersson and B. Lindgren,
Arkiv Fysik 22, 491-5

(62. 32) Franck-Condon Factors C, A-X Systems,
F. S. Ortenberg and V. B. Glasko,
"Vibrational Transition Probabilities for Band Systems of Some Diatomic Oxides,"
Astron. J. U.S.S.R. 39, 921-6

(63. 33) Wavefunctions,
K. D. Carlson and C. Moser,
J. Phys. Chem. 67, 2644-7

(64. 34) Hartree-Fock Calculations of Ground State of TiO,
K. D. Carlson and R. K. Nesbet,
J. Chem. Phys. 41, 1051-62

(65. 35) Matrix Isolation at Low Temperature,
W. Weltner, Jr., and D. MacLeod, Jr.,
J. Phys. Chem. 69, 3488-500

(65. 36) Anomalies Splitting in C, A-X Systems,
T. Kovacs,
"On the Triplet Term of the TiO Molecule,"
J. Mol. Spectrosc. 18, 229-38

(66. 37) Anamalous Triplet Splitting of A, C-X Systems,
R. Török,
"On the Anamalous Multiplet Splitting of the Triplet Terms of the TiO Molecule,"
Acta Phys. 20, 91-98

(67. 38) Hartree-Fock Calculations,
K. D. Carlson and C. R. Claydon,
Advances High Temp. Chem. 1, 43-94

(67. 39) Comparison of TiO and ScF,
C. J. Cheetham and R. F. Barrow,
Advances High Temp. Chem. 1, 7-41

TiO

(67.40) Extended-base Wavefunctions of $^3\Delta$,
K. D. Carlson and C. Moser,
J. Chem. Phys. 46, 35-46

(67.41) Dissociation Energy,
P. G. Wahlbeck and P. W. Gilles,
J. Chem. Phys. 46, 2465-73

(68.42) B. B. Laud and D. R. Kalsulkar,
"Potential Energy Curves and Dissociation Energies of some Diatomic Molecules,"
Indian J. Phys. 42, 50-5

(69.43) Dissociation Energy. Discussion,
J. Drowart, P. Coppens, and S. Smoes,
J. Chem. Phys. 50, 1046-8

(69.44) Dissociation Energy Discussion,
P. W. Gilles, P. J. Hampson, and P. G. Wahlbeck,
J. Chem. Phys. 50, 1048-9

(69.45) Excitation in Shock Waves. β System. Vibrational Analysis,
C. Linton and R. W. Nicholls,
J. Phys. B: Proc. Phys. Soc. 2, 490-8

(69.46) γ' System,
M. Kronekvist and A. Lagerqvist,
Arkiv Fysik 39, 133-7

(69.47) Prediction of Unobserved States,
L. Brewer and D. W. Green,
"The Low-lying Electronic States of ScF, TiO, ZrO,"
High Temp. Science 1, 26-45

(69.48) Spin Orbit Coupling, A, C-X System,
J. G. Phillips,
"The γ' System of the TiO Molecule,"
Astrophys. J. 157, 449-58

(70.49) Spin Orbit Coupling, A, C-X Systems,
I. Kovacs and V. M. Korwar,
"Re-Investigation of the Coupling Constants of the Revised Molecular States of TiO,"
Acta Physica Acad. 29, 399-406

(70.50) C. Linton and R. W. Nicholls,
"Measurement of Intensities of the α and β Band Systems of TiO,"
J. Quant. Spectrosc. Radiative Transfer 10, 311-4

(70.51)   D-X, g → a,d Systems,
C. M. Pathak and H. B. Palmer,
"New Electronic Transitions in TiO,"
J. Mol. Spectrosc. 33, 137-146

(71.52)   e-d System,
J. G. Phillips and S. P. Davis,
"A new $^1\Sigma - ^1\Sigma$ Transition of the TiO Molecule,"
Astrophys. J. 167, 209-212

(71.53)   Spin Orbit Coupling, A, B, C-X Systems,
J. G. Phillips,
"Satellite Bands of the γ'-System of Titanium Oxide,"
Astrophys. J. 169, 185-9

(71.54)   W. Zyrnicki and A. Czernichowski,
"The Spectroscopic Constants and the Partition Functions of the Electronic and Internal Energies of the TiO Molecule,"
Acta Physica. Polonica A39, 429-39

(71.55)   R. J. Hampson and P. W. Gilles,
"High Temperature Vaporization and Thermodynamics of the Titanium Oxides VII. Mass Spectrometry and Dissociation Energies of TiO(g) and TiO_2(g),"
J. Chem. Phys. 55, 3712-29

(71.56)   N. S. McIntyre, K. R. Thompson, and W. Weltner Jr.,
"Spectroscopy of Titanium Oxide and Titanium Dioxide Molecules in Inert Matrices at 4°K,"
J. Phys. Chem. 75, 3243-9

(71.57)   M. L. Price, K. G. P. Sultzmann, and S. S. Penner,
"Measurements of f-numbers for α- and γ bands of TiO,"
J. Quant. Spectrosc. Radiative Transfer 11, 427-442

(72.58)   f-a System,
C. Linton
"A New $^1\Delta - ^1\Delta$ Transition of the TiO Molecule,"
Can. J. Phys. 50, 312-6

(72.59)   e-d System,
B. Lindgren,
"A Search for UV Spectra of the TiO Molecule,"
J. Mol. Spectrosc. 43, 474-6

TiO

(72.60) H. Y. Wu and P. G. Wahlbeck,
"Vapor Pressures of TiO(g) in Equilibrium with $Ti_2O_3(s)$ and $Ti_3O_5(s,\beta)$: Dissociation Energy of TiO(g),"
J. Chem. Phys. 56, 4534-40

(72.61) T. Wentink Jr., and R. J. Spineller, Jr.,
"The Isoelectronic Series ScF through ThO I. Notes on the Band Spectra of TiO, HfO, and ThO,"
J. Quant. Spectrosc. Radiative Transfer 12, 1569-90

(73.62) Vibrational Lasing,
W. W. Rice and W. H. Beattie,
"Metal Atom Oxidation Lasers,"
Chem. Phys. Letters 9, 82-5

(72.L1) P. S. Dube,
"Einstein Coefficients and Oscillator Strengths for the TiO System,"
Indian J. Pure Appl. Phys. 10, 70-1

(n.p. 63) J. G. Phillips,
"Molecular Constants of the TiO Molecule,"
(to be published)

TiS

Methods of Production and Experimental Technique

Emission and absorption from a King furnace.

Band Systems

A single $^3\Delta$ - $^3\Delta$ system, degrading R, is observed in the region $9200 > \lambda > 7500$ Å.

Band heads (69.1):

v', v''	0, 1	0, 0	1, 0	2, 0
$\lambda(R_3)$	9066.22	8629.55	8287.04	7973.91
$\lambda(R_2)$	9062.44	8626.03	8283.61	7970.51
$\lambda(R_1)$	9063.96	8627.34	8284.18	7974.33

TiS

SPECTROSCOPIC CONSTANTS

State	T_o	ω_e	$x_e\omega_e$	B_o	$\alpha_e \times 10^4$	$D_e \times 10^7$	r_e	Remarks	Bibliography
$A^3\Delta_3$	11582.25 + X_3	484.12	2.55	0.18854	10.2	1.22	2.161		(69.1)
$A^3\Delta_2$	11587.05 + X_2	484.30	2.51	0.18770	9.9	1.20			
$A^3\Delta_1$	11585.31	—	—	0.18684	—	1.06			
$X^3\Delta_3$	X_3 (a)	562.07	—	0.20298	9.2	1.12	2.0825		(69.1)
$X^3\Delta_2$	X_2 (a)	562.20	1.95[b]	0.20222	9.2	1.09			
$X^3\Delta_1$	0	562.27	—	0.20135	9.0	1.03			

[a] Assuming Hund's case a, $X_2 = 90$ cm^{-1}, $X_3 = 185$ cm^{-1}; [b] Calculated using the Pekeris relation.

Dissociation energy = 4.4 ± 0.2, 35489 cm^{-1} (72.3).

Perturbations and General Information

$A^3\Delta_2$ perturbed in v = 0 level, $A^3\Delta_1$ perturbed in v = 1, 2 levels (69.1).

BIBLIOGRAPHY

(69.1)   A-X System, Rotational Analysis,
E. A. Shenyavskaya, A. A. Mal'tsev, D. I. Kataev, and
L. V. Gurvich,
Optics Spectr. U.S.S.R. 26, 509-12

(69.2)   Dissociation Energy,
S. Smoes, P. Coppens, C. Bergman, and J. Drowart,
Trans. Faraday Soc. 65, 682-7

(72.3)   M. Shafi, C. L. Beckel, and R. Engelke,
"Diatomic Molecule Ground State Dissociation Energies,"
J. Mol. Spectrosc. 42, 578-81

TlF

Methods of Production and Experimental Technique

Absorption from a saturated vapor at T = 160-700°C.
Emission from a high frequency discharge.

BAND SYSTEMS

System	Transition	Sources	Wavelength Limits	Degrading	Band Head, $\nu_{0,0}$	Remarks	Bibliography
I	$A^3\Pi_0^+ \rightleftarrows X^1\Sigma^+$	Discharge	2921-2810	R and V	35164.3		(58.5, 37.2, 34.1)
II	$B^3\Pi_1 \rightleftarrows X^1\Sigma^+$	Discharge	2808-2714	R	36805.6		(58.5, 37.2, 34.1)
III	$B' \leftarrow X^1\Sigma^+$	Absorption	~2630	Continuum	max. ~38000		(37.2, 34.1)
IV	$C(^1\Pi) \leftarrow X^1\Sigma^+$	Absorption	2244-2198	R	45480		(37.2, 34.1)
V	$C' \leftarrow X^1\Sigma^+$	Absorption	<~2200	Continuum	>45500		(37.2, 34.1)
VI	$D' \leftarrow X^1\Sigma^+$	Absorption	<~2000	Continuum	>50000		(37.2, 34.1)

TlF

I. $A^3\Pi_0^+ \rightleftarrows X^1\Sigma^+$ System

Band heads (37.2):

v', v'' (heads)	λ	Intensity
0, 2 (P)	2921.3	5
1, 2 (P)	2886.0	6
0, 1 (P)	2882.0	6
5, 5 (R)	2873.4	5
4, 4 (R)	2863.8	7
3, 3 (R)	2854.9	8
2, 2 (P)	2852.2	9
1, 1 (P)	2848.0	10
0, 0 (P)	2843.5	10
3, 2 (R)	2818.4	5
1, 0 (P)	2810.6	7

II. $B^3\Pi_1 \rightleftarrows X^1\Sigma^+$ System

Band heads (37.2):

v', v''	Heads		Intensity
2, 4	Q	2808.6	6
	R	2807.9	
1, 3	Q	2796.9	8
	R	2795.0	
1, 2	Q	2761.1	7
	R	2759.8	
0, 1	Q	2751.0	6
	R	2748.1	
2, 2	Q	2737.6	9
	R	2737.1	
1, 1	Q	2725.8	10
	R	2724.9	
0, 0	Q	2715.9	10
	R	2713.7	

IV. $C(^1\Pi) \leftarrow X^1\Sigma$ System

Band heads (R) (37.2):

(v', v'')	(0, 2)	(1, 2)	(0, 1)	(1, 1)	(0, 0)
λ(Intensity)	2244.3(5)	2227.0(8)	2221.0(9)	2203.9(8)	2198.0(10)

TlF

SPECTROSCOPIC CONSTANTS

State	T_o	ω_e	$x_e\omega_e$	B_e	$\alpha_e \times 10^3$	$D_e \times 10^6$	r_e	Remarks	Bibliography
$C^1\Pi$	45480	~345	-	-	-		-		(58.4, 58.5)
$B^3\Pi_1$	36805.6	366.6	10.2	0.2249	3.0		2.076	$y_e\omega_e = -1.15$ cm^{-1}, $\gamma = -9 \times 10^{-4}$ cm^{-1}	(58.4, 58.5)
$A^3\Pi_0^+$	35164.3	463.3	7.1	0.2309	2.7		2.049	$y_e\omega_e = 0.1$ cm^{-1}, $\gamma = -1.3 \times 10^{-4}$ cm^{-1}	(58.4, 58.5)
$X^1\Sigma^+$	0	477.3	2.3	0.22315	1.500		2.0844		(58.4, 58.5)

Dissociation energy = 4.5 ± 0.1 eV, 104 kcal/mole, 36300 cm^{-1}.

Perturbations and General Information

C-X system is diffuse in absorption and not present in emission, which indicates that the $C^1\Pi$ state predissociates. Predissociation is also observed in the $B^3\Pi_1$ state for $v > 3$ and the $A^3\Pi_0^+$ state for $v > 8$.

Potential energy curves — RKRV potential (68.15):

State	v	$U(\text{cm}^{-1})$	$r_{min}(\text{Å})$	$r_{max}(\text{Å})$	$T_e + U(\text{cm}^{-1})$
$X^1\Sigma^+$	0	238.1	2.024	2.152	238.1
	1	709.2	1.983	2.206	709.2
	2	1175.6	1.957	2.245	1175.6
	3	1643.9	1.937	2.279	1643.9
	4	2102.3	1.920	2.309	2102.3
	5	2557.3	1.906	2.338	2557.3
	6	3011.0	1.893	2.364	3011.0
	7	3457.5	1.881	2.391	3457.5
	8	3894.8	1.870	2.416	3894.8
	9	4326.7	1.861	2.439	4326.7
	10	4757.6	1.852	2.463	4757.6
	11	5185.0	1.844	2.485	5185.0
	12	5610.3	1.836	2.507	5610.3
$A^3\Pi_0$	0	177.3	1.980	2.129	35358.0
	1	512.4	1.933	2.199	35693.1
	2	823.6	1.900	2.254	36004.3
$B^3\Pi_1$	0	216.8	2.015	2.419	37086.3
	1	642.4	1.976	2.212	37511.9
	2	1047.3	1.951	2.261	37916.8
	3	1435.4	1.932	2.305	38304.9
	4	1807.0	1.916	2.347	38676.5
	5	2160.9	1.901	2.388	39030.4
	6	2497.8	1.888	2.428	39367.3
	7	2816.8	1.882	2.458	39686.3
	8	3118.8	1.875	2.490	39988.3

BIBLIOGRAPHY

(34. 1)   Absorption,
S. Boizowa and K. Butkow,
Phys. Z. U.S.S.R. <u>5</u>, 756-76

(37. 2)   Absorption and Emission,
H. G. Howell,
Proc. Roy. Soc. A <u>160</u>, 242-53

(55. 3)   J. V. R. Rao and P. T. Rao,
Ind. J. Phys. <u>29</u>, 20-6

(58. 4)   Microwave Spectra,
A. H. Barratt and M. Mandel,
Phys. Rev. <u>109</u>, 1572-89

(58. 5)   A, B → X Systems, Rotational Analysis,
R. F. Barrow, H. F. K. Cheall, P. M. Thomas, and P. B. Zeeman,
Proc. Phys. Soc. <u>71</u>, 128-30

(58. 6)   Microwave Spectra,
H. G. Fitzky,
Z. Physik <u>151</u>, 351-64

(58. 7)   Electric Resonance Studies of a Molecular Beam,
G. Gräff, W. Paul, and C. Schlier,
Z. Physik <u>153</u>, 38-63

(60. 8)   Dissociation Energy,
R. F. Barrow,
Trans. Faraday Soc. <u>56</u>, 952-8

(61. 9)   Dissociation Energy,
E. M. Bulewicz, L. F. Phillips, and T. M. Sugden,
Trans. Faraday Soc. <u>57</u>, 921-31

(64. 10)   Electric Resonance of a Molecular Beam,
R. Von Boechk, G. Gräff, and R. Ley,
Z. Physik <u>179</u>, 285-313

(65. 11)   Electric Resonance of a Molecular Beam,
R. K. Ritchie and H. Lew,
Can. J. Phys. <u>43</u>, 1701-5

(66. 12) Dissociation Energy,
E. Murad, D. L. Hildenbrand, and R. P. Main,
J. Chem. Phys. 45, 263-9

(67. 13) Dissociation Energy,
F. J. Keneshea and D. Cubicciotti,
J. Phys. Chem. 71, 1958-60

(68. 14) Dissociation Energy,
J. Berkowitz and T. A. Walter,
J. Chem. Phys. 49, 1184-9

(68. 15) S. N. Thakur, R. B. Singh, and D. K. Rai,
"Potential Energy Curves and Nature of Study in Group IIIA Monohalides,"
J. Sci. Ind. Res. 27, 389-347

(70. 16) J. Hoeft, F. J. Lovas, E. Tremann, and T. Törring,
"Microwave Spectra of AiF, GaF, InF, TiF,"
Z. Naturforsch. A25, 1029-35

(71. 17) J. Singh, K. P. R. Nair, and D. K. Rai,
"Potential Energy Curves and Dissociation Energies of Diatomic Fluorides and Chlorides of Gallium, Indium and Thallium,"
J. Quant. Spectrosc. Radiative Transfer 11, 1577-1581

(72. L1) V. S. Kushawaha, B. P. Asthana, and C. M. Pathak,
"On the Fundamental Vibrational Energy of Group III-A Monohalides,"
Spectrosc. Letters 5, 357-60

(72. L2) H. Dijkerman, W. Flegel, G. Graff, and B. Mönter,
"Report on the Stark Effect of the Molecules $^{205}Tl^{19}F$ and $^{39}K^{19}F$,"
Z. Naturforsch. 27a, 100-10

UB

Dissociation energy = 3.30 ± 0.34 eV, 76 kcal/mole, 26600 cm^{-1} (70.1).

BIBLIOGRAPHY

(70.1) K. A. Gingerich,
"Gaseous Metal Borides. II. Mass-Spectrometric Evidence for the Molecules UB_2, UB, and CeB and Predicted Stability of Gaseous Diborides of Electropositive Transition Metals,"
J. Chem. Phys. 53, 746-747

UF

Methods of Production and Experimental Technique

Exploding of a U wire in F_2 (73.1).

Perturbations and General Information

Vibrational lasing action between $10.5 \leq \lambda \leq 24$ μm has been observed by exploding a U wire in F_2 (73.1).

BIBLIOGRAPHY

(73.1) W. W. Rice and W. A. Beattie,
"Metal Atom Oxidation Lasers,"
Chem. Phys. Letters 19, 82-5

UN

UN

Dissociation energy = 5.47 ± 0.22 eV, 126.0 kcal/mole, 44100 cm^{-1} (67.1).

BIBLIOGRAPHY

(67.1)   K. A. Gingerich,
         J. Chem. Phys. 47, 2192, 3

UO

Methods of Production and Experimental Technique

Exploding of a U wire in O_2.

Isolation in an Ar matrix.

Heating in a Knudsen cell.

Band Systems

A single band at 12.2 μm has been assigned to UO (73.L2).

Spectroscopic Constants

Dissociation energy = 5.38 ± 0.10 eV, 124.0 kcal/mole, 43400 cm^{-1} (68.1).

Perturbations and General Information

Vibrational lasing action has been observed between $8.8 \leq \lambda \leq 16$ μm by exploding a U wire in O_2 (73.3).

BIBLIOGRAPHY

(68.1) R. P. Steiger,
"Mass Spectrometric Investigation of the Vaporization, Thermodynamics and Dissociation Energies of LaS, ScS, YS, ZrS, and UO,"
Disc. Abstr. B, 29, 2009

(72.2) P. E. Blackburn and P. M. Danielson,
"Electron Impact Relative Ionization Cross Sections and Fragmentation of U, UO, UO_2, and UO_3,"
J. Chem. Phys. 56, 6156-64

(73.3) W. W. Rice and W. H. Beattie,
"Metal Atom Oxidation Lasers,"
Chem. Phys. Letters 19, 82-5

UO

(71.L1) H. J. Leary, T. A. Rooney, E. D. Cater, and H. B. Friedrich,
"The Infrared Spectra of Matrix Isolated UO and UO_2,"
High Temp. Sci. 3, 433-43

(73.L2) S. D. Gabelnick, G. T. Reedy, and M. G. Chasanov,
"The Infrared Spectrum of Matrix Isolated Uranium Oxide Vapor Species,"
Chem. Phys. Letters 19, 90-3

US

Dissociation energy = 5.38 ± 0.10 eV, 124.0 kcal/mole, 43400 cm^{-1} (66.1, 68.2).

BIBLIOGRAPHY

(66.1)   E. D. Carter, E. G. Rauh, and R. J. Thorn,
J. Chem. Phys. 44, 3106-12

(68.2)   Erratum (66.1),
E. D. Carter, E. G. Rauh, and R. J. Thorn,
J. Chem. Phys. 48, 538

VCl

Methods of Production and Experimental Technique

High frequency discharge in $VCL_4 + N_2$ (70.1).

Band Systems

Nine bands are observed in the region $3800 > \lambda > 3400$ Å. The bands are complex and appear to arise, at least partially, from a $\Pi - \Sigma$ transition with large multiplicity (70.1).

BIBLIOGRAPHY

(70.1) Preliminary Observations,
D. Iacocca, A. Chatalic, P. Deschamps, and G. Pannetier,
"Absorption and Emission Spectrum of a New Radical: Vanadium Monochloride,"
C. R. Acad. Sci. C, 271, 669-72

VO

Methods of Production and Experimental Technique

Absorption in a graphite furnace at T ~ 1750°C; in matrix isolation at 4°K.

Emission from the corona of a vanadium or carbon arc.

Oxyhydrogen flame.

In astrophysics: stellar spectra.

Arc in air with VO_5 (68.24).

BAND SYSTEMS

System	Transition	Sources	Wavelength Limits	Degrading	Characteristic Bands, λ	Remarks	Bibliography
I	Infrared System	Arc	10560-10460	R	10482.2		(56.9)
II	$B^4\Pi \rightleftarrows X^4\Sigma^-$	Absorption	9000-6200	R	-		(69.25, 68.21, 57.12, 52.6)
III	$C^4\Sigma^- \rightleftarrows X^4\Sigma^-$	Absorption	7000-4400	R	5736.7(0,0) 5469.3(1,0) 5228.2(2,0)		(69.25, 68.21, 57.10, 57.12)
IV	$? \leftarrow X^4\Sigma^-$	Absorption in matrix isolation	4185-3916	(R)	4185(0,0)		-

VO

I. Infrared System

It cannot be said with certainty that the infrared bands form an independent system or whether they belong to the B-X system. Absorption studies of $V^{18}O$ in matrix isolation appear to correspond to the (2, 0) and (3, 0) bands of a system with $\nu_{00} \sim 10500$ cm^{-1}; consequently we label this system as independent of B-X.

Band heads (56.9):

λ | 10530.0 | 10509.2 | 10482.2 | 10462.6

II. $B^4\Pi \rightleftarrows X^4\Sigma^-$ System

Band heads of the (0, 0) band (69.25, 68.21, 57.12, 52.6):

Heads	$^PR_{13}$, $^PQ_{12}$	$^QR_{12}$	R_1	$^QR_{23}$, Q_2	R_2, $^RQ_{21}$	$^SR_{21}$
λ	7987.13	7977.37	7960.80	7950.29	7939.62	7919.12

Heads	R_3, $^RQ_{32}$	$^SR_{32}$	$^SR_{43}$	$^SQ_{42}$	$^TR_{42}$
λ	7910.09	7898.57	7865.47	7865.29	7851.04

III. $C^4\Sigma^- \rightleftarrows X^4\Sigma^-$ System

Band heads, λ(intensity) (69.25, 68.21, 57.10, 57.12):

v', v''	λ	Intensity	v', v''	λ	Intensity
1, 4	6976.2	4	1, 0	5469.3	9
0, 3	6919.0	4	4, 2	5324.5	6
1, 3	6532.8	6	3, 1	5275.8	6
0, 2	6477.8	6	2, 0	5228.2	6
0, 1	6086.4	8	4, 1	5057.3	4
2, 2	5837.3	4	3, 0	5010.5	4
0, 0	5736.7	10	6, 2	4904.2	4
3, 2	5567.7	4	5, 1	4858.5	4
2, 1	5517.3	5			

IV. ? ← X$^4\Sigma^-$ System

Four members of a system with v_{oo} = 23895 cm^{-1} have been observed in matrix isolation and absorption in the gas phase (69, 25).

VO

SPECTROSCOPIC CONSTANTS

State	T_e	ω_e	$x_e\omega_e$	B_e	$\alpha_e \times 10^3$	$D_e \times 10^6$	r_e	Remarks	Bibliography
$C\,^4\Sigma^-$	17420.3	865.30	6.35	0.4953	3.1	-	1.672	$\lambda = 0.53$ cm^{-1}; $\gamma_1 = -0.009$ cm^{-1}; $\gamma_2 = -0.007$ cm^{-1}	(69.25, 62.14)
$B\,^4\Pi_{5/2}$	12706.8	910.9	5.0	0.5246	4	-	1.625	$A_{43} \sim 71$ cm^{-1}; $A_{32} \sim 62$ cm^{-1}; $A_{21} \sim 55$ cm^{-1}	(69.25, 62.14)
$X\,^4\Sigma^-$	0	1012.36	5.26	0.54825	3.52	-	1.589	$\lambda = 1.371$ cm^{-1}; $\gamma_1 = 0.0112$ cm^{-1}; $\gamma_2 = 0.0111$ cm^{-1}	(69.25, 62.14)

Dissociation energy = 6.45 ± 0.11 eV, 148.8 kcal/mole, 52000 cm^{-1} (67.19).

Perturbations and General Information

Franck-Condon factors — Morse potential (63.15):

$C^4\Sigma^- - X^4\Sigma^-$

v', v''	0	1	2	3	4	5	6
0	0.329	0.363	0.206	0.078	0.022	0.005	0.000
1	0.359	0.005	0.134	0.243	0.170	0.050	-
2	0.204	0.128	0.125	0.004	0.113	-	-
3	0.080	0.232	0.003	0.176	-	-	-
4	0.024	0.167	0.121	-	-	-	-
5	0.006	0.077	-	-	-	-	-
6	0.001	-	-	-	-	-	-

Radiative lifetime: $C^4\Sigma^- - X^4\Sigma^-$ (72.27) $\tau \sim 415$ nsec.

BIBLIOGRAPHY

(27.1)   Generalities,
R. Mecke,
Z. Physik 42, 390-425

(32.2)   Vibrational Analysis,
W. F. C. Ferguson,
Bur. Stand. J. Res. 8, 381-4

(33.3)   Preliminary Note,
C. Ghosh,
Nature 132, 318

(35.4)   Excitation in Oxy-hydrogen Flames,
G. Piccardi,
Atti. Accad. Lincei Rend. Cl. Sci. Fis. Mat. Nat. 21, 836-8

(35.5)   Vibrational and Rotational Analysis,
P. C. Mahanti,
Proc. Phys. Soc. 47, 433-45

(52.6)   B-X System,
P. C. Keenan and L. W. Schroeder,
Astrophys. J. 115, 82-8

VO

(55. 7) Flame Emission Spectrum, 1.0 μm to 1.5 μm. No details,
B. Kleman and B. Liljeqvist,
Arkiv Fysik 9, 377-83

(55. 8) C-X System,
A. Lagerqvist and L. E. Selin,
"The Band Spectrum of Vanadium Oxide,"
Naturwissenschaften 3, 65

(56. 9) Emission Bands, 10500 Å,
A. Lagerqvist and L. E. Selin,
Arkiv Fysik 11, 429-30

(57. 10) C → X System, Rotational Analysis,
A. Lagerqvist and L. E. Selin,
Arkiv Fysik 12, 553-68

(57. 11) Dissociation Energy by Mass Spectra,
J. Berkowitz, W. A. Chupka, and M. G. Inghram,
J. Phys. Chem. 61, 1569-72

(57. 12) Wavelengths,
A. Gatterer, J. Junkes, E. W. Salpeter, and B. Rosen,
"Molecular Spectra of Metallic Oxides,"
Ed. Specola Vaticana 1957

(57. 13) N. R. Tawde and N. S. Murthy,
"A Note on the Validity of the Revised Rotational Structure
Constants of Green-Yellow System of VO Bands,"
Ind. J. Phys. 31, 391-4

(62. 14) Theory: Energy Levels of the $^4\Sigma$ State,
J. T. Hougen,
Can. J. Phys. 40, 598-606

(63. 15) Franck-Condon Factors, C-X,
F. S. Ortenberg and V. B. Glasko,
"Vibrational Transition Probabilities for Band Systems of
some Diatomic Oxides,"
Sov. Astron. AJ 6, 921-926

(64. 16) Franck-Condon Factors, C-X,
N. S. Murthy, T. K. S. Setty, and K. V. Sumathi,
"A Note on the Franck-Condon Factors of VO ($A^2\Delta - X^2\Delta$)
Band System,"
Ind. J. Phys. 38, 428-9

(66. 17) $^2\Delta$ and $^4\Sigma^-$ Wave Functions. LCAO-MO-SCF Calculation,
K. D. Carlson and C. Moser,
J. Chem. Phys. 44, 3259-65

(67.18) Electronic Structure: Theoretical Calculation. Review,
K. D. Carlson and C. R. Claydon,
Advances High Temp. Chem. 1, 43-94

(67.19) Dissociation Energy by Mass Spectra,
P. Coppens, S. Smoes, and J. Drowart,
Trans. Faraday Soc. 63, 2140-8

(68.20) Electronic Spin Resonance Spectra of Matrix Isolated VO. Hyperfine Nuclear Coupling Constants,
P. H. Kasai,
J. Chem. Phys. 49, 4979-84

(68.21) Hyperfine Nuclear Structure in $B^4\pi-X^4\Sigma^+$, $C^4\Sigma-X^4\Sigma^-$,
D. Richards and R. F. Barrow,
Nature 217, 842

(68.22) $\Delta F=0$ Perturbation in $X^4\Sigma^-$, Hyperfine Nuclear Coupling Constant,
D. Richards and R. F. Barrow,
Nature 219, 1244-5

(68.23) B. B. Laud and D. R. Kalsulkar,
"Potential Energy Curves and Dissociation Energies of some Diatomic Molecules,"
Ind. J. Phys. 42, 50-4

(68.24) A, C-X Systems,
B. B. Laud and D. R. Kasulkar,
"The Emission Spectrum of VO Molecule,"
Ind. J. Phys. 42, 61-71

(69.25) $B^4\pi-X^4\Sigma^-$, $C^4\Sigma^- - X^4\Sigma^-$. Rotational Analysis,
D. Richards,
Theisis, Oxford

(70.26) N. S. Murthy and B. N. Murthy,
"True Potential - Energy Curves for LaO, VO, and CP,"
J. Phys. B: Atom Mol. Phys. 3, L16-L18

(72.27) Radiative, Lifetime, C-X,
T. Wentink and G. Diebold,
"Laboratory Investigation of Absolute Intensity Constants of Metallic and Alkaline Earth Oxides by Non-Shock Tube Methods,"
Parametrics Inc. AFCRL-72-0191

(72.28) M. Farber, O. M. Uy, and R. D. Srivastava,
"Effusion-Mass Spectrometric Determination of Heats of Formation of the Gaseous Molecules V_4O_{10}, V_4O_8, VO_2 and VO,"
J. Chem. Phys. 56, 5312-5315

VS

Dissociation energy = 4.61 ± 0.15 eV, 106.4 kcal/mole, 37200 cm^{-1} (67.1).

BIBLIOGRAPHY

(67.1) J. Drowart, A. Pattoret, and S. Smoes, Proc. Brit. Ceram. Soc. 8, 67-89

WO

Band Systems

The emission spectrum of WO has been studied (63.7, 57.4, 54.3, 52.2); however, it has not been properly characterized because of the numerous perturbations. In absorption in matrix isolation seven transitions $7060 > \lambda > 3500$ Å (65.9) have been identified.

WO

SPECTROSCOPIC CONSTANTS

State	T_o	$\Delta G_{1/2}$	$x_e\omega_e$	B_e	$\alpha_e \times 10^3$	$D_e \times 10^6$	r_e	Remarks	Bibliography
G	23797	933	–						(65.9)
F	23391	937	–						(65.9)
E	21499	944	–						(65.9)
D	20799.9	980.7	4.7						(52.2)
C	19189	(922)	–						(65.9)
B	17277	(930)	–						(65.9)
A	~14160	(1000)	–						(65.9)
X	0	1055.2	3.9						(52.2)

Dissociation energy = 6.8 ± 0.4 eV, 157 kcal/mole, 54850 cm^{-1}.

BIBLIOGRAPHY

(35.1) N. S. Bayliss,
Proc. Roy. Soc. A 158, 551-61

(52.2) D System, λ
A. Gatterer and S. G. Krishnamurty,
Nature 169, 543

(54.3) Vibrational Analysis Test, λ
V. Vittalachar and S. G. Krishnamurty,
Current Science 23, 357-8

(57.4) Wavelengths,
A. Gatterer, J. Junkes, E. W. Salpeter, and B. Rosen,
"Molecular Spectra of Metallic Oxides",
Ed. Specola Vaticana 1957

(59.5) Estimated Dissociation Energy,
W. A. Chupka, J. Berkowitz, and C. F. Giese,
J. Chem. Phys. 30, 827-34

(60.6) Dissociation Energy,
G. De Maria, R. P. Burns, J. Drowart, and M. G. Inghram,
J. Chem. Phys. 32, 1373-7

(63.7) λ
R. W. B. Pearse and A. G. Gaydon,
"The Identification of Molecular Spectra, Wavelengths,"
London 1963, Ed. Chapman and Hall

(64.8) Dissociation Energy,
"JANAF Thermochemical Tables,"
Ed; Dow Chemical Corp., Midland, Mich.

(65.9) $W^{16}O$ and $W^{18}O$ Absorption Spectrum in Ne Matrix,
W. Weltner Jr., and D. MacLeod Jr.,
J. Mol. Spectrosc. 17, 276-99

YCl

Methods of Production and Experimental Technique

Emission from a microwave discharge in YCl_3 + He.

BAND SYSTEMS

	System	Transition	Sources	Wavelength Limits	Degrading	Band Head, $\nu_{0,0}$	Remarks	Bibliography
	I	$^1\Sigma^+ - X\,^1\Sigma^+$	Microwave discharge	6950-6300	R	14877.6		(66.1)

I. $^1\Sigma - X^1\Sigma^+$ System

Band heads (R) of $Y^{35}Cl$ (66.1):

(v', v'')	(0, 1)	(0, 0)	(2, 2)	(1, 0)	(2, 0)
λ(intensity)	6893.84(4)	6718.81(10)	6601.59(4)	6576.34(4)	6440.90(4)

YCl

SPECTROSCOPIC CONSTANTS

State	T_o	ω_e	$x_e\omega_e$	B_e	$\alpha_e \times 10^3$	$D_o \times 10^8$	r_e	Remarks	Bibliography
$^1\Sigma^+$	14877.60	324.5	1.14	0.1089	0.7	9.0	2.48		(66.1)
$X^1\Sigma^+$	0	380.7	1.3	0.1161	0.3	9.3	2.40		(66.1)

Dissociation energy = 3.5 ± 1 eV, 81 kcal/mole, 28230 cm^{-1}.

BIBLIOGRAPHY

(66.1)   Rotational and Vibrational Analyses of $Y^{35}Cl$ and $Y^{37}Cl$,
G. M. Janney,
J. Opt. Soc. Am. 56, 1706-11

YF

Methods of Production and Experimental Technique

Absorption of Y + AlF_3 at T ~ 2000°C.

Emission from a discharge in a hollow cathode.

Heating of YF_3 + graphite in a King furnace T ~ 2100°C in argon or helium.

BAND SYSTEMS

System	Transition	Sources	Wavelength Limits	Degrading	Characteristic Bands, λ	Remarks	Bibliography
I	$B^1\Pi - X^1\Sigma^+$	Absorption	7200-5900	R	6552.12(0, 1) 6291.91(0, 0) 6086.39(1, 0)		(67.4, 66.2, 64.1)
II	$C^1\Sigma^+ - X^1\Sigma^+$	Absorption	5600-4800	R	5208.63(0, 0) 5095.27(1, 0)		(67.4, 66.2)
III	$^1\Pi - X^1\Sigma^+$	Absorption	-	R	3947.00(0, 0)?		(67.4)
IV	$^1\Pi - X^1\Sigma^+$	Absorption	-		4025.00(0, 1)? 3925.30(0, 0)?		(67.4)
V	$^1\Sigma^+ - X^1\Sigma^+$	Absorption	3670-3500	R	3572(0, 0) 3503(1, 0)		(66.2)
VI	$^1\Pi - X^1\Sigma^+$	Absorption	3400-3000	R	3269.09(0, 1) 3203.04(0, 0) 3149.01(1, 0)		(67.4, 66.2, 64.1)
VII	$b(^3\phi) - a(^3\Delta)$	Absorption	7120-6425	R	6830.82 ⎫ 6735.42 ⎬ (0, 0) 6652.41 ⎭		(67.4, 66.2)

SPECTROSCOPIC CONSTANTS

State	T_o	$\Delta G_{1/2}$	$x_e\omega_e$	B_o	$\alpha_e \times 10^3$	$D_o \times 10^7$	r_o	Remarks	Bibliography
Singlets									
$^1\Pi$	31205.80	536.30	2.1	0.27545	2.33	2.96	–	$q_0 < 2 \times 10^{-5}$ cm^{-1}	(67.4, 66.2, 64.1)
$^1\Sigma^+$	27986.9	547.5	2.69	0.2741	–	–	–	–	(66.2)
$^1\Pi$	25464.33	(581.92)	(5.6)	0.27090	–	–	–	$q_0 = -0.0007$ cm^{-1}	(67.4)
$^1\Pi$	25324.90	–	–	0.26805	–	3.3	–	$q_0 = 0.00049$ cm^{-1}	(67.4)
C $^1\Sigma^+$	19190.35	527.20	2.45	0.26578	1.77	2.64	–	–	(67.4, 66.2)
B $^1\Pi$	15885.78	534.67	2.35	0.26631	1.56	2.61	–	$q_0 = 0.00013$ cm^{-1}	(67.4, 66.2, 64.1)
X $^1\Sigma^+$	0	631.29	2.5	0.28961	1.63	2.37	1.928	–	(67.4, 66.2, 64.1)
Triplets									
b $^3\Phi_4 + x_3$	15028.0	} 531.3	2.41	0.277					(67.4, 66.2)
b $^3\Phi_3 + x_2$	14842.8								
b $^3\Phi_2 + x_1$	14635.5								

YF

SPECTROSCOPIC CONSTANTS

State	T_o	$\Delta G_{1/2}$	$x_e\omega_e$	B_o	$\alpha_e \times 10^3$	$D_o \times 10^7$	r_o	Remarks	Bibliography
$a^3\Delta_3$	x_3	578.5	2.49	⎫					(67.4, 66.2)
$a^3\Delta_2$	x_2	577.5	2.42	⎬ 0.285			1.94		
$a^3\Delta_1$	x_1	576.4	2.39	⎭					

Dissociation energy = 6.18 ± 0.22 eV, 142.6 kcal/mole, 49880 cm^{-1} (67.3).

BIBLIOGRAPHY

(64.1) Systems I and G. Rotational Analysis,
R. F. Barrow and W. J. M. Gissane,
Proc. Phys. Soc. 84, 615-6

(66.2) Systems 1, 2, 5, 6, Rotational Analysis. Triplet System, Vibrational Analysis,
E. A. Shenyavskaya, A. A. Mal'tsev, and L. Gurvich,
Optics Spectr. U.S.S.R. 21, 374-6

(67.3) Dissociation Energy Through Mass Spectra,
K. F. Zmbov and J. L. Margrave,
J. Chem. Phys. 47, 3122-5

(67.4) Summary,
R. F. Barrow, M. W. Bastin, D. L. G. Moore, and C. J. Pott,
Nature 215, 1072-3

YO

Methods of Production and Experimental Technique

Absorption in a neon matrix at 4°K (67.15).

Emission from an arc. Oxyhydrogen flame and carbon flame + salts of Y.

In astrophysics: absorption in stellar atmospheres (47.6).

BAND SYSTEMS

System	Transition	Sources	Wavelength Limits	Degrading	Band Head, $\nu_{0,0}$	Remarks	Bibliography
I	$A^2\Pi \rightleftarrows X^2\Sigma^+$	Arc	6800-5700	R	16294.72 16722.75		(61.9, 59.8, 31.3)
II	$B^2\Sigma^+ \rightleftarrows X^2\Sigma^+$	Arc	5600-4400	R	20741.92		(61.9, 59.8, 31.2, 31.3)

I. $A^2 \rightleftarrows X^2\Sigma^+$ System

Band heads, λ(intensity) (31.3):

$^2\Pi_{3/2} - {}^2\Sigma^+ \ ({}^RQ_{21})$

(v', v'')	(0, 0)	(1, 1)	(2, 1)	(1, 0)
λ(intensity)	5972.17(10)	5987.72(10)	5713.87(6)	5697.80(5)

$^2\Pi_{1/2} - {}^2\Sigma^+ \ (Q_1)$

(v', v'')	(0, 0)	(1, 1)	(2, 2)	(2, 1)
λ(intensity)	6132.13(10)	6148.43(10)	6165.13(10)	5858.88(5)

II. $B^2\Sigma^+ \rightleftarrows X^2\Sigma^+$ System

Band heads (31.3):

(v', v'')	(1, 2)	(1, 1)	(0, 0)	(1, 0)
λ(intensity)	5050.56 / 5049.66 (5)	4842.79 / 4841.90 (7)	4818.05 / 4817.36 (10)	4650.13 / 4649.47 (9)

YO

SPECTROSCOPIC CONSTANTS

State	T_o	ω_e	$x_e\omega_e$	B_o	$\alpha_e \times 10^3$	$D_e \times 10^6$	r_o	Remarks	Bibliography
$B^2\Sigma^+$	20741.92	765.03	7.75	0.3722			1.828	$p = -0.151$ cm^{-1}	(65.13, 61.9, 31.3)
$A^2\Pi_{3/2}$	16722.75	808.9	2.96	0.3857			1.795	$\gamma = -0.148$ cm^{-1}	(65.13, 61.9, 31.3)
$A^2\Pi_{1/2}$	16294.72	812.7	2.80						
$X^2\Sigma^+$	0	852.5	2.45	0.3881			1.790	$y_e\omega_e = 0.0273$ cm^{-1}	(61.9, 31.3)

Dissociation energy = 7.39 ± 0.11 eV, 170.4 kcal/mole, 59600 cm^{-1} (67.16, 67.17).

Perturbations and General Information

Franck-Condon factors and r-centroids — Morse potential:

$A^2\Pi - X^2\Sigma^+$ (64.11)

v', v''	0	1	2	3	4	5	6
0	0.98189	0.01772	0.00038				
1	0.01806	0.94373	0.03695	0.00123	0.00002		
2	0.00005	0.03834	0.90142	0.05749	0.00264	0.00006	
3		0.00020	0.06073	0.85514	0.07911	0.00468	0.00014
4			0.00052	0.08503	0.80518	0.10153	0.00747
5				0.00109	0.11100	0.75191	0.12442
6					0.00204	0.13833	0.69573
7						0.00348	0.16650
8							0.00562

Franck-Condon factors only.

$B^2\Sigma - X^2\Sigma^+$ (67.18)

v', v''	0	1	2	3	4	5	6
0	0.792	0.185	0.021	0.002	0.000	0.000	0.003
	1.809	1.895	1.976	2.015	2.049	2.107	1.810
1	0.179	0.425	0.315	0.068	0.011	0.002	0.001
	1.734	1.818	1.898	1.977	2.020	2.057	1.860
2	0.025	0.295	0.152	0.355	0.133	0.032	0.005
	1.657	1.750	1.826	1.901	1.978	2.023	2.114
3	0.003	0.076	0.319	0.016	0.300	0.196	0.071
	1.599	1.677	1.766	1.825	1.904	1.981	2.022
4	0.001	0.015	0.139	0.255	0.014	0.183	0.232
	1.574	1.616	1.697	1.781	1.866	1.905	1.981
5	0.000	0.003	0.039	0.192	0.140	0.094	0.057
	1.569	1.582	1.634	1.717	1.798	1.866	1.901
6	0.000	0.001	0.010	0.075	0.207	0.037	0.166
	1.514	1.578	1.591	1.655	1.738	1.817	1.874

Top = Franck-Condon factors, and bottom = r-centroids.

YO

Band strengths, oscillator strengths, and Einstein coefficients $B^2\Sigma - X^2\Sigma$ System (72.21):

v', v''	$S(a_0^2 e^2)$	f	$A\ \text{sec}^{-1}$
1, 0	0.109	7.65×10^{-3}	21.98×10^5
1, 1	0.212	13.30×10^{-3}	37.85×10^5
1, 2	0.148	8.90×10^{-3}	23.29×10^5
0, 0	0.395	24.91×10^{-3}	71.60×10^5
0, 1	0.087	5.18×10^{-3}	13.70×10^5

BIBLIOGRAPHY

(30.1) Generalities,
J. Querbach,
Z. Physik 60, 109-24

(31.2) A, B → X System, Vibrational Analysis,
W. F. Meggers and J. A. Wheeler,
Bur. Stand. J. Res. 6, 239-75

(31.3) A, B → X, Vibrational Analysis,
L. W. Johnson and R. C. Johnson,
Proc. Roy. Soc. A 133, 207-19

(33.4) Excitation in Oxy-hydrogen Flames,
G. Piccardi,
Gazz. Chim. Ital. 63, 127-38

(39.5) G. Piccardi,
Spectrochim. Acta 1, 249-60

(47.6) D. N. Davis,
Astrophys. J. 106, 28-75

(57.7) A. Gatterer, J. Junkes, E. W. Salpeter, and B. Rosen,
"Molecular Spectra of Metallic Oxides,"
Ed: Specola Vaticana 1957

(59. 8)   A, B-X Systems,
          U. Uhler and L. Akerlind,
          "The Rotational Analysis of the Band Systems of Yttrium Oxides,"
          Naturwissenschaften 16, 488

(61. 9)   A-X, B-X System. Rotational Analysis,
          U. Uhler and L. Akerlind,
          Arkiv Fysik 19, 1-16

(63. 10) I. Kovacs,
          "On the Anomalous Splitting of the Multiplet States in Diatomic Molecules,"
          Bull. Sci. Fac. Chim. Ind. 21, 44-50

(64. 11) Franck-Condon Factors, A-X,
          F. S. Ortenberg, V. B. Glasko, and A. I. Dmitriev,
          "Vibrational Transition Probabilities for Band Systems of some Diatomic Oxides. II.,"
          Soviet Astronomy 8, 258-261

(64. 12) Dissociation Energy by Mass Spectra,
          R. J. Ackermann, E. G. Rauh, and R. J. Thorn,
          J. Chem. Phys. 40, 883-9

(65. 13) Pure Precession Relation Between $A^2\pi$ and $B^2\Sigma^+$,
          R. A. Berg, L. Wharton, W. Klemperer, A. Büchler, and J. L. Stauffer,
          J. Chem. Phys. 43, 2416-21

(65. 14) Dissociation Energy by Mass Spectra,
          S. Smoes, J. Drowart, and G. Verhaegen,
          J. Chem. Phys. 43, 732-6

(67. 15) Electron Spin Resonance Spectra and Optical Absorption in Ne at $4°K$,
          W. Weltner Jr., D. Macleod, Jr., and P. H. Kasai,
          J. Chem. Phys. 46, 3172-84

(67. 16) Dissociation Energy by Mass Spectra,
          L. L. Ames, P. N. Walsh, and D. White,
          J. Phys. Chem. 71, 2707-18

(67. 17) Dissociation Energy Discussion,
          P. Coppens, S. Smoes, and J. Drowart,
          Trans. Faraday Soc. 63, 2140-8

YO

(67.18) N. S. Murthy and B. N. Murthy,
"The Franck-Condon Factors and the r-Centroids of the Yttrium Oxide ($B^2\Sigma - X^2\Sigma$) Band System,"
Proc. Phys. Soc. 90, 881-3

(67.19) I. V. Veits and L. V. Gruvich,
"Absorption Spectra of the Molecules of Substances not Readily Volatile and of Radicals in Shock Waves,"
Dokl. Akad. Nauk. S. S. R. R. 173, 1325-7

(71.20) Prediction of Unobserved States,
D. W. Green,
"Low-Lying Electronic States of the Scandium Oxide, Yttrium Oxide, Lanthanum Oxide Molecules,"
J. Phys. Chem. 75, 3103-3106

(72.21) P. S. Dube,
"Einstein Coefficients and Oscillator Strenghts of the $B^2\Sigma - X^2\Sigma$ System of the YO Molecule,"
Indian J. Pure Appl. Phys. 10, 167

(72.L1) P. S. Dube, D. K. Rai, and N. L. Singh,
"Variation of the Electronic Transition Moment with Internuclear Distance and Effective Vibrational Temperature in $B^2\Sigma - X^2\Sigma$ System of YO Molecule,"
Indian J. Pure Appl. Phys. 10, 87-8

YS

Methods of Production and Experimental Technique

Absorption in a neon matrix.

Band Systems

Band heads (72.1):

$$A^2\Pi_{3/2} \leftarrow X^2\Sigma \quad \quad 7167(0,0)$$
$$6949(1,0)$$

$$A^2\Pi_{1/2} \leftarrow X^2\Sigma \quad \quad 7436(0,0)$$
$$7197(1,0)$$

Spectroscopic Constants

Dissociation energy = 5.45 ± 0.13 eV, 125.7 kcal/mole, 43960 cm^{-1} (68.1).

BIBLIOGRAPHY

(68.1) R. P. Steiger,
"Mass Spectrometric Investigation of the Vaporization, Thermodynamics, and Dissociation Energies of LaS, ScS, Ys, ZrS, and UO,"
Disc. Abstr. 29B, 2009

(72.1) N. S. McIntyre, K. C. Lin, and W. Weltner, Jr.,
"ESR and Optical Spectra of the ScS and YS Molecules,"
J. Chem. Phys. 56, 5576-5583

ZrBr

Methods of Production and Experimental Technique

Radio frequency discharge in bromine vapor over powdered zirconium metal (70.1).

Band Systems

Two systems noted: $4250 > \lambda > 4050$ Å and $3850 > \lambda > 3775$ Å. The first system appears analogous to C system of ZrCl, $^4\Pi - {}^4\Sigma$; the second system analogous to B system of ZrCl.

BIBLIOGRAPHY

(70.1) C. Sivaji and P. T. Rao,
"The Band Spectrum of ZrBr,"
<u>Proc. Roy. Irish Acad.</u> <u>70</u>, 1-7

ZrCl

Methods of Production and Experimental Technique

Emission from a high frequency discharge in $ZrCl_4$.

Band Systems

Three complex band systems have been observed. The vibrational and rotational structure degrade in opposite directions (61.1).

I. <u>4150-4040 Å Region</u>

   Appears to be $^4\Pi - {}^4\Sigma$ system. Band heads, degrade R:

   λ(intensity) - 4140.56(8) | 4138.71(10) | 4136.52(8)

II. <u>3800-3620 Å Region</u>

   Band heads, degrade R:

   λ(intensity) - 3714.98(7) | 3713.98(10) | 3713.83(9) | 3713.71(7)

III. <u>2910-2840 Å Region</u>

   Band heads, degrade R:

(v', v'')	(0, 0)	(1, 1)
λ(intensity)	2909.97(10)	2908.74(8)
	2871.37(9)	2870.09(7)
	2837.51(7)	2836.43(5)

BIBLIOGRAPHY

(61.1) P. K. Carroll and P. J. Daly,
<u>Proc. Roy. Irish Acad. A</u>, <u>61</u>, 101-7

ZrI

Methods of Production and Experimental Technique

Radio frequency discharge in iodine vapor over zirconium metal powder (70.1)

Band Systems

Three systems were observed in the region $4450 > \lambda > 3850$ Å (70.1).

I. <u>4420-4225 Å Region</u>

Four groups of line-like bands, separation ~ 285 cm^{-1}, in analogy with the C system of ZrCl and ZrBr.

II. <u>4005-3930 Å Region</u>

Four groups of bands, separation ~ 152 cm^{-1}, as in B system of ZrBr.

III. <u>3920-3840 Å Region</u>

Unclassified.

BIBLIOGRAPHY

(70.1) C. Sivaji and P. T. Rao,
"The Band Spectrum of ZrI,"
<u>Proc. Roy. Irish Acad. A</u>, <u>70</u>, 7-12

ZrN

Dissociation energy = 5.81 ± 0.26 eV, 134 kcal/mole, 46900 cm^{-1} (68.1, 68.2).

BIBLIOGRAPHY

(68.1) K. A. Gingerich,
Bull. Am. Phys. Soc. **13**, 226

(68.2) K. A. Gingerich,
J. Chem. Phys. **49**, 14-8

ZrO

Methods of Production and Experimental Technique

Absorption in an isolation matrix (65.26).

Emission from an arc (Cu electrodes + compounds of Zr). Carbon furnace.

In astrophysics: absorption in stellar atmospheres (47.12).

BAND SYSTEMS

	System	Transition	Sources	Wavelength Limits	Degrading	Band Head, $\nu_{0,0}$	Remarks	Bibliography
Infrared B.8192	I	$x' \to x''$	Zr arc	9500-8500	R[a]	10685.3[b] 10700.1 10715.3 10731.4 10750.3		(50.17, 32.2)
	II	$y' \to y''$	Zr arc	8350-8190[c]	R	12203.2[b]		(56.21, 50.17)
	III	-	Zr arc	7900-7600[d]	No heads			(50.17, 48.13)
γ	IV	$A^3\phi \rightleftarrows X'^3\Delta$	Zr arc	7600-5110	R	15442.9 15756.3 16048.4		(54.18, 32.5)
β	V	$B^3\Pi \rightleftarrows X'^3\Delta$	Zr arc	5810-5455	R	17483.5 17760.2 18007.4		(65.26, 54.19, 32.5)
α	VI	$C^3\Delta \rightleftarrows X'^3\Delta$	Zr arc	5600-4200	R	21543.7 21556.4 21640.3		(54.18, 50.17, 35.8, 32.5)
A	VII	$b^1\Sigma^+ \rightleftarrows X^1\Sigma^+$	Zr arc	4000-3390	R	27151.0		(65.26, 56.22, 50.17, 49.14)
δ	VIII	$D \to X'^3\Delta$	Zr arc	3508-3472	R	28501.7[b] 28620.1 28780.3		(50.17)

BAND SYSTEMS

	System	Transition	Sources	Wavelength Limits	Degrading	Band Head, $\nu_{0,0}$	Remarks	Bibliography
φ	IX	$E^3\Delta \to X'^3\Delta$	Zr arc	3120-2940	R	33685.2[b] 33888.4 33993.7		(50.17)
B	X	$d^1\Delta \to c^1\Delta$	Zr arc	5450-4970	R	19280.9		(56.21, 50.16)
	XI	-	Zr arc	-	R	15389.4		(65.26, 57.24)
	XII	-	Matrix absorption	-	(R)	17030.0		(65.26)
	XIII	-	Matrix absorption	-	(R)	19402.4		(65.26)

[a] Bands in this region are from two or more systems. One system degrades to V.
[b] Classification is uncertain. [c] Intense, isolated band. [d] Several, non-analyzed bands.

ZrO

I. Infrared System (x' → x'')

Band heads. Classification is uncertain (50.17):

v', v''	λ	Intensity	v', v''	λ	Intensity
0, 0	9356.12	3	1, 1	9413.75	2
	9343.19	4		9412.03	2
	9329.93	5		9401.04	2
	9315.87	5		9387.26	2
	9299.56	5		9370.74	3

II. 8192 Å System

Appears to be a singlet system (56.21).

IV. $A^3\phi \rightleftarrows X'^3\Delta$ System (γ)

Band heads (54.18, 32.5):

v', v''	λ	Intensity
2, 2	6542.98	10
	6412.29	12
	6292.79	14
1, 1	6508.15	18
	6378.32	16
	6260.89	16
0, 0	6473.67	20
	6344.91	18
	6229.40	18

V. $B^3\Pi \rightleftarrows X'^3\Delta$ System (β)

Band heads (65.26, 54.19, 32.5):

v', v''	λ	Intensity	v', v''	λ	Intensity
3, 3	5809.18	7		5724.05	11
2, 2	5778.46	10	0, 0	5718.11	20
				5629.00	14
1, 1	5748.14	16		5551.74	10
	5658.13	9			
			1, 0	5456.49	8

VI. $C^3\Delta \rightleftarrows X'^3\Delta$ System (α)

Band heads (54. 18, 50. 17).

VII. $b^1\Sigma^+ \rightleftarrows X^1\Sigma^+$ System

Band heads (65. 26, 56. 22, 50. 17):

(v', v'')	(1, 3)	(0, 1)	(0, 0)	(2, 1)	(1, 0)
λ(intensity)	3981. 45(3)	3818. 32(4)	3682. 07(10)	3589. 87(6)	3571. 96(8)

X. $d^1\Delta \rightarrow c^1\Delta$ System

Band heads (56. 21, 50. 16):

(v', v'')	(0, 0)	(1, 0)	(2, 1)
λ(intensity)	5186. 04(6)	4996. 28(2)	4969. 82(2)

XI. 6499 Å System

In absorption, what appears to be the (0, 0) band is seen at 6446 Å in an Ne matrix at 4°K (65. 26) and at 6498. 9 Å in the gas phase (57. 24). The lower state is most probably a singlet.

XII. 5872 Å System

(0, 0) band observed in absorption at 5872 Å for matrix isolation (65. 26) and at 5859. 8 Å for gas phase (57. 24).

XIII. 5154 Å System

In matrix isolation (65. 26), absorption is seen at 5154 Å; however, in gas phase absorption is seen at 5185. 0 Å. This band may actually correspond to the $d^1\Delta \rightarrow c^1\Delta$ system.

ZrO

SPECTROSCOPIC CONSTANTS

State	T_e	ω_e	$x_e\omega_e$	B_o	$\alpha_e \times 10^3$	$D_e \times 10^6$	r_e	Remarks	Bibliography
$E(^3\Delta)$	x+33993.7 x+33888.4 x+33685.2								(50.17)
D	x+28780.3 x+28620.1 x+28501.7								(50.17)
$C^3\Delta$	x+21631.5 x+21548.5 x+21536.4	820.58	3.31	0.3953 0.3926 0.3896	2.1[b]		1.775		(54.18, 32.5)
$B^3\Pi$	x+17995.9 x+17745.3 x+17466.3	845.3	3.64	0.4058 0.4032 0.3960	2.3[b]		1.755		(54.19, 32.5)
$A^3\phi$	x+16033.9 x+15741.4 x+15426.8	853.9	3.14	0.4046 0.4040 0.4027	2.1[b]		1.752		(54.18, 32.5)
$X'^3\Delta$ [a]	605.1+x 297.2+x x	936.5	3.47	0.4155 0.4147 0.4135	2.1[b]		1.728		(54.18, 54.19, 41.10, 35.8, 32.5)
y'	y''+1220								(50.17)
y''	y''								(50.17)
x'	x''+10700	862.9	8.8						(50.17)
x''	x''	945.4	8.6						(50.17)

ZrO

SPECTROSCOPIC CONSTANTS

State	T_e	ω_e	$x_e\omega_e$	B_0	$\alpha_e \times 10^3$	$D_e \times 10^6$	r_e	Remarks	Bibliography
$d^1\Delta$	c+19272.55	835.4[c]	3.15	0.3976	2.1		1.764		(56.21, 50.16, 50.17)
$c^1\Delta$	c	938.1	1.80	0.4167	1.2[b]		1.725		(56.21, 50.16)
$b^1\Sigma^+$	27144.7	843.27	3.0	0.3922	1.9		1.772		(56.22, 50.16, 50.17)
?	17025[d]	872[c]							(65.26)
?	15509[d]	854[c]							(65.26)
$X^1\Sigma^+$	0	969.76[c]	5.04	0.4229	2.3		1.711	$y_e\omega_e = 0.072$ cm^{-1}	(56.22, 50.17)

[a] Separation of x and x' state is not known, [b] calculated from the Pekeris relation,
[c] $\Delta G_{1/2}$, [d] observed in matrix isolation.

Dissociation energy = 7.8 ± 0.4 eV, 181 kcal/mole, 62900 cm^{-1}.

ZrO

Perturbations and General Information

Comparison of the spectra in gas phase to that in matrix isolation at $4°K$ (65.26):

Absorption

State	Matrix			Gas	
	$\lambda_{0,0}$	$T_{0,0}$	$\Delta G'_{1/2}$	$T_{0,0}$	$\Delta G'_{1/2}$
$b^1\Sigma^+$	3660	27315	838	27982	837.2
	5154	19397	836	?	-
	5872	17025	872	?	-
	6446	15509	854	?15383	?
$X^1\Sigma^+$		0	975	0	969.8

Emission

Transitions	Matrix		Gas	
	$\nu_{0,0}$	$\Delta G''_{1/2}$	$\nu_{0,0}$	$\Delta G''_{1/2}$
$b^1\Sigma^+ - X^1\Sigma^+$	27315	963	27982	969.8
$B^3\Pi_0 - X'^3\Delta_1$	18327	931	17996	929.6

Franck-Condon factors, r-centroids — Morse potential (67.28):

$A^3\phi - X'^3\Delta$

v', v''	0	1	2	3	4	5	6
0	8.9907^{-1} 1.7430 6229.4	9.5061^{-2} 1.8625 6613.1	5.6704^{-3} 1.9484	2.0492^{-4} 2.0700	5.1243^{-6} 2.180	8.4235^{-8} 2.280	1.0190^{-9} 2.420
1	9.5962^{-2} 1.6338 5916.4	7.1415^{-1} 1.7495 6260.9	1.7267^{-1} 1.8699 6645.3	1.6381^{-2} 1.9551	8.1901^{-4} 2.080	2.6296^{-5} 2.190	5.1322^{-7} 2.310
2	4.8455^{-3} 1.4958 5634.9	1.7586^{-1} 1.6433 5947.0	5.5235^{-1} 1.7555 6292.8	2.3341^{-1} 1.8775 6678.0	3.1423^{-2} 1.9619 7109.8	2.0407^{-3} 2.090	8.0825^{-5} 2.200

Top = Franck-Condon factor followed by a factor of ten
Middle = r-centroid
Bottom = wavelength

$A^3\phi - X'^3\Delta$ (Continued)

v', v''	0	1	2	3	4	5	6
3	1.2458^{-4} 1.4000	1.4381^{-2} 1.5101 5665.5	2.3958^{-1} 1.6525 5977.7	4.1357^{-1} 1.7609 6324.3	2.7808^{-1} 1.8850 6710.9	5.0024^{-2} 1.9686	4.0573^{-3} 2.100
4	6.7308^{-7} 1.300	5.4035^{-4} 1.410	2.8259^{-2} 1.524 5695	2.8735^{-1} 1.6615 6008.5	2.9731^{-1} 1.7653 6356.3	3.0776^{-1} 1.8927 6742.5	7.1359^{-2} 1.9754
5	6.2565^{-8} 1.160	5.0328^{-6} 1.320	1.4494^{-3} 1.420	4.5947^{-2} 1.5373	3.1979^{-1} 1.6703 6039.1	2.0269^{-1} 1.7684 6387.8	3.2370^{-1} 1.9004 6777.2
6	2.0457^{-8} 1.100	2.5546^{-7} 1.220	2.0991^{-5} 1.340	3.0794^{-3} 1.430	6.6742^{-2} 1.5503	3.3783^{-1} 1.6788 6308.7	1.2849^{-1} 1.7690 6549.7

$B^3\Pi - X'^3\Delta$

v', v''	0	1	2	3	4	5	6
0	8.7452^{-1} 1.7444 5551.7	1.1632^{-1} 1.8518 5854.3	8.7354^{-3} 1.9312 6188.7	4.1421^{-4} 2.0154	1.3908^{-5} 2.0922	3.2806^{-7} 2.1803	5.7701^{-9} 2.2572
1	1.1770^{-1} 1.6475 5304.6	6.5094^{-1} 1.7508 5580.1	2.0489^{-1} 1.8594 6070.0	2.4780^{-2} 1.9380 6423	1.6373^{-3} 2.0217	7.0978^{-5} 2.0982	2.0649^{-6} 2.1857
2	7.5214^{-3} 1.5272 5145.4	2.0966^{-1} 1.6570 5332.5	4.6448^{-1} 1.7565 5608.9	2.6752^{-1} 1.8671 5911.5	4.6589^{-2} 1.9448 6459	4.0300^{-3} 2.0281	2.1677^{-4} 2.1043
3	2.5782^{-4} 1.440	2.1960^{-2} 1.5412 5247	2.7657^{-1} 1.6662 5360.7	3.1375^{-1} 1.7611 5637.0	3.0651^{-1} 1.8748 6132.1	7.2538^{-2} 1.9517	7.9042^{-3} 2.0345
4	2.8866^{-6} 1.330	1.1016^{-3} 1.444	4.2357^{-2} 1.5547 5277	3.1981^{-1} 1.6752 5389.3	1.9665^{-1} 1.7640 5666.8	3.2453^{-1} 1.8825	1.0098^{-1} 1.9586
5	4.4732^{-8} 1.238	1.9319^{-5} 1.340	2.9080^{-3} 1.448	6.7447^{-2} 1.5677	3.4141^{-1} 1.6839	1.1039^{-1} 1.7635	3.2458^{-1} 1.8904
6	3.4953^{-8} 1.120	1.1518^{-7} 1.240	7.4169^{-5} 1.350	6.0724^{-3} 1.452	9.5717^{-2} 1.5803	3.4394^{-1} 1.6922	5.1658^{-2} 1.7556

Top = Franck-Condon factor followed by a factor of ten
Middle = r-centroid
Bottom = Wavelength

ZrO

$C^3\Delta - X'^3\Delta$

v', v''	0	1	2	3	4	5	6
0	6.7343^{-1} 1.7539 4640.4	2.6842^{-1} 1.8178 4850.0	5.1474^{-2} 1.8784 5077.6	6.1522^{-3} 1.9385 5322.0	5.0571^{-4} 1.9994	2.9960^{-5} 2.0621	1.3098^{-6} 2.1275
1	2.6241^{-1} 1.6979 4471.5	2.4269^{-1} 1.7602 4665.7	3.5085^{-1} 1.8250 4876.0	1.2055^{-1} 1.8853 5103.9	2.1040^{-2} 1.9452 5352	2.2909^{-3} 2.0058	1.6979^{-4} 2.0684
2	5.4999^{-2} 1.6403 4315.6	3.3300^{-1} 1.7055 4496.2	5.0214^{-2} 1.7643 4692.0	3.2571^{-1} 1.8323 4902.4	1.8476^{-1} 1.8922 4364.4	4.4552^{-2} 1.9518	6.1907^{-3} 2.0123
3	8.1297^{-3} 1.5794 4169.0	1.2433^{-1} 1.6488 4339.7	2.9746^{-1} 1.7128 4521.1	1.1520^{-4} 1.7750	2.4986^{-1} 1.8396 4929.0	2.3100^{-1} 1.8992	7.4703^{-2} 1.9585
4	9.3346^{-4} 1.5123	2.6900^{-2} 1.5891 4195.8	1.8265^{-1} 1.6572 4364.3	2.1647^{-1} 1.7197 4546	2.3092^{-2} 1.7948 4743	1.6181^{-1} 1.8470	2.5359^{-1} 1.9063
5	8.6170^{-5} 1.4342	4.1285^{-3} 1.5238	5.4772^{-2} 1.5986 4219.0	2.1707^{-1} 1.6654 4389.1	1.2966^{-1} 1.7260	7.2244^{-2} 1.7982	8.5089^{-2} 1.8547

$b^1\Sigma^+ - X^1\Sigma^+$

v', v''	0	1	2	3	4	5	6
0	5.1299^{-1} 1.7438 3682.1	3.5646^{-1} 1.7928 3818.3	1.0969^{-1} 1.843	1.8847^{-2} 1.8979	1.9079^{-3} 1.938	1.1020^{-4} 1.980	3.1481^{-6} 2.030
1	3.3142^{-1} 1.7014 3572.0	5.9888^{-2} 1.7531 3700.1	3.3692^{-1} 1.801 3836.4	2.0989^{-1} 1.8509 3981.5	5.4182^{-2} 1.906	7.2002^{-3} 1.940	5.0115^{-4} 2.000
2	1.1817^{-1} 1.6604 3468.9	2.8005^{-1} 1.710 3589.9	3.7186^{-3} 1.760	2.1670^{-1} 1.8100 3854.2	2.6583^{-1} 1.859 4000.7	9.7791^{-2} 1.914	1.6389^{-2} 1.980
3	3.0209^{-2} 1.6191	1.9986^{-1} 1.666 3486.7	1.5184^{-1} 1.720 3607.6	6.0552^{-2} 1.770	1.0536^{-1} 1.821	2.7830^{-1} 1.867	1.4193^{-1} 1.923

Top = Franck-Condon factor followed by a factor of ten
Middle = r-centroid
Bottom = Wavelength

ZrO

$b^1\Sigma^+ - X^1\Sigma^+$ (Continued)

v', v''	0	1	2	3	4	5	6
	6.0660^{-3}	7.7563^{-2}	2.1782^{-1}	5.4116^{-2}	1.1907^{-1}	3.4680^{-2}	2.5979^{-1}
4	1.5758	1.6255	1.6724	1.730	1.780	1.830	1.876
		3390.1	3504.2		3754.7		
	9.8870^{-4}	2.1026^{-2}	1.2296^{-1}	1.8932^{-1}	7.5642^{-3}	1.4748^{-1}	3.9140^{-3}
5	1.5277	1.5828	1.6316	1.678	1.740	1.790	1.840
			3407.9			3772.9(?)	
	1.3021^{-4}	4.3482^{-3}	4.3453^{-2}	1.5385^{-1}	1.3952^{-1}	1.2958^{-3}	1.4629^{-1}
6	1.4694	1.5200	1.5895	1.638	1.683	1.750	1.800
				3425.9			3791.6(?)

$d^1\Delta - c^1\Delta$

v', v''	0	1	2
	7.5391^{-1}	2.1101^{-1}	3.1428^{-2}
0	1.7464	1.8207	1.8854
	5185.0	5448.9	
	2.1263^{-1}	3.7803^{-1}	3.1447^{-1}
1	1.6802	1.7510	1.8259
	4969.8	5212.2	5478.4
	3.0271^{-2}	3.1937^{-1}	1.4685^{-1}
2	1.6068	1.6880	1.7534
		4996.3	
	2.9274^{-3}	7.8914^{-2}	3.4227^{-1}
3	1.526	1.6168	1.6956
		4798.9	5022.9

Top = Franck-Condon factor followed by a factor of ten
Middle = r-centroid
Bottom = Wavelength

BIBLIOGRAPHY

(29.1) Generalities,
R. Wildt,
Z. Physik 54, 856-79

ZrO

(31. 2)   Stellar Spectra,
R. S. Richardson,
Astrophys. J. 73, 216-49

(32. 3)   Infrared System,
W. F. Meggers and C. C. Kiess,
Bur. Stand. J. Res. 9, 309-26

(32. 4) α System,
L. W. Johnson,
Philos. Mag. 14, 286-91

(32. 5) α, β, γ Systems, Vibrational Analysis,
F. Lowater,
Proc. Phys. Soc. 44, 51-66

(33. 6)   Stellar Spectra,
R. S. Richardson,
Astrophys. J. 78, 354-71

(34. 7)   Stellar Spectra,
N. T. Bobrovnikoff,
Astrophys. J. 78, 354-71

(35. 8) α System, Rotational Analysis,
F. Lowater,
Philos. Trans. Roy. Soc. London A 234, 355-76

(37. 9)   Vibrational Analysis,
P. P. Dobronranin,
C. R. Acad. Sci. U. R. S. S. 17, 399-403

(41. 10) γ System,
T. Tanaka and T. Horie,
Proc. Phys. Math. Soc. 23, 464-84

(42. 11)   Generalities. Stellar Spectra,
P. Swings,
Publ. Astron. Soc. Pacific 54, 232-6

(47. 12)   Stellar Spectra,
D. N. Davis,
Astrophys. J. 106, 28-75

(48. 13)   Infrared System,
C. C. Kiess,
Publ. Astron. Soc. Pacific 60, 252-3

(49.14)   A System,
          G. H. Herbig,
          Astrophys. J. 109, 109-15

(49.15)   Preliminary Note to (50.17),
          M. Afaf,
          Nature 164, 752-3

(50.16)   B System,
          M. Afaf,
          Proc. Phys. Soc. A 63, 674-5

(50.17)   Table of Band System Between 10,000 and 3000 A,
          M. Afaf,
          Proc. Phys. Soc. A 63, 1156-70

(54.18) α and γ Systems. Rotational Analysis, Isotope Effect,
          A. Lagerqvist, U. Uhler, and R. F. Barrow,
          Arkiv Fysik 8, 281-93

(54.19) β System. Rotational Analysis. Isotope Effect,
          U. Uhler,
          Arkiv Fysik 8, 295-304

(54.20)   Comparison of the Triplet States of ZrO and TiO,
          U. Uhler,
          Thesis, Stockholm

(56.21)   B Singlet System. Rotational Analysis. Isotope Effect,
          L. Akerlind,
          Arkiv Fysik, Sverige 11, 395-404

(56.22)   A Singlet System. Rotational Analysis. Isotope Effect,
          U. Uhler and L. Akerlind,
          Arkiv Fysik 10, 431-46

(57.23)   Dissociation Energy by Mass Spectra,
          W. A. Chupka, J. Berkowitz, and M. G. Inghram,
          J. Chem. Phys. 26, 1207-10

(57.24)   Wavelengths, Vibrational Analysis,
          A. Gatterer, J. Junkes, E. W. Salpeter, and B. Rosen,
          "Molecular Spectra of Metallic Oxides,"
          Ed: Specola Vaticana 1957

ZrO

(62.25)   N. S. Murthy,
"On the Validity of the Revised Rotational Constants of α(C-X) Band Systems of ZrO,"
Ind. J. Phys. 36, 101-104

(65.26)   ZrO in an Isolation Matrix. Emission and Optical Absorption Bands. Ground State Determination,
W. Weltner, Jr., and D. MacLeod, Jr.,
J. Phys. Chem. 69, 3488-500

(67.27)   P. D. Singh and A. N. Pathak,
"Franck-Condon Factors and r-Centroids for the C-X Band System of ZrO,"
Proc. Phys. Soc. 90, 543-4

(67.28)   Franck-Condon Factors, r-Centroids, A, B, C-X′, b-X′, c-d,
R. W. Nicholls and D. C. Tyte,
"Franck-Condon Factors and r-Centroids for Band Systems of ZrO,"
Proc. Phys. Soc. 91, 489-96

(67.29)   P. D. Singh and A. N. Pathak,
"Vibrational Transition Probabilities and r-Centroids of the A → X System of ZrO,"
Proc. Phys. Soc. 91, 497-8

(68.30)   B. B. Land and D. A. Kalsulkar,
"Potential Energy Curves and Dissociation Energies of some Diatomic Molecules,"
Ind. J. Phys. 42, 50-5

(69.31)   Predictions of Unobserved Levels,
L. Brewer and D. W. Green,
"The Low-Lying Electronic States of ScF, TiO, ZrO,"
High Temp. Science 1, 26-45

(71.32)   Franck-Condon Factors, C-X′, b-X
H. L. Liszt and W. H. Smith,
"RKR Franck-Condon Factors for Blue and Ultraviolet Transitions of some Metal Oxides,"
J. Quant. Spectrosc. Radiative Transfer 11, 1043-62

ZrS

Dissociation energy = 5.96 ± 0.15 eV, 137.4 kcal/mole, 48070 cm^{-1} (68.1, 68.2).

BIBLIOGRAPHY

(68.1) R. P. Steiger,
"Mass Spectrometric Investigation of the Vaporization, Thermodynamics, and Dissociation Energies of LaS, ScS, YS, ZrS and UO,"
Disc. Abstr. 29B, 2009

(68.2) R. P. Steiger and E. D. Cater,
TR #C00-1182-23